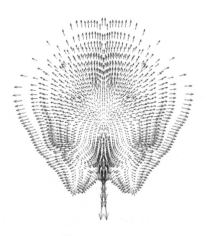

Smoothed Particle Hydrodynamics

a meshfree particle method

G R Liu

National University of Singapore

M B Liu

Smoothed Particle Hydrodynamics

a meshfree particle method

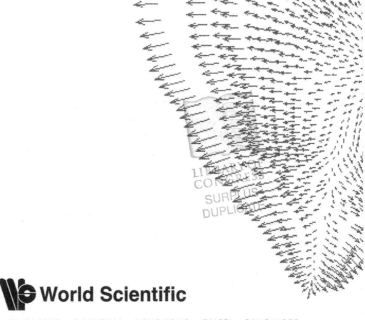

W‌e World Scientific

NEW JERSEY · LONDON · SINGAPORE · SHANGHAI · HONG KONG · TAIPEI · BANGALORE

Published by

World Scientific Publishing Co. Pte. Ltd.

5 Toh Tuck Link, Singapore 596224

USA office: 27 Warren Street, Suite 401-402, Hackensack, NJ 07601

UK office: 57 Shelton Street, Covent Garden, London WC2H 9HE

Library of Congress Cataloging-in-Publication Data
Liu, G. R. (Gui-Rong)
   Smoothed particle hydrogynamics : a meshfree particle method / G.R. Liu and M.B. Liu
   p. cm.
   Includes bibliographical references and index.
   ISBN-13 978-981-238-456-0
   ISBN-10 981-238-456-1
   1. Meshfree methods (Numerical analysis)   2. Hydrodynamics.   I. Liu, M. B.
   QA297 .L566 2003
   532.5--dc21

                               2005295550

British Library Cataloguing-in-Publication Data
A catalogue record for this book is available from the British Library.

First published 2003
Reprinted 2005, 2007, 2009

Printed in Singapore by World Scientific Printers

Preface

Background

Numerical simulation using computers has increasingly become a very important approach for solving problems in engineering and science. It plays a valuable role in providing tests and examinations for theories, offering insights to complex physics, and assisting in the interpretation and even the discovery of new phenomena. Grid or mesh based numerical methods such as the finite difference methods (FDM) and the finite element methods (FDM) have been widely applied to various areas of computational fluid dynamics (CFD) and computational solid mechanics (CSM), and are currently the dominant methods in numerical simulations for solving problems in engineering and science.

Despite the great success, grid-based numerical methods suffer from difficulties in some aspects, which limit their applications in many complex problems. The major difficulties are inherited from the use of grid or mesh. Because the entire formulation is based on the grid/mesh, a time-consuming and costly process of generating/regenerating a quality grid/mesh is necessary. The use of grid/mesh can lead to difficulties in dealing with problems with free surface, deformable boundary, moving interface (for FDM), and extremely large deformation (for FEM).

A recent strong interest is focused on the next generation computational methods — meshfree methods, which are expected to be superior to conventional grid-based FDM and FEM in many applications. A large number of meshfree methods have been proposed for different applications, as detailed in the recent monograph on meshfree methods by G. R. Liu (2002). These meshfree methods share some common features, but are different in the function approximation, the approaches for creating discretized system equations and the detailed implementation process.

This book will be focusing on a class of meshfree particle methods (MPMs), in particular, the smoothed particle hydrodynamics (SPH) method and its variations. The reasons for devoting this volume to the SPH method are listed as follows.

1. The SPH method is capable of dealing with problems with free surface, deformable boundary, moving interface as well as extremely large deformation.
2. The SPH method, as one of the oldest MPMs, is quickly approaching its mature stage.
3. With the continuing improvements and modifications, the accuracy, stability and adaptivity of the SPH method have reached an acceptable level for practical engineering applications.
4. Applications of the SPH method are very wide, ranging from micro-scale to macro-scale and even to astronomical scale, and from discrete systems to continuum systems.
5. Some commercial codes have incorporated the SPH processor into their software packages with many successful practical applications.

The authors and their research teams started to work on SPH since 1997, when they were searching for an alternative numerical approach for simulating explosion of high explosives, underwater explosion, etc. They were using the FEM, FVM, and ALE for explosion related projects (Lam and Liu et al. 1996; Liu and Lam et al. 1998; Zhao et al., 1998; Chong et al. 1998a; b; 1999; etc.). It was very often to encounter unexpected terminations during the computation due to mesh distortion related problems. It had been a constantly painful daily struggle to adjust all kinds of parameters in the codes that they used, and then re-run the simulation, hoping to get a successful run at the end. The research works using these methods were indeed not very productive, but there were no other choices. An alternative method without using a mesh is definitely necessary for these problems. The quest of the authors' teams for such an alternative method has led to their active venturing into the meshfree methods.

The SPH method was then chosen as one of the alternatives due to the above-mentioned reasons. The successful applications of the SPH method to many explosion related problems have encouraged the authors and their team members to further extend the SPH method for high velocity impact and penetration problems. These applications will be presented in detail in this book after a systematic presentation on the basic idea, theory, and the formation of the SPH method of various versions.

To Zuona

Yun, Kun, Run,

and my family

for the time and support they gave to me

G. R. Liu

To Huiqi

for her love, support and encouragement

M. B. Liu

This book

For the very first time, the SPH method is presented in detail in a book form. The book aims to provide an introduction to the SPH method for readers who are interested in learning, using or further developing the method. It covers the theoretical background, numerical techniques, code implementation issues and many different applications.

The book is written for senior university students, graduate students, researchers and professionals both in engineering and science. The presented techniques, intriguing applications and sample code will be useful to mechanical, civil, and aeronautical and astronautically students, engineers, researchers and professionals in both CFD and CSM areas.

Substantial efforts have been made in presenting the materials in an easy-to-understand manner, so that the 3rd year undergraduates in engineering or science schools can understand most of the materials discussed in this book. Background knowledge on the grid-based numerical methods such as FEM (see, e.g., Liu and Quek, 2003) and FDM is not required but would be helpful in understanding the procedure and the method, and in reading the book. Basic knowledge on CFD will also help in reading the book, and a section is therefore provided to prepare the students for the knowledge on fundamental equations that governs the fluid flows.

In theories, the kernel approximation in continuous form and particle approximation in discretized form are described, as these two approximations form the foundation for the SPH methods. The reproducibility of the SPH approximation is investigated, which is accompanied with systematically derived consistency conditions in both continuous and discretized forms. The continuous consistency conditions result in a generalized approach to construct analytical smoothing functions that play a key role in the SPH formulation. The discretized consistency conditions provide means to restore the particle inconsistency in the conventional SPH, and lead to the development of the corrective smoothed particle method (or CSPM). A discontinuous SPH (or DSPH) formulation is presented to simulate discontinuous phenomena such as shock waves. The DSPH formulation improves the boundary deficiency and restores consistency in the discontinuous region. An adaptive SPH (or ASPH) that provides better directional smoothing features is also introduced.

In numerical techniques, some detailed treatments such as artificial viscosity, artificial heat, physical viscosity, variable smoothing length, zero-energy mode problem, artificial compressibility, solid boundary treatment, choice of time step are addressed for enhancing the stability in the computational process and the accuracy of the results.

In applications, many interesting and practical examples are presented. The applications are dispersed in different chapters when addressing the corresponding topics. Except for some routine cases for benchmarking the SPH method, most of the presented applications are quite difficult for the grid-based numerical methods. These includes incompressible flows, free surface flows, high compressible flows, high explosive (HE) detonation and explosion, underwater explosion and water mitigation of explosive shocks, high velocity impact and penetration, and multiple scale simulations coupled with the molecular dynamics method. The presented numerical examples demonstrate the powerful capabilities of the SPH method and offer ample opportunities to further improve and extend the SPH method for more complex and practical applications in engineering and science.

In implementation, this book also discusses some important issues in the computer coding of the SPH method, including basics of parallel coding. A source code in FORTRAN 77 is provided with detailed descriptions.

Numerical simulations using the SPH method are a relatively new area of research, and are still under continuing development. There are problems awaiting further improvements in the SPH method. These problems in turn offer ample opportunities for researchers to develop more advanced particle methods as the next generation of numerical methods. The authors hope that the formulation and the source code provided in this book can serve the purpose to have a smoother start to efficiently learn, test, practise and further develop the SPH methods.

Outline of the book

This book provides a comprehensive introduction to the SPH method and its variations such as the CSPM, DSPH, and ASPH. It is organized in a total of ten chapters that are briefed as follows.

Chapter 1 introduces some background knowledge of numerical simulation. The features and limitations of the grid-based numerical methods are discussed. The basic ideas of meshfree and particle methods are briefed. Some general features of the meshfree particle methods, especially smoothed particle hydrodynamics, are described. The invention, development, applications and extensions of the SPH method are briefly addressed.

Chapter 2 provides the fundamentals of the SPH method such as the basic concepts and the essence of SPH formulations. The essential formulations will be useful in the later Chapters.

Chapter 3 presents a general approach to construct analytical smoothing functions for the meshfree methods including SPH. The approach not only systematically derives the constructing conditions for the smoothing function, but also addresses the related particle consistency problem in the SPH method. The efficiency of the approach has been demonstrated by developing various existing smoothing functions including a newly constructed smoothing function.

Chapter 4 describes the implementation of the SPH method on the Navier-Stokes equation, and presents some applications to some general fluid dynamic problems. Some numerical aspects such as artificial viscosity, artificial heat, physical viscosity, variable smoothing length, zero-energy mode problem, artificial compressibility, solid boundary treatment, choice of time step are discussed. Nearest neighbor particle searching (NNPS) algorithms and pair interaction technique for particle interaction are described. The implemented three-dimensional SPH code is applied to solve different flow problems.

Chapter 5 proposes a smoothed particle hydrodynamics formulation, which is superior to the traditional SPH and the corrective smoothed particle method (CSPM) in simulating discontinuous phenomena. With this formulation, the solution accuracy near the boundaries and the discontinuities are both considerably improved.

Chapter 6 presents the application of SPH to high explosive (HE) detonations process, and the later expansion process. An adaptive procedure to evolve the smoothing length is employed to meet the needs of simulating large deformation and explosion events. Besides some basic numerical examples, explosions of shaped charge are also simulated using the SPH method with some revealing results.

Chapter 7 introduces the application of SPH to the simulation of the early time phenomena such as shock waves in underwater explosions. A particle-to-particle interface treat technique is employed, which allows the kernel and particle approximations among particles from different materials, and applies a special penalty force to penetrating particles. A comparison investigation is also carried out on the real and the artificial detonation models of HE as well as their influences to the entire underwater explosion shock simulations. This chapter also presents an investigation on the water mitigation problems using the SPH method. Contact and non-contact water mitigation simulations have been carried out and are compared with the case without water to study the water mitigation effects.

Chapter 8 investigates the hydrodynamics with material strength using both the SPH method and the adaptive smoothed particle hydrodynamics (ASPH) that is a modified version of SPH. A comparison study of SPH and ASPH and their applications in hydrodynamics with material strength is presented. The constitutive model and equation of state for the material are incorporated into the SPH and ASPH equations. Two numerical tests of impact and penetration are presented.

Chapter 9 presents an approach for coupling length scale (CLS) via combining SPH with molecular dynamics (MD). The molecular dynamics is applied to the atomic-sized regions for accuracy, whereas SPH is applied to other peripheral regions for efficiency. Handshaking MD/SPH is implemented by allowing interaction between SPH particles and MD atoms. The validity of this novel particle-particle CLS approach is preliminarily demonstrated by the simulation of the Poiseuille flow and Couette flow of simple fluids.

Chapter 10 addresses issues related to the computer implementation of the SPH method. The general procedure for Lagrangian particle simulation is described. Computer implementations of the SPH method on serial and parallel computers are also addressed. A sample SPH code is presented. The main features of the sample code, more detailed descriptions and the source code of the related subroutines are given. The programs demonstrate most of the concepts and techniques related to the MPMs related to SPH.

Chapters 1 to 4 are essential for the SPH method, and may be read orderly before reading other chapters. Chapters 5 to 9 are mainly applications and special topics of the SPH method. They may be read independently in any order depending on the interests of the readers. Proper cross-references are provided when substantial materials and issues are shared between these chapters. Chapter 10 provides a SPH code and the detailed descriptions. It may be read after Chapter 4 for the beginners and any time for the experienced users who are interested only in using the source code.

Acknowledgement

The authors' work in the area of meshfree particle methods has been profoundly influenced by the works of Prof. J. J. Monaghan, L. D., Libersky, L. Lucy, W. Benz, J. K. Chen and many others working in this area. Without their significant contributions, writing this book would not be possible.

In preparing this book, many colleagues and students have supported and contributed to the writing of this book. The authors would like to express their sincere thanks to all of them. Special thanks are given to Prof. K. Y. Lam, Dr. H. F. Qiang, Mr. Z. R. Li, Mr. G. L. Chin, Dr. X., Han, Dr. Zong Z., etc. They have helped either in preparing some examples, or in reading the drafts of the book, or have been involved in carrying out a number of projects related to the SPH methods. Special thanks also go to Y. Liu for reading the drafts of this thick volume and providing very useful editorial comments.

Finally, the authors would also like to thank A*STAR, Singapore, and the National University of Singapore for their partial financial sponsorship for the research projects undertaken by the authors and their teams related to the topic of this book.

The Authors

Dr. G. R. Liu received his PhD from Tohoku University, Japan in 1991. He was a Postdoctoral Fellow at Northwestern University, U. S. A. He is currently the Director of the Centre for Advanced Computations in Engineering Science (ACES), National University of Singapore, and the President of the Association for Computational Mechanics (Singapore). He is also an Associate Professor at the Department of Mechanical Engineering, National University of Singapore. He has provided consultation services to many national and international organizations. He authored more than **300** technical publications including more than 180 international journal papers and five authored books, including the popular book entitled "Mesh Free Method: moving beyond the finite element method". He serves as an editor and a member of editorial boards of five scientific journals. He is the recipient of the **Outstanding University Researchers Awards**, the **Defence Technology Prize** (National award), and the **Silver Award** at **CrayQuest** (Nationwide competition). His research interests include Computational Mechanics, Meshfree Methods, Nano-scale Computation, Micro bio-system computation, Vibration and Wave Propagation in Composites, Mechanics of Composites and Smart Materials, Inverse Problems and Numerical Analysis.

M. B. Liu received his B.E. and M. E. degrees from Xi'an JiaoTong University (XJTU), China in 1993 and 1996, respectively, and received his PhD from the National University of Singapore (NUS) in 2003. He was a lecturer in the Northern JiaoTong University (NJTU), China, from July 1996 to Nov. 1998. He is currently a research fellow at the Department of Mechanical Engineering in NUS. His areas of interests include meshfree particle methods, computational fluid dynamics, computational aerodynamics, computational mechanics, and high performance computing techniques. He has conducted extensive research projects including turbo-machinery aerodynamic, vibration and strength analysis, free surface flows, incompressible flows, highly compressible flows, high explosive (HE) detonation, HE explosion, underwater explosion, water mitigation of shocks, high velocity impact, penetration, and multiple scale simulations coupled with atomistic method.

Contents

Chapter 1

Introduction

1.1 Numerical simulation

1.1.1 Role of numerical simulation

Numerical simulation using computers or computational simulation has increasingly become a very important approach for solving complex practical problems in engineering and science. Numerical simulation translates important aspects of a physical problem into a discrete form of mathematical description, recreates and solves the problem on a computer, and reveals phenomena virtually according to the requirements of the analysts. Rather than adopting the traditional theoretical practice of constructing layers of assumptions and approximations, this modern numerical approach attacks the original problems in all its detail without making too many assumptions, with the help of the increasing computer power.

Numerical simulation provides an alternative tool of scientific investigation, instead of carrying out expensive, time-consuming or even dangerous experiments in laboratories or on site. The numerical tools are often more useful than the traditional experimental methods in terms of providing insightful and complete information that cannot be directly measured or observed, or difficult to acquire via other means.

Numerical simulation with computers plays a valuable role in providing a validation for theories, offers insights to the experimental results and assists in the interpretation or even the discovery of new phenomena. It acts also as a bridge between the experimental models and the theoretical predictions as schematically shown in Figure 1.1.

1

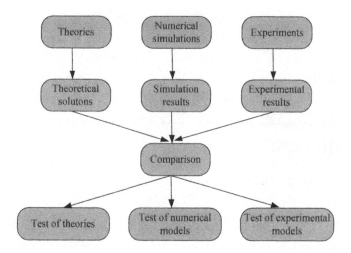

Figure 1.1 Connection between the numerical simulations, theories and experiments. The connection is very important in conducting scientific investigations. In the connection, the numerical simulation is playing an increasingly important role.

1.1.2   Solution procedure of general numerical simulations

Numerical simulations follow a similar procedure to serve a practical purpose. There are in principle some necessary steps in the procedure, as shown in Figure 1.2.   From the physical phenomena observed, mathematical models are established with some possible simplifications and assumptions.   These mathematical models are generally expressed in the form of governing equations with proper boundary conditions (BC) and/or initial conditions (IC).   The governing equations may be a set of ordinary differential equations (ODE), partial differential equations (PDE), integration equations or equations in any other possible forms of physical laws[1].   Boundary and/or initial conditions are necessary for determining the field variables in space and/or time.

To numerically solve the governing equations, the involved geometry of the problem domain needs to be divided into discrete components.   The domain discretization techniques may be different for different numerical methods. Domain discretization usually refers to representing a continuum problem domain with a finite number of components, which form the computational

---

[1] The formulation of a problem could be well-posed or ill-posed.   This book discusses problems that are well-posed.   For ill-posed problems, special techniques are required (see, e.g., Liu and Han, 2003).

frame for the numerical approximation. The computational frame is traditionally a set of mesh or grid, which consists of a lattice of points, or grid nodes to approximate the geometry of the problem domain. The grid nodes are the locations where the field variables are evaluated, and their relations are defined by some kind of *nodal connectivity*. Grid nodes are connected to form a *mesh* following the connectivity. Accuracy of the numerical approximation is closely related to the mesh cell size and the mesh patterns.

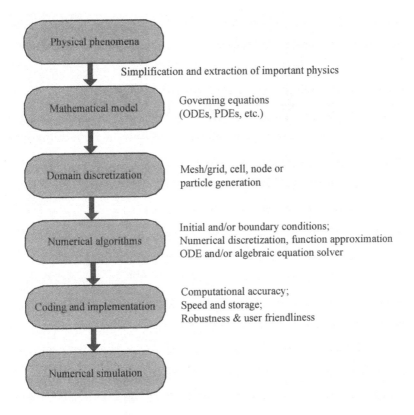

Figure 1.2 Procedure of conducting a numerical simulation.

Numerical discretization provides means to change the integral or derivative operations in the governing equations in continuous forms to discrete representations. It is closely related to the domain discretization technique applied. The numerical discretization is based on a theory of function approximations (Liu, 2002). After domain discretization and numerical discretization, the original physical equations can be changed into a set of algebraic equations or ordinary differential equations, which can be solved using the existing numerical routines. In the process of establishing the algebraic or ODE equations, the so-called strong or weak form formulation can be used (see, e.g., Liu, 2002). As recently reported by Liu and Gu (2002, 2003), these two forms of formulation can also be combined together to take the fullest advantages of both weak and strong form formulations. This combination led to the development of the meshfree weak-strong (MWS) form methods.

Implementation of a numerical simulation involves translating the domain decomposition and numerical algorithms into a computer code in some programming language(s). In coding a computer program, the accuracy, and efficiency (speed and storage) are two very important considerations. Other considerations include robustness of the code (consistency check, error trap), user-friendliness of the code (easy to read, use and even to modify), etc. Before performing a practical numerical simulation, the code should be verified to reproduce sets of experimental data, theoretical solutions, or the exact results from other established methods for benchmark problems or actual engineering problems.

For numerical simulations of problems in fluid mechanics, the governing equations can be established from the conservation laws, which states that a certain number of system field variables such as the *mass*, *momentum* and *energy* must be conserved during the evolution process of the system (see, Chapter 4 for details). These three fundamental principles of conservation, together with additional information concerning the specification of the nature of the material/medium, conditions at the boundary, and conditions at the initial stage completely determine the behavior of the fluid system[2]. These fundamental physical principles can be expressed in the form of some basic mathematical equations called governing equations. In most circumstances the governing equations are partial differential equations (PDE).

Except for a few circumstances, it is very difficult to obtain analytical solution of these integral equations or partial differential equations. Computational fluid dynamics (CFD) deals with the techniques of spatially approximating the integral or the differential operations in the integral or differential equations with the discretized counterparts that are usually in a form of simple algebraic summation. This approximation leads to a set of algebraic equations (or ODEs with respect to time only) that can be solved to obtain numerical values for field functions (such as density, pressure, velocity, etc.) at discrete points in space and/or time (Figure 1.3). A typical numerical simulation of a CFD problem involves the following factors.

---

[2] Engineering problems can be formulated in a deterministic manner that establishes one-to-one relationships between the causes and results of the problems. They can also be formulated in a probabilistic manner that establishes relations between the causes and results in a probabilistic sense. This book focuses on the deterministic problems.

1. governing equations,
2. proper boundary conditions and/or initial conditions,
3. domain discretization technique,
4. numerical discretization technique,
5. numerical technique to solve the resultant algebraic equations or ordinary differential equations (ODE).

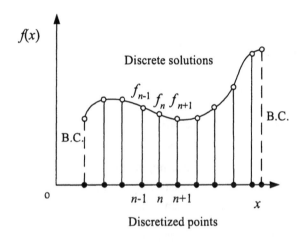

Figure 1.3 Domain and numerical discretization for a numerical simulation of a field function $f(x)$ defined in one-dimensional space.

1.2 Grid-based methods

There are two fundamental frames for describing the physical governing equations: the Eulerian description and the Lagrangian description. The Eulerian description is a spatial description, and is typically represented by the finite difference method (FDM) (see, e.g., Anderson, 1995; Hirsch, 1988; Wilkins, 1999). The Lagrangian description is a material description, and is typically represented by the finite element method (FEM) (see, e.g., Zienkiewicz and Taylor, 2000; Liu and Quek, 2003, etc.). For example, in fluid mechanics, if the viscosity and the heat conduction as well as the external forces are neglected, the conservation equations in PDE form for these two descriptions are very much different, as listed in Table 1.1.

Table 1.1 Conservation equations in PDE form in the Lagrangian and Eulerian descriptions

Conservation	Lagrangian description	Eulerian description
Mass	$\dfrac{D\rho}{Dt} = -\rho \dfrac{\partial v^\beta}{\partial x^\beta}$	$\dfrac{\partial \rho}{\partial t} + v^\beta \dfrac{\partial \rho}{\partial x^\beta} = -\rho \dfrac{\partial v^\beta}{\partial x^\beta}$
Momentum	$\dfrac{Dv^\beta}{Dt} = -\dfrac{1}{\rho} \dfrac{\partial p}{\partial x^\beta}$	$\dfrac{\partial v^\beta}{\partial t} + v^\alpha \dfrac{\partial v^\beta}{\partial v^\alpha} = -\dfrac{1}{\rho} \dfrac{\partial p}{\partial x^\beta}$
Energy	$\dfrac{De}{Dt} = -\dfrac{p}{\rho} \dfrac{\partial v^\beta}{\partial x^\beta}$	$\dfrac{\partial e}{\partial t} + v^\beta \dfrac{\partial e}{\partial x^\beta} = -\dfrac{p}{\rho} \dfrac{\partial v^\beta}{\partial x^\beta}$

In Table 1.1, ρ, e, v and x are density, internal energy, velocity and position vector respectively. The Greek superscripts α and β are used to denote the coordinate directions, while the summation in the equations is taken over repeated indices. It is seen that the differences between the two sets of equations are inherited in the definition of the *total time derivative* as the combination of the *local derivative* and the *convective derivative*, i.e.,

$$\frac{D}{Dt} = \frac{\partial}{\partial t} + v^\alpha \frac{\partial}{\partial v^\alpha} \qquad (1.1)$$

where D/Dt is the total time derivative (or *substantial derivative, material derivative*, or *global derivative*) that is physically the time rate of change following a moving fluid elements; $\partial/\partial t$ is the local derivative that is physically the time rate of change at a fixed point; $v^\alpha \, \partial/\partial v^\alpha$ is the convective derivative that is physically the change due to the movement of the fluid element from one location to another in the flow field where the flow properties are spatially different. Therefore, the total time derivative describes that the flow property of the fluid element is changing, as a fluid element sweeps passing a point in the flow. This is because 1) at that point, the flow field property itself may be fluctuating with time (the local derivative); 2) the fluid element is on its way to another location in the flow field where the flow property may be different (the convective derivative).

The Eulerian and Lagrangian descriptions correspond to two disparate kinds of grid of domain discretization: the Eulerian grid and the Lagrangian grid. Both of them are widely used in numerical methods with preferences on types of problems, and hence are briefed in the following.

1.2.1 Lagrangian grid

In the Lagrangian grid-based methods such as the well-known and widely used FEM (see, e.g., Zienkiewicz and Taylor, 2000; Liu and Quek, 2003; etc.), the grid is fixed to or attached on the material in the entire computation process, and therefore it moves with the material as illustrated in Figure 1.4.

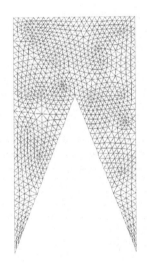

Figure 1.4 Lagrangian mesh/cells/grids for a shaped charge detonation simulation. The triangular cells and the entire mesh of cell move with the material.

Since each grid node follows the path of the material at the grid point, the relative movement of the connecting nodes may result in expansion, compression and deformation of a mesh cell (or element). Mass, momentum and energy are transported with the movement of the mesh cells. Because the mass within each cell remains fixed, no mass flux crosses the mesh cell boundaries. When the material deforms, the mesh deforms accordingly.

The Lagrangian grid-based methods have several advantages.

1. Since no convective term exists in the related partial differential equations, the code is conceptually simpler and should be faster as no computational effort is necessary for dealing with the convective terms.
2. Since the grid is fixed on the moving material, the entire time history of all the field variables at a material point can be easily tracked and obtained.
3. In the Lagrangian computation, some grid nodes can be placed along boundaries and material interfaces. The boundary conditions at free surfaces, moving boundaries, and material interfaces are automatically imposed, tracked and determined simply by the movement of these grid nodes.
4. Irregular or complicated geometries can be conveniently treated by using an irregular mesh.
5. Since the grid is required only within the problem domain, no additional grids beyond the problem domain is required, and hence the Lagrangian grid-based methods are computationally efficient.

Due to these advantages, Lagrangian methods are very popular and successful in solving computational solid mechanics (CSM) problems, where the deformation is not as large as that in the fluid flows.

However, Lagrangian grid-based methods are practically very difficult to apply for cases with extremely distorted mesh, because their formulation is always based on mesh. When the mesh is heavily distorted, the accuracy of the formulation and hence the solution will be severely affected. In addition, the time step, which is controlled by the smallest element size, can become too small to be efficient for the time marching, and may even lead to the breakdown of the computation.

A possible option to enhance the Lagrangian computation is to rezone the mesh or re-mesh the problem domain. The mesh rezoning involves overlaying of a new, undistorted mesh on the old, distorted mesh, so that the following-up computation can be performed on the new undistorted mesh. The physical properties in the new mesh cells are approximated from the old mesh cells through calculating the mass, momentum and energy transport in a Eulerian description. Adaptive rezoning techniques are quite popular for simulations of impact, penetration, explosion, fragmentation, turbulence flows, and fluid-structure interaction problems. The rezoning procedure in Lagrangian computations can be tedious and very time-consuming. Moreover, with each rezoning, some material diffusion occurs and material histories may be lost. In addition, the Lagrangian codes under frequent remesh turn to resemble a Eulerian code in an overall sense. Therefore, even though there are some very good advantages in Lagrangian grid-based methods, the disadvantages can result in numerical difficulties when simulating events of extremely large deformation (Benson, 1992; Charles, 1987; Hans, 1999).

A Lagrangian numerical method, whose solution does not depend on a mesh and hence is not affected by the heavy movement of the nodes, is indeed desirable.

1.2.2 Eulerian grid

Contrary to the Lagrangian grid, the Eulerian grid is fixed on the space, in which the simulated object is located and moves across the fixed mesh cells in the grid (illustrated in Figure 1.5). Therefore, all grid nodes and mesh cells remain spatially fixed in space and do not change with time while the materials are flowing across the mesh. The flux of mass, momentum and energy across mesh cell boundaries are simulated to compute the distribution of mass, velocity, energy, etc. in the problem domain. The shape and volume of the mesh cell remain unchanged in the entire process of the computation.

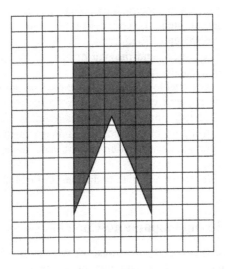

Figure 1.5 Eulerian mesh/cells/grids for a shaped charge detonation simulation. The mesh/grid is fixed in space and does not move or deform with time. The material moves/flows across the fixed mesh cells.

Since the Eulerian grid is fixed in space and with time, large deformations in the object do not cause any deformations in the mesh itself and therefore do not cause the same kind of numerical problems as in the Lagrangian grid-based methods. Eulerian methods are therefore dominant in the area of computational fluid dynamics, where the flow of the material dominates. In principle, all hydrodynamic problems can be numerically solved using a multi-material Eulerian method that calculates the mass, momentum and energy flux across the fixed Eulerian mesh cell boundaries. Early simulations of problems with large deformation such as explosion and high velocity impacts were usually performed using some kind of Eulerian methods (Benson, 1992; Charles, 1987; Hans, 1999). However, there are many disadvantages associated with the Eulerian methods.

1. It is very difficult to analyze the time history of field variables at a fix point on the material, because the movement of the material cannot be tracked using a fixed mesh. One can only have the time history of field variables at fixed-in-space Eulerian grid.
2. It is not easy to treat the irregular or complicated geometries of material/media in the Eulerian grid-based methods. A complicated mesh generation procedure to convert the irregular geometry of problem domain into a regular computational domain is usually necessary. Sometimes, expensive numerical mapping is required.
3. Since the Eulerian methods track the mass, momentum and energy flux across the mesh cell boundaries, the position of free surfaces, deformable boundaries, and moving material interfaces are difficult to be determined accurately.
4. Since the Eulerian methods require a grid over a computational domain, which should be large enough to cover the entire area to which the material can possibly flow. It sometimes requires the modeler to use a much coarse grid for computational efficiency at the expenses of the resolution of domain discretization and the accuracy of the solution.

The features of both the Lagrangian and Eulerian methods are summarized in Table 1.2.

1.2.3 Combined Lagrangian and Eulerian grids

The different but complementary features of the Lagrangian and Eulerian descriptions suggest that it would be computationally beneficial to combine these two descriptions so as to strengthen their advantages and to avoid their disadvantages. This idea has led to the development of two complicated approaches that apply both the Lagrangian and Eulerian descriptions: the Coupled Eulerian Lagrangian (CEL) (Hans, 1999) and the Arbitrary Lagrange Eulerian (ALE) (Benson, 1992; Hirt et al., 1974; Belytschko et al., 2000).

The CEL approach employs both the Eulerian and Lagrangian methods in separate (or with some overlap) regions of the problem domain. One of the most common practices is to discretize solids in a Lagrangian frame, and fluids (or materials behaving like fluids) in a Eulerian frame. The Lagrangian region and Eulerian region continuously interact with each other through a coupling module in which computational information is exchanged either by mapping or by special interface treatments between these two sets of grid.

Table 1.2 Comparisons of Lagrangian and Eulerian methods.

	Lagrangian methods	**Eulerian methods**
Grid	Attached on the moving material	Fixed in the space
Track	Movement of any point on materials	Mass, momentum, and energy flux across grid nodes and mesh cell boundary
Time history	Easy to obtain time-history data at a point attached on materials	Difficult to obtain time-history data at a point attached on materials
Moving boundary and interface	Easy to track	Difficult to track
Irregular geometry	Easy to model	Difficult to model with good accuracy
Large deformation	Difficult to handle	Easy to handle

The ALE is closely related to the rezoning techniques for Lagrangian mesh, and aims to move the mesh independently of the materials so that the mesh distortion can be minimized. In an ALE, Lagrangian motion is computed at every time step in the beginning, followed by a possible rezoning stage in which the mesh is not rezoned (pure Lagrangian description), or rezoned to the original shape (Eulerian description), or rezoned to some more advantageous shape (somewhat between the Lagrangian and Eulerian description).

These two approaches of combining Eulerian and Lagrangian descriptions receive much research interest and have achieved a lot in obtaining more stable solutions. Many commercial hydrocodes such as MSC/Dytran (MSC/Dytran, 1997; Liu and Lam et al. 1998; Lam and Liu et al., 1996), DYNA2D and DYNA3D (Hallquist, 1988; 1986), LS-DYNA (Hallquist, 1998), and AUTODYN (Century dynamics, 1997) have incorporated CEL or/and ALE for coupled analyses of dynamic phenomena with fluid solid interaction behavior. Unfortunately, even with the CEL and ALE formulations a highly distorted mesh can still introduce severe errors in numerical simulations (Benson, 1992; Hirt et al., 1974; Belytschko et al., 2000).

In our earlier study using the FEM, FVM, and ALE for explosion related projects (Lam and Liu et al. 1996; Liu and Lam et al. 1998; Zhao et al., 1998; Chong et al. 1998a; b; 1999; etc.), it was very often to encounter unexpected terminations in the computation process. The reasons were mainly the prohibitively very small time step resulted from very small Lagrangian elements or Euler cells or negative density that created by the extremely large deformation or movement of the material/media. Small Euler cells can also be created in using the ALE when the movement of the Lagrangian elements cut through the Euler cells and leave extremely small volume in the Euler cells. It had been a constantly painful daily struggle to adjust all kinds of parameters in the codes that we used, and then re-run the simulation hoping to get through a successful run at the end. The research works using these methods were indeed not very productive, but there were not other choices. An alternative method is definitely required for these types of problems. The quest of the authors' teams for such an alternative method has led to their active venturing into the meshfree methods.

1.2.4   Limitations of the grid-based methods

Conventional grid-based numerical methods such as FDM and FEM have been widely applied to various areas of CFD and CSM, and currently are the dominant methods in numerical simulations of domain discretization and numerical discretization. Despite the great success, grid-based numerical methods suffer from some inherent difficulties in many aspects, which limit their applications to many problems.

In grid-based numerical methods, mesh generation for the problem domain is a prerequisite for the numerical simulations. For the Eulerian grid methods like FDM, constructing a regular grid for irregular or complex geometry has never been an easy task, and usually requires additional complex mathematical transformation that can be even more expensive than solving the problem itself. Determining the precise locations of the inhomogeneities, free surfaces, deformable boundaries and moving interfaces within the frame of the fixed Eulerian grid is also a formidable task. The Eulerian methods are also not well suited to problems that need monitoring the material properties in fixed volumes, e.g. particulate flows. For the Lagrangian grid methods like FEM, mesh generation is necessary for the objects being simulated, and usually occupies a significant portion of the computational effort. Treatment of large deformation is an important issue in a Lagrangian grid-based method. It usually requires special techniques like rezoning. Mesh rezoning, however, is tedious and time consuming, and may introduce additional inaccuracy into the solution.

The difficulties and limitations of the grid-based methods are especially evident when simulating hydrodynamic phenomena such as *explosion* and *high velocity impact* (HVI). In the whole process of an explosion, there exist special features such as large deformations, large inhomogeneities, moving material interfaces, deformable boundaries, and free surfaces. These special features pose great challenges to numerical simulations using the grid-based methods. High velocity impact problems involve shock waves propagating through the colliding or impacting bodies that behave like fluids. Analytically, the equations of motion and a high-pressure equation of state are the key descriptors of material behavior. In HVI phenomena, there exist large deformations, moving material interfaces, deformable boundaries, and free surfaces, which are, again, very difficult for grid-based numerical methods. As can be seen from the next chapters, simulation of hydrodynamic phenomena such as explosion and HVI by methods without using a mesh is a very promising alternative.

The grid-based numerical methods are also not suitable for situations where the main concern of the object is a set of discrete physical particles rather than a continuum, e.g., the interaction of stars in astrophysics, movement of millions of atoms in an equilibrium or non-equilibrium state, dynamic behavior of protein molecules, and etc. Simulation of such discrete systems using the continuum grid-based methods may not always be a good choice.

1.3 Meshfree methods

A recent strong interest is focused on the development of the next generation of computational methods — meshfree methods, which are expected to be superior to the conventional grid-based FDM and FEM for many applications. The key idea of the meshfree methods is to provide accurate and stable numerical solutions for integral equations or PDEs with all kinds of possible boundary conditions with a set of arbitrarily distributed nodes (or particles) without using any mesh that provides the connectivity of these nodes or particles. Details on many existing meshfree methods can be found in a recent monograph by Liu (2002). One important goal of the initial research is to modify the internal structure of the grid-based FDM and FEM to become more adaptive, versatile and robust. Much effort is concentrated on problems to which the conventional FDM and FEM are difficult to apply, such as problems with free surface, deformable boundary, moving interface (for FDM), large deformation (for FEM), complex mesh generation, mesh adaptivity, and multi-scale resolution (for both FDM and FEM). Recently, a number of meshfree methods have been proposed for analyzing solids and structures as well as fluid flows. These meshfree methods share some common features, but are different in the means of function approximation and the implementation process.

Smoothed particle hydrodynamics (SPH) (Lucy, 1977; Gingold and Monaghan, 1977), as a meshfree and particle method, was originally invented for modeling astrophysical phenomena, and later widely extended for applications to problems of continuum solid and fluid mechanics. The SPH method and its different variants are the major type of particle methods, and have been incorporated into many commercial codes.

Liszka and Orkisz (1980) proposed a generalized finite difference method that can deal with arbitrary irregular grids. Nayroles et al. (1992) are the first to use moving least square approximations in a Galerkin method to formulate the so-called diffuse element method (DEM). Based on the DEM, Belytschko et al. (1994) advanced remarkably the element free Galerkin (EFG) method. The EFG is currently one of the most popular meshfree methods, and applied to many solid mechanics problems (Krysl and Belytschko, 1996a; b; Noguchi, et al. 2000; etc.) with the help of a background mesh for integration.

Atluri and Zhu (1998) have originated the Meshless Local Petrov-Galerkin (MLPG) method that requires only local background cells for the integration. Because the MLPG does not need a global background mesh for integration, it has been applied to the analysis of beam and plate structures (Atluri, et al., 1999; Gu and Liu, 2001e, Long and Atluri, 2002), fluid flows (Lin and Atluri, 2001; Wu and Liu, 2003a,b; etc.), and other mechanics problems. Detailed descriptions of the MLPG and its applications can be found in the monograph by Atluri and Shen (2000).

W. K. Liu and his co-workers (Chen and Liu, 1995, Liu et al., 1995a, b; 1996a, b, c; 1997), through revisiting the consistency and reproducing conditions in SPH, proposed a reproducing kernel particle method (RKPM) which improves the accuracy of the SPH approximation especially around the boundary. There are comprehensive literatures available on RKPM and its applications (see, W. K. Liu et al., 1996b; Li and Liu, 2002).

G. R. Liu and his colleagues in a series of papers developed the point interpolation method (PIM) and some variants (Liu 2002; Gu and Liu, 2001a, c; 2002; Liu and Gu, 1999; 2001a-d). Their struggle has been on the singularity issue in the polynomial PIMs, and different ways to solve the problem have been attempted (Liu, 2002). The use of radial basis function (or together with the polynomials) has well resolved the problem for both the local Petrov-Galerkin weak-form (Liu and Gu, 2001b) and the global Galerkin weak-form (Wang and Liu, 2000; 2001a; 2002). Recently, a meshfree weak-strong (MWS) form formulation based on a combined weak and strong forms (Liu and Gu, 2002; 2003c; 2003; Wu and Liu 2003b) has been proposed. The MWS method uses both MLS and the radial PIM shape functions, and needs only a local background mesh for nodes that is near the natural boundaries of the problem domain.

Table 1.3 Some typical meshfree methods in chronological order

Methods	References	Methods of approximation
Smoothed particle hydrodynamics (SPH)	Lucy, 1977; Gingold and Monaghan, 1977, etc.	Integral representation
Finite point method	Liszka and Orkisz, 1980; Onate et al., 1996, etc.	Finite difference representation
Diffuse element method (DEM)	Nayroles et al., 1992	Moving least square (MLS) approximation Galerkin method
Element free Galerkin (EFG) method	Belytschko et al., 1994, 1996; 1998; etc.	MLS approximation Galerkin method
Reproduced kernel particle method (RKPM)	Liu et al., 1995; 1996, etc.	Integral representation Galerkin method
HP-cloud method	Duarte and Oden, 1996, etc.	MLS approximation, Partition of unity
Free mesh method	Yagawa and Yamada, 1996; 1998, etc.	Galerkin method
Meshless local Petrov-Galerkin (MLPG) method	Atluri and Zhu, 1998; 1999; Atluri and Shen, 2002; etc.	MLS approximation Petrov-Galerkin method
Point interpolation method (PIM)	Liu and Gu, 1999; 2001a-d; Gu and Liu, 2001a,c; Liu,2002; Wang and Liu, 2000; 2001; 2002	Point interpolation, (Radial and Polynomial basis), Galerkin method, Petrov-Galerkin method
Meshfree weak-strong form (MWS)	Liu and Gu, 2002; 2003c; 2003; etc.	MLS, PIM, radial PIM (RPIM), Collocation plus Petrov-Galerkin

Other notable representatives of meshfree methods include the HP-cloud method by Duarte and Oden (1996), free mesh method (FMM) by Yagawa and Yamada (1996; 1998), the point assembly method by Liu (1999), etc. Some typical meshfree methods either in strong or weak form are listed in Table 1.3.

Comprehensive investigations on meshfree methods are closely related to the applications to complex computational solid and fluid mechanics problems. Since the computational frame in the meshfree methods is a set of arbitrarily distributed nodes rather than a system of pre-defined mesh/grid, the meshfree methods are attractive in dealing with problems that are difficult for traditional grid-based methods. The interesting applications of meshfree methods include large deformation analyses in solids (Chen et al., 1996; 1997; 1998; Jun et al., 1998; Li et al., 2000a, b; 2002; etc.), vibration analyses especially for plates and shells (Gu and Liu, 2001d,e; Liu and Gu, 2000a; Liu and Chen, 2001; Liu and Tan, 2002; L. Liu et al., 2002; Dai et al., 2002; etc.), structure buckling problems (Liu and Chen, 2002), piezoelectric structure simulations (Liu, Dai et al., 2002a, b), non-linear foundation consolidation problems (Wang et al. 2000; 2001a,b; 2002), incompressible flows (Lin and Atluri, 2001; Wu and Liu, 2003a,b; etc.), singular boundary-value problems (X. Liu et al., 2002), and impact and explosion simulation that will be discussed in the following chapters.

Meshfree methods have also been developed for boundary integral equations to develop boundary meshfree methods, in which only the boundary of the problem domain needs to be represented with nodes. The formulation developed by Mukherjee, et al. (1997a; b) was based on the formulation of EFG using the MLS approximation. Boundary point interpolation methods (BPIM) were developed by Gu and Liu using polynomial PIM and radial PIM interpolations (Gu and Liu, 2001; 2002; Liu and Gu, 2003b), which give much a set of much smaller discretized system equations due to the delta function property of the PIM shape functions.

In practical applications, a meshfree method can be coupled with other meshfree methods or a conventional numerical method to take the full advantages of each method. Examples include SPH coupling with FEM (Attaway et al., 1994; Johnson, 1994; Century Dynamics, 1997), EFG coupling with boundary element method (BEM) (Gu and Liu, 2001b; Liu and Gu, 2000b), and MPLG coupling with BEM or FEM (Liu and Gu, 2000d). A meshfree method can also be coupled with another meshfree method for particular applications (Liu and Gu, 2000c). An adaptive stress analysis package based on the meshfree technology, MFree2D$^\copyright$, has been developed (Liu and Tu, 2001; 2002; Liu, 2002).

There are basically three types of meshfree methods: methods based on strong form formulations, methods based on weak form formulations, and particle methods. The *strong form method* such as the *collocation method* has attractive advantages of being simple to implement, computationally efficient and "truly" meshfree, because no integration is required in establishing the discrete system equations. However, they are often unstable and less accurate, especially when irregularly distributed nodes are used for problems governed by partial differential equations with Neumann (derivative) boundary conditions, such as solid mechanics problems with stress (natural) boundary conditions. On the other hand, *weak form method* such as the EFG, MLPG and PIM has the advantages of excellent stability, accuracy. The Neumann boundary conditions can be naturally satisfied due to the use of the weak form that involves smoothing (integral) operators[3]. However, the weak form method is said not to be "truly" meshfree, as a background mesh (local or global) is required for the integration of the weak forms.

Recently, a novel meshfree weak-strong (MWS) form method is proposed by Liu and Gu (2002; 2003c; 2003) based on a combined formulation of both the strong form and the local weak form. In the MWS method, the strong form formulation is used for all the internal nodes and the nodes on the essential boundaries. The local weak form (Petrov-Galerkin weak form) is used only for nodes near the natural boundaries. Hence, there is no need for numerical integrations for all the internal nodes and the nodes on the essential boundaries. The numerical integration is performed locally only for the nodes on the natural boundaries and thus only local background cells for the nodes near the natural boundaries are required. The locally supported radial point interpolation and the moving least squares (MLS) approximation have been used to construct the meshfree shape functions for the MWS. The final system matrices were sparse and banded for computational efficiency. The MWS method is, so far, the meshfree method that uses least meshes in the entire computation and produce stable solutions even for solid mechanics problems using irregularly distributed nodes. It is one more step close to realize the dream of the "truly" meshfree method that is capable of producing stable and accurate solutions for solid mechanics problems using irregularly distributed nodes.

Some excellent reviews on the meshfree and particle methods can be found in papers by Belytschko et al. (1996), Li and Liu (2002), etc. In the monograph on meshfree methods, Liu (2002) has comprehensively addressed the history, development, theory and applications of the major existing meshfree methods. According to Liu (2002), the theories of function approximations used in the meshfree methods can be classified into three major categories: integral representation methods, series representation methods, and differential representation methods.

This book focuses on meshfree particle methods that in fact are the earliest class of meshfree methods, particularly on the smoothed particle hydrodynamics (SPH) method that uses the integral representation method for field function approximation. As mentioned by Liu (2002, page 54), the SPH method is

---

[3] The integral operator is know as a smoothing operator that can smear out the errors contained in the function been operated on. (c.f., e.g., Chapter 2 of a monograph by Liu and Han (2003)).

actually very much similar to the weak form method. The difference is that the weak form operation (integral operation) is implemented in the stage of function approximation in the SPH, rather than in the stage of creating the discrete system equation as in the usual weak form method (EFG, MLPG, RKPIM, PIM, etc.). The use of the integral representation of field functions passes the differentiation operations[4] on the field function to the smoothing (weight) function. It reduces the requirement on the order of continuity of the approximated field function. Therefore, the SPH has been proven stable for arbitrarily distributed nodes for many problems with extremely large deformations. The accuracy of the solution in the SPH method depends, naturally, very much on the choice of the smoothing function (see Chapter 3).

1.4   Meshfree particle methods (MPMs)

A meshfree particle method (MPM) in general refers to the class of meshfree methods that employ a set of finite number of discrete particles to represent the state of a system and to record the movement of the system. Each particle can either be directly associated with one discrete physical object, or be generated to represent a part of the continuum problem domain. The particles can range from very small (nano or micro) scale, to meso scale, to macro scale, and even to astronomical scale. For CFD problems, each particle possesses a set of field variables such as mass, momentum, energy, position etc, and other variables (e.g., charge, vorticity, etc.) related to the specific problem. The evolution of the physical system is determined by the conservation of mass, momentum and energy. Some typical particle methods or particle-like methods are listed in Table 1.4.

Based on the *length scale*, the meshfree particle methods can be roughly divided into three classes: atomistic/microscopic scale meshfree particle methods, mesoscopic meshfree particle methods, and macroscopic meshfree particle methods.

A typical atomistic MPM is the molecular dynamics (MD) method, either *ab inito* or classic that uses force potential functions. Mesoscopic MPMs include dissipative particle dynamics (DPD) (Hoogerbrugge and Koelman, 1992; Español, 1998), lattice gas Cellular Automata (CA) (Wolfram, 1983; Kandanoff et al. 1989) etc. Macroscopic MPMs includes SPH (Lucy, 1977; Gingold and Monaghan, 1977), Particle-in-Cell (PIC) (Harlow, 1963; 1964), Marker-and-Cell (MAC) (Harlow, 1964), Fluid-in-Cell (FLIC) (Gentry et al., 1966), MPS

---

[4] The differential operator is know as a harshening operator that can magnify the errors contained in the function been operated on. (see, e.g., Chapter 2 of a monograph by Liu and Han (2003)).

(Moving Particle Semi-implicit) (Koshizuka et al., 1998), particle-particle (PP), particle-mesh (PM), particle-particle-mesh (P^3M) (Hockney and Eastwood, 1988), and other various meshfree particle models.

Table 1.4 Some typical particle methods.

Methods	References
Molecular dynamics (MD)	Alder and Wainright ,1957; Rahman, 1964; Stillinger and Rahman, 1974; etc.
Monte Carlo (MC)	Metropolis and Ulam, 1949; Binder, 1988, 1992; etc.
Direct simulation Monte Carlo (DSMC)	Bird, 1994; Pan et al., 1999, 2000, 2002; etc.
Dissipative particle dynamics (DPD)	Hoogerbrugge and Koelman, 1992; Español, 1998; etc.
Lattice gas Cellular Automata (CA)	Wolfram, 1983; Kandanoff et al. 1989; etc.
Lattice Bolztmann equation (LBE)	Chen and Doolen, 1998; Qian et al., 2000; etc.
Particle-in-Cell (PIC)	Harlow, 1963; 1964; etc.
Marker-and-Cell (MAC)	Harlow, 1964; etc.
Fluid-in-Cell (FLIC)	Gentry et al., 1966; etc.
Moving Particle Semi-implicit (MPS)	Koshizuka et al., 1998; etc.
Discrete element method (DEM)	Cundall, 1987; Owen, 1996; etc.
Vortex methods	Chorin, 1973; Leonard, 1980; etc.
Smoothed particle hydrodynamics (SPH)	Lucy, 1977; Gingold and Monaghan, 1977; etc.

Many MPMs were initially developed for systems with discrete particles. Examples include SPH simulation of the interaction of stars in astrophysics, MD simulation of movement of millions of atoms in an equilibrium or non-equilibrium state, discrete element method (DEM) simulation of soils and sands (Cundall, 1987), etc. MPMs have also been modified, extended, and applied to system of continuum media. In such cases, an additional operation is required to generate a set of particles to represent the continuum media. Each particle represents a part of the problem domain, with some attributes such as mass, position, momentum and energy concentrated on the mass or geometric center of the sub-domain. Examples include SPH, PIC, MAC, FLIC, MPS, vortex methods (Chorin, 1973; Leonard, 1980), corrective smoothed particle method (CSPM) (Chen et al., 1999a-c; 2000), and many others.

According to the mathematical model used, the meshfree particle methods can be deterministic or probabilistic. The deterministic MPMs deal directly with the governing system equations of physical law. In the deterministic MPMs, once the initial and boundary conditions are given, the particle evolution in the later time stages theoretically can be precisely determined based on the physical laws that govern the problem. Many meshfree particle methods are inherent with probabilistic nature based on statistical principles. Main representatives of the probabilistic MPMs include MD, MD based Monte Carlo (MC) (Metropolis and Ulam, 1949; Binder, 1988; 1992), DPD, direct simulation Monte Carlo (DSMC) (Bird, 1994; Pan et al., 1999; 2000; 2002); lattice gas Cellular Automata, and lattice Bolztmann equation (LBE) (Chen and Doolen, 1998; Qian et al., 2000), etc.

It is noted that some meshfree particle methods can have mixed features. One typical example is the SPH method. It was originally invented to simulate astrophysical problems, and is currently being applied to macroscopic continuum problems of computational solid and fluid mechanics, and to atomistic scale simulations (Nitsche and Zhang, 2002; Liu et al., 2002b). The SPH method is now widely used for both discrete particle systems and continuum systems. It was actually initially developed as a probabilistic meshfree particle method (Lucy, 1977), and was later modified and applied as a deterministic meshfree particle method.

There are some other points that need to be further clarified. Firstly, most meshfree particle methods are inherently Lagrangian methods, in which the particles represent the physical system moving in the Lagrangian frame according to the internal interactions and external force, and thus evolve the system in time. There are also examples in which the particles are fixed in the Eulerian space as interpolation points rather than moving objects (Laguna, 1995). Secondly, most meshfree particle methods use explicit methods for the time integration. Some exceptions, however, use implicit or semi-implicit procedures such as the MPS (Koshizuka et al.; 1998), in which the pressure term in the momentum equations are implicitly determined. Thirdly, most of the particle methods are basically meshfree methods, in which the particles form the computational frame for the field variable approximation. However, there are some particle methods that still use some kind of mesh (e.g. PIC, LBE etc.) for background interpolation, or for other purposes. Lastly, in using particles to represent a continuum domain, some kind of mesh may be needed to generate the initial distribution of particles.

The advantages of the MPMs over conventional grid-based numerical methods can be roughly summarized as follows:

1. In the MPMs, the problem domain is discretized with particles without a fixed connectivity. Treatment of large deformation is relatively much easier;
2. Discretization of complex geometry for the MPMs is relatively simpler as only an initial discretization is required;
3. Refinement of the particles is expected much easier to perform than the mesh refinement;
4. It is easy to obtain the features of the entire physical system through tracing the motion of the particles. Therefore, identifying free surfaces, moving interfaces and deformable boundaries is no longer a tough task. Time history of field variables at any point on the material can also be naturally obtained.

1.5   Solution strategy of MPMs

The MPMs aim to perform numerical analyses for complex problems without the use of a mesh pre-defined using a connectivity of nodes. Similar to the simulations with the conventional grid-based numerical methods, a typical simulation using MPMs also involves

1) governing equations with proper boundary conditions (BC) and/or initial conditions (IC),
2) domain discretization technique for creating particles,
3) numerical discretization technique (weak form, strong form, particle methods),
4) numerical technique to solve the resultant algebraic equations or ordinary differential equations (ODE).

It is seen that the differences between the grid-based methods and MPMs lie in:

- The problem domain is discretized with or represented by particles (spatial discretization with *particle representation*);

- Functions, derivatives and integrals in the governing equations are approximated using the particles rather than over a mesh (numerical discretization with *particle approximation*).

1.5.1 Particle representation

In the MPMs, there is no need to prescribe the connectivity between the particles. All one needs is an initial distribution or generation of the particles that represent the problem domain, if the problem domain is not initially in discrete particle form. Different ways of generating particles for continuous domains can be employed. Since mesh generation algorithms (e.g. triangulation algorithm) are readily available for both the 2D and 3D space, one simple approach is to deploy particles in the mass or geometric centers of the mesh cells, as shown in Figure 1.6 and Figure 1.7.

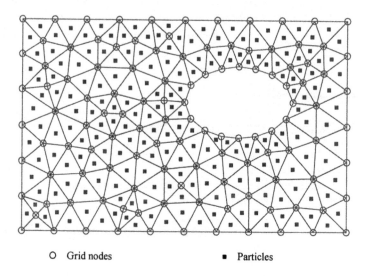

O   Grid nodes   ■   Particles

Figure 1.6 Initial particle generation for a continuum using a triangular mesh in two-dimensional space.

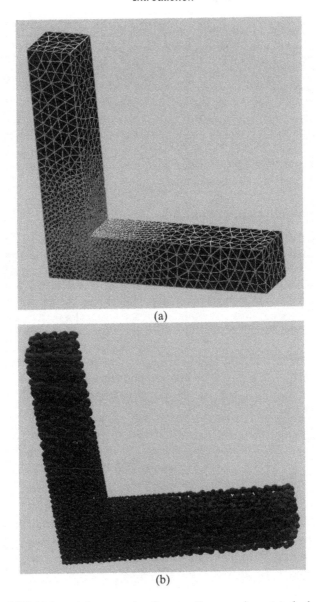

(a)

(b)

Figure 1.7 Initial particle generation for a continuum using a tetrahedron mesh in three-dimensional space. (a) tetrahedron mesh, (b) generated particles.

The triangular or tetrahedron meshes are preferred, because they can always be generated automatically for domains of complex geometry using existing mesh generation methods that are even commercially available. For domains of simple geometry, quadrilateral or hexahedron meshes can also be used for particle generation.

Note that when the particles are placed at the mass centers of the corresponding cells, the problem domain is approximated, and the original smoothing surface becomes a rough surface. Locating the particles at the nodes of the mesh can provide an apparently smoother surface. Place particles at both the centers and the nodes of the cells may be another alternative to provide a good representation for the continuum.

1.5.2   Particle approximation

Numerical discretization in the MPMs involves approximating the values of functions, derivatives and integrals at a particle with particle approximations using the information at all the neighboring particles that have influence on the particle. The area of influence of a particle is determined by the so-called influence domain or support domain (refer to the next chapter for more details). The neighboring particles within the support domain of a particle provide all the necessary and sufficient information for the field variable approximations at the particle.

For example, the field variable (e.g., a component of the velocity) u for a particle located at $x = (x, y, z)$ within the problem domain can be approximated using the information on the particles within the support domain of the particle at x (Figure 1.8):

$$u(x) = \sum_{i=1}^{N} \phi_i(x) u_i \qquad (1.2)$$

where N is the number of particles within the support domain of the particle at x; u_i is the field variable at particle i, ϕ_i is the shape function at particle i constructed using the information on all particles within the support domain of the particle at x. The shape functions can then be used for establishing a set of discretized system equations using weak form, strong form, or both.

1.5.3   Solution procedure of MPMs

The procedure in a MPM simulation is mainly similar to that in a grid-based numerical simulation, except for the particle representation and the particle approximation. A typical MPM simulation procedure for dynamic fluid flow problems can be briefed as follows:

Meshfree particles

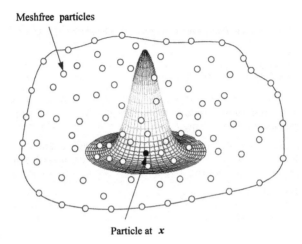

Particle at *x*

Figure 1.8 The support domain of a particle at *x* .

1. Represent the problem domain with particles so that the computational information is known at the discrete particles at an initial instant *t* with a proper treatment on the boundary conditions.
2. Discretize the derivatives or integrals in the governing equations with proper particle approximations;
3. From the given velocity and/or position, calculate the strain rate and/or strain, and then calculate the stress at each discrete particles at the instant *t* ;
4. Calculate the acceleration at each discrete particles using the calculated stress;
5. Use the acceleration at the instant *t* to calculate the new velocities and the new positions at time instant $t + \Delta t$, where Δt is the incremental time step;
6. From new velocities and/or new positions, calculate the new strain rate and/or new strain at time instant $t + \Delta t$, and then calculate the new stress at time instant $t + \Delta t$. Repeat step 4, 5 and 6 to march forward in time until the final specified time instant.

1.6   Smoothed particle hydrodynamics (SPH)

1.6.1   The SPH method

In the SPH method, the state of a system is represented by a set of particles, which possess individual material properties and move according to the governing conservation equations (see, Figure 1.9). Since its invention to solve astrophysical problems in three-dimensional open space (Lucy, 1977; Gingold and Monaghan, 1977), SPH has been extensively studied and extended to dynamic response with material strength as well as dynamic fluid flows with large deformations.

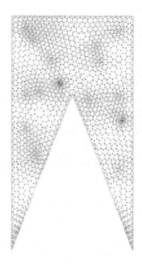

Figure 1.9 SPH particles used in a shaped charge detonation simulation. The particles are irregularly distributed.

Smoothed particle hydrodynamics, as a meshfree, Lagrangian, particle method, has its particular characteristics. It has some special advantages over the traditional grid-based numerical methods, the most significant one among which is the *adaptive* nature of the SPH method. This adaptability of SPH is achieved at the very early stage of the field variable approximation that is performed at *each time step* based on a *current local set* of *arbitrarily distributed* particles. Because of this *adaptive* nature of the SPH approximation, the formulation of SPH is not affected by the arbitrariness of the particle distribution. Therefore, it can naturally handle problems with extremely large deformation. This is, therefore, the most attractive feature of the SPH method.

The meshfree nature of the SPH method is also due to the above-mentioned adaptive formulation and the use of particles to represent the problem domain and to act as the computational frame for field variable approximations. The SPH approximation does not require a pre-defined mesh to provide any connective of the particles in the process of computation, and it works well even without any particle refinement operation. This meshfree nature is very attractive for problems where the traditional FEM or FDM encounters difficulties mentioned earlier.

Besides the meshfree and adaptive nature, another exciting and attractive feature of the SPH method is the harmonic combination of the Lagrangian formulation and particle approximation. Unlike the meshfree nodes in other meshfree methods, which are only used as interpolation points, the SPH particles also carry material properties, and are allowed to move in light of the internal interactions and external forces. Functioning as both approximation points and material components, the SPH particles seem to be endowed with life.

The pith and marrow of the method are fully embodied in the three terms of *SMOOTHED PARTICLE HYDRODYNAMICS*. The first term SMOOTHED represents the *smoothed* approximation nature by using the weighted average over the neighboring particles for stability; the third term HYDRODYNAMICS is the right niche of the method in the application to hydrodynamics problems. It is this harmonic combination of the adaptive, Lagrangian and particle nature in the SPH method that leads to various practical applications in different areas in engineering and science.

1.6.2   Briefing on the history of the SPH method

Smoothed particle hydrodynamics (SPH) is a meshfree, Lagrangian particle method for modeling fluid flows. SPH was first invented to solve astrophysical problems in three-dimensional open space, in particular polytropes (Lucy, 1977; Gingold and Monaghan, 1977), since the collective movement of those particles is similar to the movement of a liquid or gas flow, and it can be modeled by the governing equations of the classical Newtonian hydrodynamics.

Although the traditional grid-based numerical methods such as the finite difference methods (FDM) and finite element methods (FEM) exist, and in some circumstances, are better developed methods than the SPH method, they have difficulties in handling some complex phenomena as discussed earlier. This motivated researchers to seek for alternatives to solve these kinds of problems, and the SPH method has then become a good choice. Several review papers on the SPH method have been published, including those by Benz (1989; 1990) and Monaghan (1992).

Unlike the famous particle-in-cell (PIC) method developed in early 1960s (Harlow, 1963; 1964; Brackbill et al., 1988; Munz et al., 1999; Cushman et al., 2000), the SPH method does not need a grid/mesh to calculate the spatial derivatives. These particles are capable of moving in the space, carry all the computational information, and thus form the computational frame for solving the partial differential equations describing the conservation laws of the continuum fluid dynamics.

Invention

Smoothed particle hydrodynamics (SPH) is a meshfree, adaptive, Lagrangian particle method for modeling fluid flows. SPH was first invented to solve astrophysical problems in three-dimensional open space, in particular polytropes (Lucy, 1977; Gingold and Monaghan, 1977), since the collective movement of those particles is similar to the movement of a liquid or gas flow, and it can be modeled by the governing equations of the classical Newtonian hydrodynamics.

Extensions

Today, the SPH method is being used in many areas, such as the simulations of binary stars and stellar collisions (Benz, 1988; 1990; Monaghan, 1992; Frederic et al., 1999), supernova (Hultman and Pharayn, 1999), collapse as well as the formation of galaxies (Monaghan and Lattanzio, 1991; Berczik and Kolesnik, 1993; 1998; Berczik, 2000), coalescence of black holes with neutron stars (Lee, 1998; 2000), single and multiple detonation of white dwarfs (Senz et al., 1999), even the evolution of the universe (Monaghan, 1990). The SPH method has also been applied extensively to a vast range of problems in either computational fluid or solid mechanics because of relatively strong ability to incorporate complicated physical effects into the SPH formulations. In some sense, the term hydrodynamics may be interpreted as mechanics in general. When the SPH approximation is used to create point-dependent shape functions (see, Chapter 3), it can be applied to other areas of mechanics rather than classical hydrodynamics. Hence in some literatures (Kum et al., 1995, Posch et al. 1995), it is called Smoothed Particle Mechanics.

Applications

The earliest applications of SPH were mainly focused on fluid dynamics related areas. These include elastic flow (Swegle, 1992), magneto-hydrodynamics (Morris, 1996), multi-phase flows (Monaghan and Kocharyan, 1995), quasi-incompressible flows (Monaghan, 1994; Morris et al., 1997), gravity currents (Monaghan, 1995a), flow through porous media (Morris et al. 1999; Zhu et al. 1999), heat conduction (Monaghan, 1995b; Chen et al., 1999a), shock simulations (Monaghan and Gingold, 1983; Monaghan, 1987; 1989; Morris and Monaghan, 1997), heat transfer and mass flow (Cleary, 1998), ice and cohesive grains (Gutfraind and Savage, 1998; Oger and Savage, 1999). Benz and Asphaug (1993; 1994; 1995) extended SPH to the simulation of the fracture of brittle solids. Bonet and Kulasegaram (2000) applied SPH to the simulation of metal forming. The SPH method has been very attractive in simulating large deformation and impulsive loading events. One significant application area is high (or hyper) velocity impact (HVI) problems concerning the effects of projectiles impacting upon space assets (satellites, space stations, shuttles). In HVI problems, shock waves propagate through the colliding bodies, which behave like fluids (Zukas, 1982, 1990). Libersky and his co-workers (Libersky et al., 1991; 1993; 1995; Randles and Libersky, 1996; Randles et al., 1995a, b) and Johnson et al. (1993; 1996a, b) have made outstanding contributions in the application of SPH to impact problems. Another important application of the SPH method is the explosion phenomena arising from the detonation of high explosive (HE). Swegle and Attaway (1995) have investigated the feasibility of using the SPH method for underwater explosion calculations. Recently, Liu and his co-workers have applied the SPH method to model a series of explosion phenomena including high explosive detonation, explosion, underwater shock, and water mitigation of shocks (Liu, et al. 2000; 2002a; 2003a-f).

The application of SPH to a wide range of problems has led to significant extensions and improvements of the original SPH method. The numerical aspects have been gradually improved, some inherent drawbacks of SPH were identified, and modified techniques or corrective methods were also proposed. Swegle et al. (1995) identified the tensile instability problem that can be important for materials with strength; Morris (1996) identified the particle inconsistency problem that can lead to poor accuracy in the solution. Over the past years, different modifications or corrections have been tried to restore the consistency and to improve the accuracy of the SPH method. These modifications lead to various versions of the SPH methods and corresponding formulations. Monaghan (1988; 1982; 1985) proposed symmetrization formulations that were reported to have better effects. Johnson and Beissel (Johnson et al., 1996; Johnson and Beissel, 1996) gave an axis symmetry normalization formulation so that, for velocity fields that yield constant values of normal velocity strains, the normal velocity strains can be exactly reproduced. W. K. Liu et al. (Liu and Chen, 1995) presented the reproducing kernel particle method (RKPM) that can result in better accuracy in the particle approximation. Chen et al. (1999a; b; c) proposed a corrective smoothed particle method (CSPM) which improves the simulation accuracy both inside the problem domain and around the boundary area. Randles and Libersky (2000) extended the stress point method (Dyka and Ingel, 1995; Dyka et al., 1997) to multi-dimensional space to improve the tensile instability and zero energy mode problems (Vignjevic et al., 2000). Other notable modifications or corrections of the SPH method include the moving least square particle hydrodynamics (MLSPH) by Dilts (1999; 2000), the integration kernel correction by Bonet and Kulasegaram (2000), and the correction by Belytschko et al. (1998). Presently, the SPH is a method that can simulate general fluid dynamic problems fairly well.

Challenges

Though the SPH method has been extensively applied to different areas, there are still a lot of issues that need to be further investigated. This is especially true in the numerical analyses of the method. Due to the meshfree particle nature of the method, it is not always straightforward to directly apply the techniques that were developed for grid-based Eulerian methods or Lagrangian methods to the SPH method. Some authors have tried to perform theoretical and numerical analyses on SPH (Gingold and Monaghan, 1982; Monaghan, 1982; Morris, 1994; 1996; Balsara, 1995; Meglicki, 1995; Ben et al., 1996a, b; Fulk, 1994; Swegle, et al., 1994; 1995). Through these studies, issues related to the stability, accuracy and convergence properties of the SPH method are gradually becoming understood. However, most of the analyses are based on uniformly distributed particles, and sometimes only for one-dimensional cases, the results obtained by such analyses are often limited to idealized circumstances. For more general cases especially those with large deformations and impulsive loadings where the particles are usually highly disordered, the obtained results may not always be reliable, as it is not yet very clear how the particle irregularity affects the accuracy of the solutions.

There is still a long way for the method to become extensively applicable, practically useful and robust as the traditional grid-based methods such as FEM and FDM. This is because much work needs to be done to consolidate the theoretical foundations of the SPH method, and to remedy its inherent numerical drawbacks. Moreover, there should be a necessary process for any numerical technique to develop, advance, improve, and to be validated so as to be more efficient, robust in practical applications.

1.6.3   The SPH method in this book

As can be seen in the previous discussions, there are so many kinds of meshfree methods for different applications. The purpose of this book is not to provide a comprehensive record of all the emerged meshfree methods. For such purpose, the readers are suggested to refer to the monograph on meshfree methods by Liu (2002) and some other review papers (Belytschko et al., 1996; 1998; Li and Liu, 2002). The emphasis of this book will be on the meshfree particle methods, specifically, the smoothed particle hydrodynamics and its different modifications and variations. Some novel applications of the SPH method will also be addressed in detail. Devoting this volume entirely to the SPH method is based on the following reasons.

1. The SPH method, as the oldest MPM, is quickly approaching to its mature stage;
2. With the continuing improvements and modifications, the accuracy, stability and adaptively of the SPH method have reached an acceptable level for practical engineering applications;
3. Applications of the SPH method are very wide, ranging from CFD to CSM, from micro-scale to macro-scale and to astronomical scale, from discrete systems to continuum systems;
4. Some commercial codes have incorporated the SPH processor into their software packages with many successful practical applications.

There are a number of versions of the SPH method proposed so far. This book provides an introduction to the traditional SPH method and its variations such as the CSPM, DSPH, and ASPH. The theories related to the SPH method will be systematically discussed. A comprehensive but concise collection of numerical techniques is presented. Some important implementation issues are discussed.

A SPH source code in FORTRAN is provided. The SPH code consists of most of the standard SPH techniques, and can be easily extended to other variations of SPH with modifications either on the continuous integral representation or the discretized particle approximation. Releasing the sample source code is to suit the needs of readers for an easy comprehension, understanding, quick implementation, practical applications and further development of the MPMs.

Many novel and interesting applications in the areas related to CFD will be presented. These include

- incompressible flows,
- free surface flows,
- high compressive flows,
- high explosive (HE) detonation,
- HE explosion,
- underwater explosion,
- water mitigation of shocks,
- high velocity impact,
- penetration, and
- multiple scale simulations coupled with atomistic method.

As mentioned in the preceding discussions, a simulation using the SPH method involves two major steps: particle representation and particle approximation. The particle representation is an issue related to only the initial creation of the particles, and it can be solved using the existing software packages commercially available. Therefore, the effort of this book is mainly on the central issue of the SPH particle approximation. The next Chapter will be focusing on the basic ideas and essential formulations of the SPH method that are useful for all different versions of the SPH methods.

Chapter 2

SPH Concept and Essential Formulation

In Chapter 1, it is seen that as a meshfree particle method, the smoothed particle hydrodynamics (SPH) shares many similarities with other particle methods and meshfree methods while possessing its unique features. In this chapter, the basic concept and the essential formulations of the SPH method are introduced. The SPH approximations are discussed, which include the strategy of the SPH method, the continuous integral representation (kernel approximation) and the discretized particle approximation. Some tricks to be used in deriving the SPH formulations for complex PDEs are presented. Concepts on the support domain and the influence domain are described. Differences between the SPH method and other particle methods such as the particle-in-cell (PIC) method are discussed. These concepts, tricks and essential formulations discussed in this chapter are very useful in the development of various variants of the SPH method.

2.1 Basic ideas of SPH

As mentioned in detail in Chapter 1, the SPH method was developed for hydrodynamics problems that are basically in the form of partial differential equations (PDE) of *field variables* such as the density, velocity, energy, etc. Obtaining analytical solutions for such a set of PDEs is not usually possible, except for very few simple cases. Therefore, efforts have been made in seeking for numerical solutions. In doing so, one needs to first discretize the *problem domain* where the PDEs are defined. Next, a method is needed to provide an approximation for the values of the field functions and their derivatives at any point. The *function approximation* is then applied to the PDEs to produce a set of ordinary differential equations (ODE) in a discretized form with respect only to time. This set of discretized ODE can then be solved using one of the standard integration routines of the conventional finite difference method.

In the SPH method the following key ideas are employed to achieve the above-mentioned task.

1. The problem domain is represented by a set of *arbitrarily distributed particles*, if the domain is not yet in the form of particles. No connectivity for these particles is needed. (*Meshfree*)

2. The *integral representation method* is used for field function approximation. This is termed in the SPH method as the *kernel approximation*. (*Integral function representation*)

3. The kernel approximation is then further approximated using particles. This is termed in SPH as *particle approximation*. It is done by replacing the integration in the integral representation of the field function and its derivatives with summations over all the corresponding values at the neighboring particles in a local domain called the *support domain*. (*Compact support*)

4. The particle approximation is performed at every time step, and hence the use of the particles depends on the current local distribution of the particles. (*Adaptive*)

5. The particle approximations are performed to all terms related to field functions in the PDEs to produce a set of ODEs in discretized form with respect to time only. (*Lagrangian*)

6. The ODEs are solved using an *explicit* integration algorithm to achieve fast time stepping, and to obtain the time history of all the field variables for all the particles. (*Dynamic*)

Item 1 determines the meshfree nature of the SPH method. It is not difficult to formulate a numerical method without using a mesh. But the key problem is how to ensure the stability of the numerical solution, especially when the irregular nodes or particles in *compactly supported* domains are used for problems with Neumann (derivative) boundary conditions (Liu and Gu, 2002; 2003).

Item 2 mathematically provides the necessary stability to the SPH method, as the integral representation has a smoothing effect that behaves as a weak form formulation. The weak form formulations are usually very stable as long as the numerical integration is accurately performed.

Item 3 produces banded or sparse discretized system matrices, which are extremely important as far as the computation effort is concerned. This is because problems with large deformation require a huge number (could be millions) of particles to represent the problem domain. It could take unacceptably long CPU time to solve such a large system equations, if the matrices are full.

The adaptability of SPH is achieved by performing item 4 at *each time step* based on particles *arbitrarily distributed* in the *current* support domain. Because of this *adaptive* SPH approximation performed at the very early stage of field variable approximation, the formulation of SPH is not affected by the arbitrariness of the particle distribution that changes with time. Therefore, it can naturally handle problems with extremely large deformation. Note that this particle approximation is, in fact, a method to perform the numerical integration required at item 2. To ensure the accuracy of the integration and hence the numerical stability, sufficient particles (the support domain should be sufficiently large) have to be used in the summation. The particles are now assigned a mass after the particle approximation, meaning that they are actually the physical material particles.

The use of the Lagrangian description at item 5 allows the SPH method to have all the features of any Lagrangian method that were mentioned in Section 1.2.1.

Item 6 is a conventional means of time marching to solve dynamic problems. All one needs to do in the SPH method is to figure out a proper way to determine the time step to ensure stable time integration.

The combination of the 6-point strategy makes the SPH method be a meshfree, adaptive, stable and Lagrangian solver for dynamic problems. The detailed formulation of the SPH method will begin from the next Section, and these features of SPH will be demonstrated in many examples in the later chapters.

2.2   Essential formulation of SPH

2.2.1   Integral representation of a function

The formulation of SPH is often divided into two key steps. The first step is the *integral representation* or the so-called *kernel approximation* of field functions. The second one is the *particle approximation*.

In the first key step, the integration of the multiplication of an arbitrary function and a *smoothing kernel function* gives the kernel approximation in the form of integral representation of the function. More details on the methods of function approximation can be found in a monograph by Liu (2002). The integral representation of the function is then approximated by summing up the values of the nearest neighbor particles, which yields the particle approximation of the function at a discrete point or particle. Though some other names such as *kernel estimate* or *particle estimate* are also used in other literatures, in this book, the term of integral representation or kernel approximation and particle approximation will be used.

The concept of integral representation of a function $f(x)$ used in the SPH method starts from the following identity.

$$f(x) = \int_\Omega f(x')\delta(x-x')dx' \qquad (2.1)$$

where f is a function of the three-dimensional position vector x, and $\delta(x-x')$ is the Dirac delta function given by

$$\delta(x-x') = \begin{cases} 1 & x = x' \\ 0 & x \neq x' \end{cases} \qquad (2.2)$$

In equation (2.1), Ω is the volume of the integral that contains x. Equation (2.1) implies that a function can be represented in an integral form. Since the Dirac delta function is used, the integral representation in equation (2.1) is exact or rigorous, as long as $f(x)$ is defined and continuous in Ω.

If the Delta function kernel $\delta(x-x')$ is replaced by a smoothing function $W(x-x',h)$, the integral representation of $f(x)$ is given by

$$f(x) \doteq \int_\Omega f(x')W(x-x',h)dx' \qquad (2.3)$$

where W is the so-called *smoothing kernel function*, or *smoothing function*, or *smoothing kernel*, or *kernel function* or simply *kernel* in many SPH literatures. This book chooses to use *smoothing function* or *kernel* for short. In the smoothing function, h is the smoothing length defining the influence area of the smoothing function W. Note that as long as W is not the Dirac function, the integral representation in equation (2.3) can only be an approximation. This is the origination of the term of *kernel approximation*.

In the SPH convention, the *kernel approximation operator* is marked by the angle bracket $\diamond$ (see, e.g., Fulk, 1994.), and therefore equation (2.3) is rewritten as

$$< f(x) > = \int_\Omega f(x')W(x-x',h)dx' \qquad (2.4)$$

Note that the equal sign is used in equation (2.4) that is the standard expression of the kernel approximation of a function.

The smoothing function W is *usually* chosen to be an *even function* for reasons given later. It should also satisfy a number of conditions. The first one is the *normalization condition* that states

$$\int_{\Omega} W(x - x', h)dx' = 1 \tag{2.5}$$

This condition is also termed as *unity condition* since the integration of the smoothing function produces the unity.

The second condition is the *Delta function property* that is observed when the smoothing length approaches zero

$$\lim_{h \to 0} W(x - x', h) = \delta(x - x') \tag{2.6}$$

The third condition is the *compact condition*

$$W(x - x', h) = 0 \quad \text{when } |x - x'| > \kappa h \tag{2.7}$$

where κ is a constant related to the smoothing function for point at x, and defines the effective (non-zero) area of the smoothing function. This effective area is called the support domain for the smoothing function of point x (or the support domain of that point). More detailed discussions on the support domain can be found in the next section. Using this compact condition, integration over the entire problem domain is localized as integration over the support domain of the smoothing function. Therefore, the integration domain Ω can be, and usually is the same as the support domain.

In the SPH literatures, the kernel approximation is often said to have h^2 accuracy or second order accuracy (Monaghan, 1982; 1992; Hernquist and Katz; 1989; Fulk, 1994). The reason is given as follows.

Note from equation (2.7) that the support domain of the smoothing function is $|x' - x| \leq \kappa h$, the errors in the SPH integral representation can be roughly estimated using the Taylor series expansion of $f(x')$ around x, where $f(x)$ is differentiable. Using equation (2.4) leads to

$$< f(x) >= \int_\Omega [f(x) + f'(x)(x' - x) + r((x' - x)^2)] W(x - x', h) dx'$$

$$= f(x) \int_\Omega W(x - x', h) dx' \tag{2.8}$$

$$+ f'(x) \int_\Omega (x' - x) W(x - x', h) dx' + r(h^2)$$

where r stands for the residual. Note that W is an even function with respect to x, hence $(x' - x)W(x - x', h)$ should be an odd function, hence we should have

$$\int_\Omega (x' - x) W(x - x', h) dx' = 0 \tag{2.9}$$

Using equations (2.5) and (2.9), equation (2.8) becomes

$$< f(x) >= f(x) + r(h^2) \tag{2.10}$$

From the above equation, it can be seen that, in the SPH method, the integral representation or kernel approximation of a function is of second order accuracy. However, this kernel approximation is not necessarily of second order accuracy if the smoothing function is not an even function, or if the normalization condition is not satisfied. This will be further discussed in Chapter 3 in greater detail.

2.2.2   Integral representation of the derivative of a function

The approximation for the spatial derivative $\nabla \cdot f(x)$ is obtained simply by substituting $f(x)$ with $\nabla \cdot f(x)$ in equation (2.4), which gives

$$< \nabla \cdot f(x) >= \int_\Omega [\nabla \cdot f(x')] W(x - x', h) dx' \tag{2.11}$$

where the divergence in the integral is operated with respect to the primed coordinate. Since

$$[\nabla \cdot f(x')] W(x - x', h) =$$
$$\nabla \cdot [f(x') W(x - x', h)] - f(x') \cdot \nabla W(x - x', h) \tag{2.12}$$

From equation (2.11), the following equation is obtained,

$$< \nabla \cdot f(x) >=$$
$$\int_{\Omega} \nabla \cdot [f(x')W(x-x',h)]dx' - \int_{\Omega} f(x') \cdot \nabla W(x-x',h)dx' \qquad (2.13)$$

The first integral on the right hand side (RHS) of equation (2.13) can be converted using the divergence theorem into an integral over the surface S of the domain of the integration, Ω.

$$< \nabla \cdot f(x) >= \int_{S} f(x')W(x-x',h) \cdot \bar{n}dS - \int_{\Omega} f(x') \cdot \nabla W(x-x',h)dx' \qquad (2.14)$$

where $\bar{n}$ is the unit vector normal to the surface S.

Since the smoothing function W is defined to have compact support, when the support domain is located within the problem domain (Figure 2.1), the surface integral on the right hand side of equation (2.14) is zero. If the support domain overlaps with the problem domain (Figure 2.2), the smoothing function W is truncated by the boundary and the surface integral is no longer zero. Under such circumstances, modifications should be made to remedy the boundary effects if the surface integration is treated as zero in equation (2.14).

Therefore, for those points whose support domain is inside the problem domain, (2.14) is simplified as follows.

$$< \nabla \cdot f(x) >= - \int_{\Omega} f(x') \cdot \nabla W(x-x',h)dx' \qquad (2.15)$$

From the above equation, it can be seen that the differential operation on a function is transmitted to a differential operation on the smoothing function. In other words, the SPH integral representation of the derivative of a field function allows the spatial gradient to be determined from the values of the function and the derivatives of the smoothing function W, rather than from the derivatives of the function itself. This feature is very similar to that in the weak form methods that reduce the consistency requirement on the assumed field functions and produce stable solutions for PDEs (Liu, 2002).

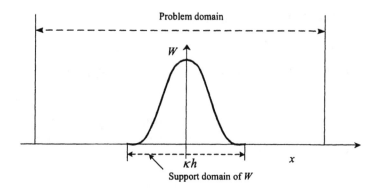

Figure 2.1 The support domain of the smoothing function W and problem domain. The support domain is located within the problem domain. Therefore, the surface integral on the right hand side of equation (2.14) is zero.

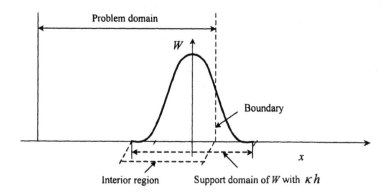

Figure 2.2 The support domain of the smoothing function W and problem domain. The support domain intersects with the problem domain. Therefore, the smoothing function W is truncated by the boundary, and the surface integral on the right hand side of equation (2.14) is on longer zero.

2.2.3 Particle approximation

In the SPH method, the entire system is represented by a finite number of particles that carry individual mass and occupy individual space. This is achieved by the following *particle approximation*, which is another key operation in the SPH methods.

The continuous integral representations concerning the SPH kernel approximation (expressed in equations (2.4) and (2.15)) can be converted to discretized forms of summation over all the particles in the support domain shown in Figure 2.3. The corresponding discretized process of summation over the particles is commonly known as particle approximation in the SPH literatures. This process is carried out as follows.

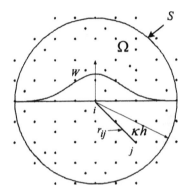

Figure 2.3 Particle approximations using particles within the support domain of the smoothing function W for particle i. The support domain is circular with a radius of κh.

If the infinitesimal volume dx' in the above integrations at the location of particle j is replaced by the finite volume of the particle ΔV_j that is related to the mass of the particles m_j by

$$m_j = \Delta V_j \rho_j \qquad (2.16)$$

where ρ_j is the density of particle j (=1, 2, ..., N) in which N is the number of particles within the support domain of particle j.

The continuous SPH integral representation for $f(x)$ can be written in the following form of discretized *particle approximation*.

$$f(x) = \int_{\Omega} f(x')W(x-x',h)dx'$$

$$\cong \sum_{j=1}^{N} f(x_j)W(x-x_j,h)\Delta V_j$$

$$= \sum_{j=1}^{N} f(x_j)W(x-x_j,h)\frac{1}{\rho_j}(\rho_j\Delta V_j) \qquad (2.17)$$

$$= \sum_{j=1}^{N} f(x_j)W(x-x_j,h)\frac{1}{\rho_j}(m_j)$$

or

$$f(x) = \sum_{j=1}^{N} \frac{m_j}{\rho_j} f(x_j)W(x-x_j,h) \qquad (2.18)$$

Note that the particle approximation is performed at the second step in the above derivation.

The particle approximation for a function at particle i can finally be written as

$$<f(x_i)> = \sum_{j=1}^{N} \frac{m_j}{\rho_j} f(x_j) \cdot W_{ij} \qquad (2.19)$$

where

$$W_{ij} = W(x_i - x_j, h) \qquad (2.20)$$

Equation (2.19) states that the value of a function at particle i is approximated using the average of those values of the function at all the particles in the support domain of particle i weighted by the smoothing function.

Following the same argument, the particle approximation for the spatial derivative of the function is

$$<\nabla \cdot f(x)> = -\sum_{j=1}^{N} \frac{m_j}{\rho_j} f(x_j) \cdot \nabla W(x-x_j,h) \qquad (2.21)$$

where the gradient ∇W in the above equation is taken with respect to the particle j. The particle approximation for a function at particle i can finally be written as

$$< \nabla \cdot f(x_i) >= -\sum_{j=1}^{N} \frac{m_j}{\rho_j} f(x_j) \cdot \nabla_i W_{ij} \qquad (2.22)$$

where

$$\nabla_i W_{ij} = \frac{x_i - x_j}{r_{ij}} \frac{\partial W_{ij}}{\partial r_{ij}} = \frac{x_{ij}}{r_{ij}} \frac{\partial W_{ij}}{\partial r_{ij}} \qquad (2.23)$$

Equation (2.22) states that the value of the gradient of a function at particle i is approximated using the average of those values of the function at all the particles in the support domain of particle i weighted by the gradient of the smoothing function.

It can be seen that the particle approximation in equation (2.19 and (2.21) actually converts the continuous integral representations of a function and its derivatives to the discretized summations based on an arbitrarily set of particles. This use of particle summations to approximate the integral is, in fact, a key approximation that makes the SPH method simple without using a background mesh for numerical integration[1].

Note that the particle approximation introduces the *mass* and *density* of the particle into the equations. This can be conveniently applied to hydrodynamic problems in which the density is a key field variable. This is probably one of the major reasons for the SPH method being particularly popular for dynamic fluid flow problems. If the SPH particle approximation is applied to solid mechanics problems, special treatments are required. One of the ways is to use the SPH approximation to create shape functions, and to establish the discrete system equations (see, e.g., Liu 2002).

The particle approximation is, however, related to some numerical problems inherent in the SPH method, such as the particle inconsistency and the tensile instability. It may be mentioned here that the number of the sampling points for integration should be more than the field nodes (particles). This is at least true for meshfree methods based on weak forms for solid mechanics problems (see, Example 6.2 in the monograph by Liu, 2002). Otherwise, it may (not always) lead to some kind of instability problems. For meshfree methods based on weak forms, the stability may be restored using stabilization terms in the weak form

---

[1] A background mesh is often needed for many meshfree methods for performing the integration resulted from the use of weak forms (see, e.g., Liu 2002).

for solid mechanics (Beissel and Belytschko, 1996). Some possible ways of solving the tensile instability is mentioned in Section 8.4.

In summary, for a given particle i, according to the particle approximation, the value of a function and its derivative for particle i are approximated as

$$< f(x_i) > = \sum_{j=1}^{N} \frac{m_j}{\rho_j} f(x_j) W_{ij} \tag{2.24}$$

$$< \nabla \cdot f(x_i) > = \sum_{j=1}^{N} \frac{m_j}{\rho_j} f(x_j) \cdot \nabla_i W_{ij} \tag{2.25}$$

$$W_{ij} = W(x_i - x_j, h) = W(|x_i - x_j|, h) \tag{2.26}$$

$$\nabla_i W_{ij} = \frac{x_i - x_j}{r_{ij}} \frac{\partial W_{ij}}{\partial r_{ij}} = \frac{x_{ij}}{r_{ij}} \frac{\partial W_{ij}}{\partial r_{ij}} \tag{2.27}$$

where r_{ij} is the distance between particle i and j. It should be noted that $\nabla_i W_{ij}$ is taken with respect to particle i, so the negative sign in equation (2.21) is removed in equation (2.25).

It should be noted that if substituting the function $f(x)$ with the density function ρ in equation (2.24), the SPH approximation for the density is obtained as

$$\rho_i = \sum_{j=1}^{N} m_j W_{ij} \tag{2.28}$$

Equation (2.28) is one of the most popular forms of obtaining density in SPH, and is generally referred to as *summation density approach*.

Note that W_{ij} has a unit of the inverse of volume. Equation (2.28) states that the density of a particle is a weighted average of those of all the particles in its support domain. More detailed discussions on the density evolution in the SPH method can found in Chapter 4.

2.2.4   Some techniques in deriving SPH formulations

By using the above-described procedure of kernel approximation and particle approximation, SPH formulations for partial differential equations (PDEs) can always be derived. There are in fact a number of ways to derive SPH formulation of PDEs. One approach (Benz, 1990) to derive the SPH equations for PDEs is to multiply each term with the smoothing function, integrate over the volume with the use of integration by parts and Taylor expansions. Monaghan (1992) used a more straightforward approach of directly using equations (2.24) and (2.25). In that approach, the following two identifies are employed to place the density inside the gradient operator,

$$\nabla \cdot f(x) = \frac{1}{\rho}[\nabla \cdot (\rho f(x)) - f(x) \cdot \nabla \rho] \tag{2.29}$$

$$\nabla \cdot f(x) = \rho[\nabla \cdot (\frac{f(x)}{\rho}) + \frac{f(x)}{\rho^2} \cdot \nabla \rho] \tag{2.30}$$

The above two identities may be substituted into the integral in equation (2.11). The same procedure of the particle approximation to obtain equation (2.25) is applied to each gradient term on the right hand side of equations (2.29) and (2.30). Note that each expression at the outside of every gradient term is evaluated at the particle itself, the results from equations (2.29) and (2.30) for the divergence of $f(x)$ at particle i are obtained as

$$\nabla \cdot f(x_i) = \frac{1}{\rho_i}\left[\sum_{j=1}^{N} m_j \left[f(x_j) - f(x_i)\right] \cdot \nabla_i W_{ij}\right] \tag{2.31}$$

$$\nabla \cdot f(x_i) = \rho_i\left[\sum_{j=1}^{N} m_j \left[(\frac{f(x_j)}{\rho_j^2}) + (\frac{f(x_i)}{\rho_i^2})\right] \cdot \nabla_i W_{ij}\right] \tag{2.32}$$

One of the good features for the above two equations is that the field function $f(x)$ appears in the form of paired particles.

Besides the above-mentioned two identities, some other rules of operation can be convenient to use in deriving the SPH formulations for a complex system equations. For two arbitrary functions of field variables f_1 and f_2, the following rules exist.

$$\langle f_1 + f_2 \rangle = \langle f_1 \rangle + \langle f_2 \rangle \tag{2.33}$$

$$\langle f_1 f_2 \rangle = \langle f_1 \rangle \langle f_2 \rangle \tag{2.34}$$

Hence, an SPH approximation of the sum of functions equals to the sum of the SPH approximations of the individual function, and an SPH approximation of a product of functions equals to the product of the SPH approximations of the individual function.

If f_1 is a constant denoted by c, we should have

$$\langle cf_2 \rangle = c \langle f_2 \rangle \qquad (2.35)$$

Equations (2.33) and (2.35) state that the SPH approximation operator is a linear operator. It is also easy to see that the SPH approximation operator is commutative, i.e.,

$$\langle f_1 + f_2 \rangle = \langle f_2 + f_1 \rangle \qquad (2.36)$$

and

$$\langle f_1 f_2 \rangle = \langle f_2 f_1 \rangle \qquad (2.37)$$

2.3   Other fundamental issues

2.3.1   Support and influence domain

In the above discussion, we used the concept of support domain. There is another concept of influence domain widely used. However, the concepts of support and influence domains are somewhat confusing and may need to be further clarified. The following is the definition given by Liu (2002) in the general context of meshfree methods.

By definition, the support domain for a field point at $x = (x, y, z)$ is the domain where the information for all the points inside this domain is used to determine the information at the point at x. The influence domain is defined as a domain where a node exerts its influences. Hence, the influence domain is associated with a node in the meshfree methods, and the support domain goes with any field point x, which can be, but does not necessarily have to be a node. From the definition, it is seen that

- When the concept of support domain is used, the consideration is based on a field point x; when the concept of influence domain is used, the consideration is based only on the nodes.

- If a node i is within the support domain of point x, then node i exert an influence on point x, and thus point x is within the influence domain of node i.

- If the field point x happens to be a node i, then this node will have a support domain and an influence domain (such as the particles in the SPH method). For such circumstances, the support domain of the node i at point x can be the same as its influence domain.

The concepts of support and influence domain can lead to different ways of implementation and coding in meshfree methods.

The support domain in meshfree methods for a point can be local as a sub-region of the entire problem domain, or be global as the entire problem domain. In the later case, the solution of that point is related to all nodes or particles in the problem domain, though the importance of different nodes or particles can be different. In order to save computational expense, a local support domain is usually preferred, in which only the nodes or particles that are within a local region of finite dimensions of a point are used for the approximations of the field variables at that point. The dimension and shape of the support domain for different points may be different. As illustrated in Figure 2.4, the most commonly used support domain shapes are elliptic (or strictly circular) and rectangular (or strictly square). The support domain is usually taken to be symmetric, and sometimes non-symmetric, especially for points near the boundary, or for points with special considerations.

The influence domain for a node can also be global as the entire problem domain or local as a sub-region of the entire problem domain with finite dimension and shape. The dimension and shape of the influence domains for different nodes may also be different as shown in Figure 2.5. Point x is within the influence domain of nodes 1, 3 and 4. It is however, not within the influence domain of node 2, though the distance between x and node 2 is shorter than that between x and node 1.

As a particle method, the SPH method approximates field variables only on the particles. In other words, a point x is always on a particle (node). Therefore, a particle in the SPH method has both the support domain and influence domains.

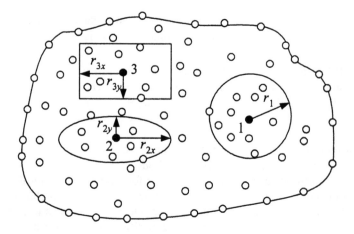

Figure 2.4 Different dimensions and shapes of the support domains for different points.

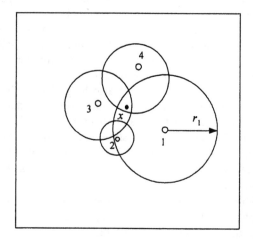

Figure 2.5 Influence domains for nodes 1, 2, 3 and 4 when approximating the field variables at point x. Nodes 1, 3 and 4 are within the support domain of point x. The circles represent the influence domain for the corresponding nodes. Note that compared with node 1, although node 2 is closer to point x, it does not have any influence on point x.

For the SPH method, the concepts of support and influence domains for a particle are closely related to the smoothing length h of that particle. The smoothing length h multiplied by a factor κ determines the support domain or influence domain in which the smoothing function applies. As can be seen in the later chapters, the smoothing length can vary both temporarily or spatially. In some cases, the smoothing length can also vary in dimensions and therefore is a scalar in 1D, a vector in 2D and a tensor in 3D (Owen, 1998). The support domain and influence domain of a particle in SPH are therefore directly related to the region of κh for that particle.

We discuss the differences of these two domains because historically they lead to two different particle approximation models in the SPH method. The use of the concept of support and influence domains in the SPH method can lead to, respectively, the scatter model (Figure 2.6a), and gather model (Figure 2.6b) in approximating the field variables at particle i (Hernquist and Katz, 1989). In the scatter model, approximations on particle i are carried out on the particles whose influence domain covers particle i. In the gather model, approximations on particle i are carried out on the particles within the support domain of particle i.

Since the smoothing length for two particles may not necessarily be the same, it may happen that a particle i that falls within the influence domain of particle j does not influence on the solution at particle j. In other words, particle j influences particle i while particle i does not influence particle j. This is a usual case in meshfree methods which does not involve particle nature, and the domains are used only for numerical interpolation of field variables.[2] However, for meshfree particle methods such as SPH, this unbalanced or non-symmetric influence can lead to severe nonphysical solution (Fulk, 1994; Monaghan, 1992; Roberto and Roberto, 2000). In the SPH method, if h_i is smaller than h_j then the influencing domain of particle j may cover particle i but not necessarily vice versa. Therefore, it is possible for particle j to exert a force on particle i without i exerting the corresponding reaction on j. This is an obvious violation of Newton's Third Law. In order to overcome this problem, some kind of mean value (see, Section 4.4.5) of the smoothing length for the two interacting particles is usually applied to associate with the support and influence domains of these two particles. For such circumstances, the above mentioned scatter model will be the same as the gather model. Therefore the support domain and influence domains for a SPH particle are practically the same.

---

[2] Note that the use of different sizes of support domains in the weak-form meshfree methods, such as the EFG, may lead to compatibility problems. A sudden change in the size of the support domain can cause the similar jumping phenomena discussed by Liu (2002, Section 5.11), which could produce incompatible shape functions. However, this issue at this detail has not yet been well studied.

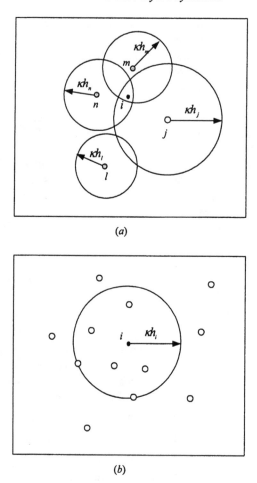

(a)

(b)

Figure 2.6 Scatter and gather models in SPH approximations. (a) Scatter model that uses the concept of influence domain. The circles represent the influence domain of the corresponding particles. (b) Gather model that uses the concept of support domain. The circle represents the support domain of particle i.

2.3.2 Physical influence domain

Partial differential equations (PDE) either in hyperbolic, parabolic or elliptic forms are also physically associated with the concept of influence domain. As can be seen in Figure 2.7, for hyperbolic PDE, information at point P influences only the shaded region, which is the downstream area between the two left- and right advancing characteristic lines passing through point P. The solution at point P depends on the boundary that is intercepted by and included between the two characteristic lines through point P, i.e. interval ab. For a parabolic PDE, information at point P influences the entire region of the plane to one side of P as shown in Figure 2.8. The solution at point P depends on the boundary information along the entire y axis, as well as along the portion of the x axis from a to b. For an elliptic PDE, the information at point P influences all points in the entire closed domain, and in turn the solution at point P is dependent on the information along the entire closed boundary $abcd$ (Figure 2.9).

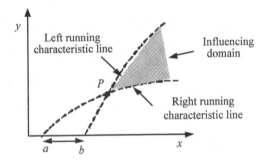

Figure 2.7 Physical influence domain for the solution of a 2D hyperbolic PDE.

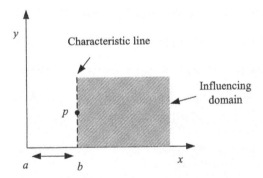

Figure 2.8 Physical influence domain for the solution of a 2D parabolic PDE.

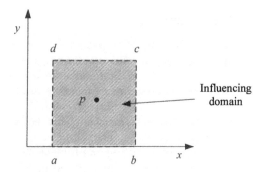

Figure 2.9 Physical influence domain for the solution of a 2D elliptic PDE.

Note that the support and influence domains discussed in Section 2.3.1 are used in a numerical treatment in meshfree methods, and are different from the above-discussed physical influence domains. Choosing a numerical support or influence domain by considering the physical influence domain in a proper manner should lead to more accurate solution more efficiently. This concept of using the physical influence domain has been well practiced in the conventional finite different method in designing different schemes, such as the up-wind scheme, down-wind scheme, etc.

2.3.3 Particle-in-Cell (PIC) method

Particle-in-Cell method is another widely applied particle method, and has a longer history than the SPH method. The PIC method was initiated in the Fluid Dynamics Group at the Las Alamos National Laboratory (LANL) in the late 50s and 60s to solve complex computational fluid dynamic problems including reactive flows, multi-material flows, multi-phase flows and flows with spatial discontinuities (Johnson, 1996). The motivation is to invent a new approach, which can effectively avoid the disadvantages of mesh entanglement in the Lagrangian simulation of multi-material under high pressure, in which solids behave like fluids. Harlow and his co-workers in LANL proposed and developed the PIC method (Harlow, 1957; Evans and Harlow, 1957), where Lagrangian particles are used to carry material mass, position and other information on a 2D uniform Eulerian mesh. The PIC method is therefore a dual description method with both Lagrangian and Eulerian features: Lagrangian description to move the mass particles, while Eulerian descriptions to interpolate information between mass particles and Eulerian nodes. Due to its special advantages and success, the PIC method was comprehensively investigated and widely applied to various areas due to its special advantages. Different variants were developed including: (1) Fluid-in-Cell (FLIC) (Gentry et al., 1966) to address the particle fluctuations areas due to its special advantages. Different variants were developed including: (1) Fluid-in-Cell (FLIC) (Gentry et al., 1966) to address the particle fluctuations and large memory requirements of PIC, (2) Vorticity and Stream Function Method (Fromm and Harlow, 1963) for incompressible flows, (3) Marker-and-Cell (MAC) (Harlow and Welch, 1965) for free surface flow, etc. Some recent developments of PIC include the work of Brackbill and his colleagues (Brackbill, 1988), and the material point method by Sulsky et al. (1995). In the followings, a brief formulation of the PIC method is provided (Franz, 2001).

Consider a general fluid dynamic problem in 2D space, discretize the time as $t_{n+1} = t_n + \Delta t$ and discretize the problem domain with an Eulerian mesh with a number of particles enclosed in each mesh cell, as shown in Figure 2.10. Each particle represents a fluid element and carries properties such as position and velocity. For mesh cell (i, j), the density and velocity can be obtained by a summation over the particles within the cell.

$$\rho_{i,j}^n = \frac{1}{\Delta l^2} \sum_k^N m_k \delta[x_k^n(i, j)]$$ (2.38)

$$v_{i,j}^n = \frac{1}{\Delta l^2} \sum_k^N m_k v_k^n \delta[x_k^n(i, j)]$$ (2.39)

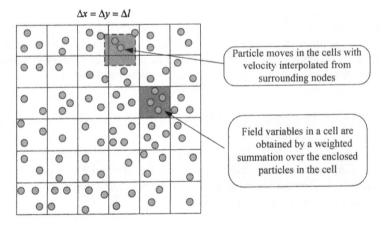

Figure 2.10 Domain discretization in the particle-in-cell (PIC) method.

where m is the mass of the particles, and

$$\delta[x(i, j)] = \delta[\text{int}(\frac{x}{\Delta l} - i)]\delta[\text{int}(\frac{y}{\Delta l} - j)] \tag{2.40}$$

For the momentum equation, we have (see, Chapter 4)

$$\frac{\partial(\rho v)}{\partial t} + \nabla \cdot (\rho v v) = -\nabla p \tag{2.41}$$

It can be rewritten in the following form

$$\rho \frac{\partial v}{\partial t} = -\nabla p - [v \frac{\partial \rho}{\partial t} + \nabla \cdot (\rho v v)] \tag{2.42}$$

The contribution from pressure term in RHS of equation (2.42) can be approximated as

$$\begin{cases} vx_{i,j}^{n+1} = vx_{i,j}^{n} - \dfrac{\Delta t}{2\Delta l \rho_{i,j}^{n}} (p_{i+1,j}^{n} - p_{i-1,j}^{n}) \\[4mm] vy_{i,j}^{n+1} = vy_{i,j}^{n} - \dfrac{\Delta t}{2\Delta l \rho_{i,j}^{n}} (p_{i,j+1}^{n} - p_{i,j-1}^{n}) \end{cases} \tag{2.43}$$

where vx and vy are the velocity component in x and y directions respectively. Pressures in the RHS of equation (2.43) can be obtained from an equation of state.

It is the second Lagrangian part in equation (2.42) that involves particle movement. Its treatment is the essence of the PIC method, in which the velocity for particle k is accumulated by the following weighted summation

$$v_{k}^{n+\frac{1}{2}} = \frac{1}{2} \sum_{m} \frac{S_{m}}{\Delta l^{2}} [v_{m}^{n+1} + v_{m}^{n}] \tag{2.44}$$

The summation is based on the mth mesh cells overlapped with a square of width Δl centered at particle k considered. s is the overlapped area of the neighboring cells with the measure square.

After determining the particle velocity, the particle position in the next time step is

$$x_k^{n+1} = x_k^{n+1} + \Delta t x_k^{n+\frac{1}{2}} \tag{2.45}$$

Therefore a general implementation procedure for the PIC method is

1. Introduce an Eulerian grid with a number of fluid particles within each grid cell;
2. Solve the momentum equation by a standard FDM scheme to obtain the velocity at each grid cell (e.g. equation (2.43));
3. Accumulate the velocity by a weighted summation to obtain the particle velocity (equation (2.44));
4. Move the particle to a new position using the obtained particle velocity (equation (2.45));
5. Calculate the cell properties over summation of the enclosed particles (equation (2.38), (2.39)).
6. Repeat stage 2-5 until the stopping criteria are satisfied.

It is clear that the PIC method is a dual description method with Lagrangian particles to move in the Eulerian grid cells. The inherent background Eulerian mesh determines that PIC method is not a meshfree particle method.

Table 2.1 Comparison of SPH and PIC

	SPH	**PIC**
Method description	Lagrangian	Dual (Eulerian and Lagrangian)
Mesh/grid	No	Yes
Function of the Particles	Material particles and field function approximation	Material particles
Particle information	Directly obtained by solving the conservation equation using the SPH approximations.	Interpolated from or to the Eulerian mesh cells.
Interaction between particles	Yes	No

As summed up in Table 2.1, except that the particles in SPH and PIC both represent material blocks, the SPH method and the PIC method are quite different. The SPH method is a Lagrangian description, while the PIC method is an Eulerian-Lagrangian description. The SPH method is a truly meshfree method, while the PIC method is inherent with an Eulerian mesh for calculating the pressure gradient. The particles in SPH method not only represent material particles, but also act as computational frame for the approximations of both the field functions and their derivatives. In the SPH method, the particle information is obtained by solving the conservation equation using the SPH approximations. While in the PIC method, the particle information is interpolated from the background Eulerian mesh. The particles within the influence domain in the SPH method interact with each other in the form of particle approximations. The particles in the PIC method do not interact with each other, but exchange information with a background mesh.

2.4   Concluding remarks

SPH is a particle-based meshfree approach, which is attractive in many applications especially in hydrodynamic simulations in which the density is a field variable in the system equations. Similar to other meshfree particle methods, the computational frames in SPH are neither grid cells as in the finite difference methods, nor mesh elements as in the finite element methods, but the moving particles in space. The particle approximation in the SPH method is performed at every time step with particles in the current support domain, and it is done for the governing PDEs in Lagrangian description. This adaptive, meshfree, particle, and Lagrangian natures of the SPH method avoid problems such as the mesh deformation, and thus it is very attractive in treating large deformation and impulsive loading events.

In summarizing the contents in this chapter, the following remarks can be made.

1. The SPH method employs particles to represent material and form the computational frame. There is no need for predefined connectivity between these particles. All one needs is the initial particle distribution.

2. The SPH approximation consists of kernel approximation and particle approximation. The kernel approximation of a function and its derivative are carried out in the continuum domain, and the particle approximations of a function and its derivative are carried out using discretized particles in the support domain at the current time step.

3. Each particle in the SPH method is associated with a support domain and influence domain. For most practical applications, the support domain of a particle can be equal to its influence domain.
4. The SPH method is different from the PIC method since SPH is a purely meshfree method, while PIC is a dual approach with both Eulerian and Lagrangian descriptions.

Though the SPH method sometimes suffers from some numerical problems such as boundary inaccuracy and tensile instability, and in some circumstances may produce results with errors larger than those obtained using other methods tailored for specific problems, its accuracy can be improved by various correction schemes. The balance between accuracy and efficiency as well as adaptivity should be taken into account when applying the SPH method and its corrective variants.

The applications of the SPH method range from very small scale to very large scale, from astrophysics to fluid and solid mechanics problems, from discrete physical systems to continuum systems. SPH is an extremely versatile method, and can easily handle problems with highly irregular and even dynamic geometry as well as problems with large deformation and high impulsive loadings. There are many applications where the SPH method can be exploited to give accurate results without requiring the use of complicated grid/mesh refinement algorithms. The following chapters will provide detailed SPH formulations for many practical applications.

Chapter 3

Construction of Smoothing Functions

In Chapter 2, the basic ideas and essential formulations of the SPH method have been presented. It has been shown that the *smoothing function* plays a very important role in the SPH approximations, as it determines the accuracy of the function representation and efficiency of the computation.

In this chapter, a generalized approach to construct the smoothing functions for the SPH method is introduced. The approach uses the integral form of function representation with the help of the Taylor series expansion. A set of conditions are derived systematically, which can then be utilized to construct both analytical smoothing functions and point-dependent smoothing functions that can only be given in numerical forms. These conditions not only ensure the consistency in the SPH approximations, but also describe the compact support requirements for the smoothing function. Examples of SPH smoothing functions constructed include many existing ones used so far in the SPH literature, and a new quartic smoothing function derived recently by Liu, Liu and Lam (2002). The new quartic smoothing function is then applied to the simulation of a one-dimensional shock problem and a two-dimensional heat conduction problem. Particle inconsistency problem in the discretized form of particle approximation is also discussed with an approach for the consistency restoration.

3.1   Introduction

One of the central issues for the meshfree methods is how to effectively perform function approximation based on a set of nodes scattered in an arbitrary manner without using a predefined mesh or grid that provides the connectivity of the nodes. The methods for function approximation are classified in the monograph by Liu (2002) into three categories: 1) integral representation, 2) series representation, and 3) differential representation. The SPH method employs the integral representation using a *smoothing function*. The smoothing function (also called *smoothing kernel function, smoothing kernel* or simply *kernel* in many literatures) is of utmost importance since it not only determines the pattern for the function approximation, defines the dimension of the support domain of particles, but also determines the consistency and hence the accuracy of both the kernel and particle approximations.

Different smoothing functions have been used in the SPH method as shown in the published literatures. Various requirements or properties for the smoothing functions are discussed in different literatures. Some of them are listed in Chapter 2. All the major properties are now summarized and described in the following discussion.

1. The smoothing function must be normalized (*Unity*) over its support domain as discussed in Chapter 2.

$$\int_\Omega W(x - x', h)dx' = 1 \qquad (3.1)$$

2. The smoothing function should be compactly supported (*Compact support*), i.e.,

$$W(x - x') = 0, \text{ for } |x - x'| > \kappa h \qquad (3.2)$$

The dimension of the compact support is defined by the smoothing length h and a scaling factor κ, where h is the smoothing length, and κ determines the spread of the specified smoothing function. $|x - x'| \le \kappa h$ defines the support domain of the particle at point x (see, Chapter 2).

3. $W(x - x') \ge 0$ for any point at x' within the support domain of the particle at point x (*Positivity*).

4. The smoothing function value for a particle should be monotonically decreasing with the increase of the distance away from the particle (*Decay*).

5. The smoothing function should satisfy the Dirac delta function condition as the smoothing length approaches to zero (*Delta function property*).

$$\lim_{h \to 0} W(x - x', h) = \delta(x - x') \qquad (3.3)$$

6. The smoothing function should be an even function (*Symmetric property*).
7. The smoothing function should be sufficiently smooth (*Smoothness*).

The first normalization property ensures that the integral of the smoothing function over the support domain to be unity. It can be shown later that it also ensures the zero-th order consistency (C^0) of the integral representation of a continuum function.

The second property transforms a SPH approximation from global operation to a local operation. This will lead to a set of sparse discretized system matrices, and therefore is of significance as far as the computational efforts are concerned.

The third property states that the smoothing function should be non-negative in the support domain. It is not mathematically necessary as a convergent requirement, but important to ensure a physically meaningful (or stable) representation of some physical phenomena. Some smoothing functions used in the practice are negative in parts of the support domain. However in hydrodynamic simulations, negative value for the smoothing function can have serious consequences that result in some unphysical parameters such as negative density and energy.

The fourth property is based on the physical consideration in that nearer particle should have a bigger influence on the concerned particle. In other words, with the increase of the distance of two interacting particles, the interaction force decreases.

The fifth property makes sure that as the smoothing length tends to be zero, the approximation value approaches the function value, i.e. $\langle f(x) \rangle = f(x)$. This property is naturally met since using the requirements 1-4 will lead to a function that approaches the Dirac delta function when its support domain approaches zero.

The sixth property means that particles from the same distance but different positions should have equal effect on a given particle. This is not a very rigid condition, and it is sometimes violated in some meshfree particle methods that provide higher consistency.

The seventh property aims to obtain better approximation. For the approximations of a function and its derivatives, the smoothing function needs to be sufficiently continuous to obtain good results. A smoothing function with smoother value of the function and derivatives would usually yield better results. This is because the smoothing function will not be sensitive to particle disorder, and the errors in approximating the integral interpolants are small provided the particle disorder is not too extreme (Monaghan, 1992; Fulk, 1994).

Any function having the above properties can be employed as SPH smoothing function functions. Many researchers and practitioners have tried different kinds of smoothing functions. The following lists some of the most frequently used ones in the SPH literatures.

In the original SPH paper, Lucy (1977) used the following bell-shaped function (Figure 3.1) as the smoothing function.

$$W(x-x',h) = W(R,h) = \alpha_d \begin{cases} (1+3R)(1-R)^3 & R \le 1 \\ 0 & R > 1 \end{cases} \tag{3.4}$$

where α_d is $5/4h$, $5/\pi h^2$ and $105/16\pi h^3$ in one-, two- and three-dimensional space, respectively, so that the condition of unity can be satisfied for all the three dimensions. In equation (3.4), R is the relative distance between two points (particles) at points x and x', $R = \dfrac{r}{h} = \dfrac{|x-x'|}{h}$, where r is the distance between the two points.

Monaghan (1992) stated that to find a physical interpretation of an SPH equation, it is always best to assume the smoothing function to be a Gaussian. Gingold and Monaghan (1977) in their original paper selected the following Gaussian kernel (Figure 3.2) to simulate the non-spherical stars,

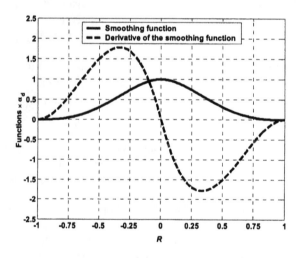

Figure 3.1 The smoothing function and its first derivative used by Lucy (1977). α_d is $5/4h$, $5/\pi h^2$ and $105/16\pi h^3$ in one-, two- and three-dimensional space, respectively.

$$W(R,h) = \alpha_d e^{-R^2} \tag{3.5}$$

where α_d is $1/\pi^{1/2}h$, $1/\pi h^2$ and $1/\pi^{3/2}h^3$, respectively, in one-, two- and three-dimensional space, for the unity requirement.

The Gaussian kernel is sufficiently smooth even for high orders of derivatives, and is regarded as a golden selection since it is very stable and accurate especially for disordered particles. It is, however, not really compact, as it never goes to zero theoretically, unless R approaches to infinity. Because it approaches to zero very fast numerically, it is practically compact. Note that it is computationally more expensive since it can take a longer distance for the kernel to approach zero, especially for higher order derivatives of the smoothing function. This can result in a large support domain with an inclusion of more particles for the particle approximation, and therefore leads to a larger bandwidth in the discrete system matrix.

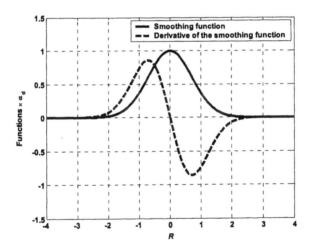

Figure 3.2 The Gaussian kernel and its first derivative. α_d is $1/\pi^{1/2}h$, $1/\pi h^2$ and $1/\pi^{3/2}h^3$ in one-, two- and three-dimensional space, respectively.

Monaghan and Lattanzio (1985) devised the following smoothing function (Figure 3.3) based on the cubic spline functions known as the B-spline function.

$$W(R,h) = \alpha_d \times \begin{cases} \frac{2}{3} - R^2 + \frac{1}{2}R^3 & 0 \le R < 1 \\ \frac{1}{6}(2-R)^3 & 1 \le R < 2 \\ 0 & R \ge 2 \end{cases} \qquad (3.6)$$

In one-, two- and three-dimensional space, $\alpha_d = 1/h$, $15/7\pi h^2$ and $3/2\pi h^3$, respectively. The cubic spline function has been, so far, the most widely used smoothing function in the emerged SPH literatures since it resembles a Gaussian function while having a narrower compact support. However, the second derivative of the cubic spline is piecewise linear functions, and accordingly, the stability properties can be inferior to those of smoother kernels. In addition, the smoothing function is in pieces, which is slightly more difficulty to use compared to one piece smoothing functions.

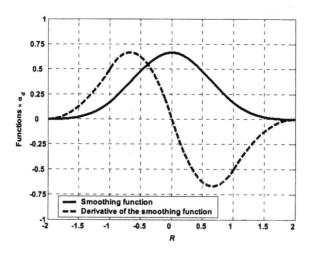

Figure 3.3 The cubic spline kernel and its first derivative. α_d is $1/h$, $15/7\pi h^2$ and $3/2\pi h^3$ in one-, two- and three-dimensional space, respectively.

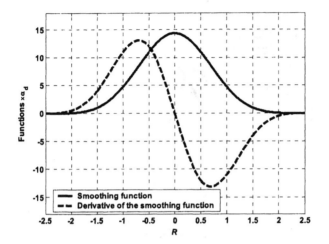

Figure 3.4 The quartic smoothing function and its first derivative. α_d is $1/24h$ in one-dimensional space.

Morris (1994, 1996) introduced higher order (quartic and quintic) splines that are more closely approximating the Gaussian and more stable. The quartic spline (Figure 3.4) is

$$W(R,h) = \alpha_d \times \begin{cases} (R+2.5)^4 - 5(R+1.5)^4 + 10(R+0.5)^4 & 0 \le R < 0.5 \\ (2.5-R)^4 - 5(1.5-R)^4 & 0.5 \le R < 1.5 \\ (2.5-R)^4 & 1.5 \le R < 2.5 \\ 0 & R > 2.5 \end{cases} \quad (3.7)$$

where α_d is $1/24h$ in one-dimensional space.
The quintic spline (Figure 3.5) is

$$W(R,h) = \alpha_d \times \begin{cases} (3-R)^5 - 6(2-R)^5 + 15(1-R)^5 & 0 \le R < 1 \\ (3-R)^5 - 6(2-R)^5 & 1 \le R < 2 \\ (3-R)^5 & 2 \le R < 3 \\ 0 & R > 3 \end{cases} \quad (3.8)$$

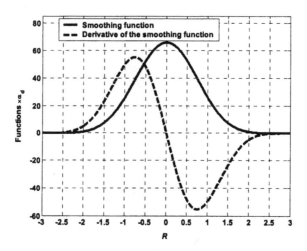

Figure 3.5 The quintic smoothing function and its first derivative. α_d is $120/h$, $7/478\pi h^2$ and $3/359\pi h^3$ in one-, two- and three-dimensional space, respectively.

where α_d is $120/h$, $7/478\pi h^2$ and $3/359\pi h^3$ in one-, two- and three-dimensional space, respectively.

Johnson et al. (1996b) used the following quadratic smoothing function (Figure 3.6) to simulate the high velocity impact problem

$$W(R,h) = \alpha_d(\frac{3}{16}R^2 - \frac{3}{4}R + \frac{3}{4}) \quad 0 \le R \le 2 \qquad (3.9)$$

where in one-, two- and three-dimensional space, $\alpha_d = 1/h$, $2/\pi h^2$ and $5/4\pi h^3$, respectively. Unlike other smoothing functions, the derivative of this quadratic smoothing function always increases as the particles move closer, and always decreases as they move apart. This is regarded by Johnson as an important improvement over the cubic spline function, and it is reported to relieve the problem of compressive instability.

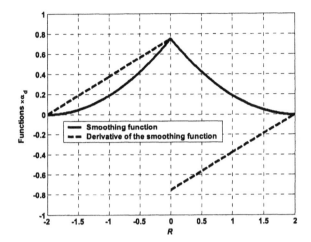

Figure 3.6 The quadratic smoothing function and its first derivative used by Johnson et al. (1996b). α_d is $1/h$, $2/\pi h^2$ and $5/4\pi h^3$ in one-, two- and three-dimensional space, respectively.

Some higher order smoothing functions that are devised from lower order forms were constructed (Monaghan and Lattanzio, 1985), such as the super-Gaussian kernel

$$W(R,h) = \alpha_d(\frac{3}{2} - R^2)e^{-R^2} \qquad 0 \le R \le 2 \tag{3.10}$$

where α_d is $1/\sqrt{\pi}$ in one-dimensional space.

One disadvantage of the high order smoothing function is that the kernel is negative in some region of its support domain. This may lead to unphysical results for hydrodynamic problems (Fulk, 1994).

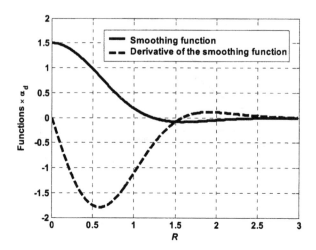

Figure 3.7 The super-Gaussian smoothing function and its first derivative. α_d

is $1/\sqrt{\pi}$ in one-dimensional space.

3.2   Conditions for constructing smoothing functions

Section 3.1 presented a number of smoothing functions used so far.   The
question now is whether there is a standard procedure to develop the smoothing
functions. This section discusses conditions that a smoothing function has to
satisfy in order to ensure a certain order of accuracy in the SPH approximations.
These conditions can then be used to construct smoothing functions in a
systematical manner.

Any numerical approximation should represent as closely as possible the
corresponding field function of the physical problem. In the traditional finite
difference methods (FDM), the concept of consistency defines how well the
discretized system equations model the partial differential equations of physical
laws (Hirsch, 1988; Anderson, 1995). A numerical interpolation scheme (in
FDM) is consistent if it has the ability to exactly represent the differential
equations in the limit as the number of the grid points approaches infinity and the
maximal mesh size approaches zero. On one hand, consistency is a basic
requirement to construct a finite difference scheme; on the other hand,
consistency is a prerequisite for convergence. This is due to the Lax-Richtmyer

equivalence theorem, which states that a consistent finite difference scheme for a well-posed partial differential equation is convergent if and only if it is stable.

Similarly, by using the Taylor series expansion, analysis can be carried out on how well the SPH approximations represent the physical equations in the limit as the particle spacing approaches zero. This analysis is carried out in the stage of the SPH kernel approximation for a function and its derivatives. The analysis shows that, to exactly approximate a function and its derivatives, certain conditions need to be satisfied. These conditions can, in turn, be used to construct the smoothing functions. The following development is performed in reference to the work by Liu, Liu and Lam (2002).

3.2.1 Approximation of a field function

In the SPH method, for a field function f, multiplying f with the smoothing function W, and then integrating over the support domain Ω of a point can approximate the field function value at that point using the following form of integral representation.

$$f(x) = \int_{\Omega} f(x')W(x-x',h)dx' \qquad (3.11)$$

If $f(x)$ is sufficiently smooth, applying the Taylor series expansion of $f(x')$ in the vicinity of x yields

$$f(x') = f(x) + f'(x)(x'-x) + \frac{1}{2}f''(x)(x'-x)^2 + \dots$$
$$= \sum_{k=0}^{n} \frac{(-1)^k h^k f^{(k)}(x)}{k!} (\frac{x-x'}{h})^k + r_n(\frac{x-x'}{h}) \qquad (3.12)$$

where $r_n(\frac{x-x'}{h})$ is the remainder of the Taylor series expansion.

Substituting equation (3.12) into equation (3.11) leads to

$$f(x) = \int_\Omega \sum_{k=0}^n \frac{(-1)^k h^k f^{(k)}(x)}{k!} (\frac{x-x'}{h})^k W(x-x',h)dx' + r_n(\frac{x-x'}{h})$$

$$= \sum_{k=0}^n \frac{(-1)^k h^k f^{(k)}(x)}{k!} \int_\Omega (\frac{x-x'}{h})^k W(x-x',h)dx' + r_n(\frac{x-x'}{h}) \qquad (3.13)$$

$$= \sum_{k=0}^n A_k f^{(k)}(x) + r_n(\frac{x-x'}{h})$$

where

$$A_k = \frac{(-1)^k h^k}{k!} \int_\Omega (\frac{x-x'}{h})^k W(x-x',h)dx' \qquad (3.14)$$

Comparing the LHS with the RHS of equation (3.13), in order for $f(x)$ to be approximated to n-th order, the coefficients A_k must equal to the counterparts for $f^{(k)}(x)$ in the LHS of equation (3.13). Therefore, the following conditions for the smoothing function W can be obtained.

$$\left.\begin{aligned}
A_0 &= \int_\Omega W(x-x',h)dx' = 1 \\[2mm]
A_1 &= -h \int_\Omega (\frac{x-x'}{h})W(x-x',h)dx' = 0 \\[2mm]
A_2 &= \frac{h^2}{2!} \int_\Omega (\frac{x-x'}{h})^2 W(x-x',h)dx' = 0 \\
&\;\;\vdots \\
A_n &= \frac{(-1)^n h^n}{n!} \int_\Omega (\frac{x-x'}{h})^n W(x-x',h)dx' = 0
\end{aligned}\right\} \qquad (3.15)$$

These conditions can be further written in the following simplified expressions in terms of the k-th moments M_k of the smoothing function.

$$M_0 = \int_\Omega W(x - x', h)dx' = 1$$

$$M_1 = \int_\Omega (x - x')W(x - x', h)dx' = 0$$

$$M_2 = \int_\Omega (x - x')^2 W(x - x', h)dx' = 0$$ (3.16)

$$\vdots$$

$$M_n = \int_\Omega (x - x')^n W(x - x', h)dx' = 0$$

Note that the first equation in (3.16) is, in fact, the unity condition expressed in equation (3.1), and the second equation in (3.16) represents the symmetric property given below equation (3.3). Satisfaction of these two conditions ensures the first order consistency for the SPH kernel approximation for a function.

3.2.2 Approximation of the derivatives of a field function

In the computational fluid dynamics (CFD), since the highest derivative of the filed function in the governing PDE (see, Chapter 4) is second order. Hence, in the following discussions, the approximation of first and second order derivatives of a field function is primarily considered. This is also because the procedure can be easily extended to approximate higher derivatives and obtain similar results. In addition, one can regard a higher derivative as the derivative of the lower derivative (e.g., second derivative is the derivative of the first derivative, and so on).

First derivative

The approximation of the first derivative can be obtained by replacing the function $f(x)$ in equation (3.11) with its derivative $f'(x)$, i.e.,

$$f'(x) = \int_\Omega f'(x')W(x - x', h)dx'$$ (3.17)

Integrating by parts, the above equation can be rewritten as,

$$f'(x) = \int_S f(x')W(x - x', h) \cdot \bar{n}dS - \int_\Omega f(x')W'(x - x', h)dx'$$ (3.18)

where the first integral is over the surface S of the support domain Ω; $\bar{n}$ is the unit vector normal to the surface S. Substituting (3.12) into the second integral on the RHS of equation (3.18) yields

$$f'(x) = \int_S f(x')W(x-x',h) \cdot \bar{n}dS - \int_\Omega [(\sum_{k=0}^n \frac{(-1)^k h^k f^{(k)}(x)}{k!}(\frac{x-x'}{h})^k$$

$$+r_n(\frac{x-x'}{h}))W'(x-x',h)]dx'$$

$$= \int_S f(x')W(x-x',h) \cdot \bar{n}dS$$

$$\qquad (3.19)$$

$$-[\sum_{k=0}^n \frac{(-1)^k h^k f^{(k)}(x)}{k!} \int_\Omega (\frac{x-x'}{h})^k W'(x-x',h)dx' + r_n(\frac{x-x'}{h})]$$

$$= \int_S f(x')W(x-x',h) \cdot \bar{n}dS + \sum_{k=0}^n A_k' f^{(k)}(x) + r_n(\frac{x-x'}{h})$$

where

$$A_k' = \frac{(-1)^{k+1} h^k}{k!} \int_\Omega (\frac{x-x'}{h})^k W'(x-x',h)dx' \qquad (3.20)$$

It is clear that, if the following equations are satisfied, $f'(x)$ can be approximated to n-th order,

$$\left.\begin{array}{l} M_0' = \int_\Omega W'(x-x',h)dx' = 0 \\[2mm] M_1' = \int_\Omega (x-x')W'(x-x',h)dx' = 1 \\[2mm] M_2' = \int_\Omega (x-x')^2 W'(x-x',h)dx' = 0 \\[2mm] \quad\vdots \\[2mm] M_n' = \int_\Omega (x-x')^n W'(x-x',h)dx' = 0 \end{array}\right\} \qquad (3.21)$$

and

$$W(x - x', h)\mid_S = 0 \qquad (3.22)$$

Equation (3.22) requires the smoothing function to vanish on the surface of the support domain, which naturally leads the surface integration $\int_S f(x')W(x - x', h) \cdot \bar{n}dS$ to vanish for any arbitrarily function $f(x)$. The first expression in equation (3.21) is actually another representation of equation (3.22), as we can easily confirm from the following development.

$$\int_\Omega W'(x - x', h)dx' = \int_S 1 \cdot W(x - x', h) \cdot \bar{n}dS - \int_\Omega (1)' \cdot W(x - x', h)dx'$$

$$= \int_S W(x - x', h) \cdot \bar{n}dS = 0 \qquad (3.23)$$

Note that equation (3.22) is in fact the compact support condition give in equation (3.2).

Second derivative

The approximation of the second derivative can be obtained by directly substituting the function $f(x)$ in (3.11) with its second derivative $f''(x)$, which gives

$$f''(x) = \int_\Omega f''(x')W(x - x', h)dx' \qquad (3.24)$$

Using integration by parts, the above equation can be rewritten as,

$$f''(x) = \int_S f'(x')W(x - x', h) \cdot \bar{n}dS - \int_\Omega f'(x')W'(x - x', h)dx'$$

$$= \int_S f'(x')W(x - x', h) \cdot \bar{n}dS - [\int_S f(x')W'(x - x', h) \cdot \bar{n}dS \qquad (3.25)$$

$$- \int_\Omega f(x')W''(x - x', h)dx']$$

Substituting (3.12) into the third integral on the RHS in the above equation, we have

$$
\begin{aligned}
f''(x) &= \int_S f'(x')W(x-x',h)\cdot\bar{n}dS - \int_S f(x')W'(x-x',h)\cdot\bar{n}dS \\
&\quad + \int(\sum_{k=0}^{n}\frac{(-1)^k h^k f^{(k)}(x)}{k!}(\frac{x-x'}{h})^k + r_n(\frac{x-x'}{h}))W''(x-x',h)dx' \\
&= \int_S f'(x')W(x-x',h)\cdot\bar{n}dS - \int_S f(x')W'(x-x',h)\cdot\bar{n}dS \\
&\quad + [\sum_{k=0}^{n}\frac{(-1)^k h^k f^{(k)}(x)}{k!}\int_\Omega(\frac{x-x'}{h})^k W''(x-x',h)dx' + r_n(\frac{x-x'}{h})] \\
&= \int_S f'(x')W(x-x',h)\cdot\bar{n}dS - \int_S f(x')W'(x-x',h)\cdot\bar{n}dS \\
&\quad + \sum_{k=0}^{n}A_k'' f^{(k)}(x) + r_n(\frac{x-x'}{h})
\end{aligned}
\tag{3.26}
$$

where

$$
A_k'' = \frac{(-1)^k h^k}{k!}\int_\Omega(\frac{x-x'}{h})^k W''(x-x',h)dx'
\tag{3.27}
$$

It can be seen that, if the following equations are satisfied, $f''(x)$ can be approximated to n-th order accuracy.

$$
\left.
\begin{aligned}
M_0'' &= \int_\Omega W''(x-x',h)dx' = 0 \\
M_1'' &= \int_\Omega (x-x')W''(x-x',h)dx' = 0 \\
M_2'' &= \int_\Omega (x-x')^2 W''(x-x',h)dx' = 2 \\
&\vdots \\
M_n'' &= \int_\Omega (x-x')^n W''(x-x',h)dx' = 0
\end{aligned}
\right\}
\tag{3.28}
$$

$$W(x - x', h)|_S = 0 \tag{3.29}$$

and

$$W'(x - x', h)|_S = 0 \tag{3.30}$$

Equations (3.29) and (3.30) determine the surface integral terms in equation (3.26) to vanish for an arbitrarily function $f(x)$ and its first derivative $f'(x)$. The first expression in equation (3.28) is actually another representation of equation (3.30) as we can see from the following expression.

$$\begin{aligned}
\int_\Omega W''(x - x', h)dx' &= \int_S 1 \cdot W'(x - x', h) \cdot \bar{n} dS - \int_\Omega (1)' \cdot W'(x - x', h)dx' \\
&= \int_S W'(x - x', h) \cdot \bar{n} dS = 0
\end{aligned} \tag{3.31}$$

If equations (3.29) and (3.30) are satisfied, equations (3.21) and (3.28) can be derived from the equation (3.16) (except the first expression in equation (3.21) and first two expressions in equation (3.28)) by using the following integration by parts.

$$\begin{aligned}
\int_\Omega (x - x')^k W(x - x', h)dx' &= -\frac{1}{(k+1)} \int_\Omega [(x - x')^{k+1}]'W(x - x', h)dx' \\
&= -\frac{1}{(k+1)} [\int_S (x - x')^{k+1} W(x - x', h) \cdot \bar{n} dS - \int_\Omega (x - x')^{k+1} W'(x - x', h)dx'] \\
&= \frac{1}{(k+1)} \int_\Omega (x - x')^{k+1} W'(x - x', h)dx' \\
&= \frac{1}{(k+1)}(-\frac{1}{(k+2)}) \\
&\left(\int_S (x - x')^{k+2} W'(x - x', h) \cdot \bar{n} dS - \int_\Omega (x - x')^{k+2} W''(x - x', h)dx'\right) \\
&= \frac{1}{(k+1)(k+2)} \int_\Omega (x - x')^{k+2} W''(x - x', h)dx'
\end{aligned} \tag{3.32}$$

In summary, if a function and its first two derivatives are to be reproduced to n-th order accuracy, then the smoothing function should satisfy:

$$
\left.\begin{aligned}
M_0 &= \int_\Omega W(x-x',h)dx' = 1 \\[6pt]
M_1 &= \int_\Omega (x-x')W(x-x',h)dx' = 0 \\[6pt]
M_2 &= \int_\Omega (x-x')^2 W(x-x',h)dx' = 0 \\[4pt]
&\;\;\vdots \\[4pt]
M_n &= \int_\Omega (x-x')^n W(x-x',h)dx' = 0
\end{aligned}\right\} \tag{3.33}
$$

and

$$
\left.\begin{aligned}
W(x-x',h)\big|_S &= 0 \\
W'(x-x',h)\big|_S &= 0
\end{aligned}\right\} \tag{3.34}
$$

These conditions can be used to construct the smoothing functions. It can be seen that the conditions of the smoothing functions can be classified into two groups. The first group shows the ability of the smoothing function to reproduce polynomials, which will be further discussed later. Satisfying the first group, the function can be approximated to n-th order accuracy. The second group defines the surface values of the smoothing function as well as its first derivatives, and is the requirements of the property of compact support for the smoothing function and its first derivative. Satisfying these conditions, the first two derivatives of the function can be exactly approximated to the n-th order.

Following the same procedure, the conditions for producing higher order derivatives of a function can also be obtained, and classified into two similar groups. Except those from the conditions expressed in equations (3.33) and (3.34), the higher order derivatives of the smoothing function should also be compactly supported (vanish on the boundary of the support domain). In general, the compact support conditions for the k-th order derivative to be approximated to the n-th order accuracy are:

$$
\left.\begin{aligned}
W(x - x', h)\,|_S &= 0 \\
W'(x - x', h)\,|_S &= 0 \\
&\vdots \\
W^{k-1}(x - x', h)\,|_S &= 0
\end{aligned}\right\}
\tag{3.35}
$$

Revisiting the previously listed properties of the smoothing function, the normalization (unity) property expressed in equation (3.1) is actually a constituent of equation (3.33). In other words, for the function to be reproduced to the zero-th order accuracy the averaged value of the smoothing function over the support domain should be unity. For the function to be reproduced to the n-th order accuracy, the moments of the smoothing function up to n-th order should vanish, in addition to the unity condition.

The compact support property of the smoothing function is also a constituent of the surface equations (3.34). In other words, for the first derivative of a function to be exactly reproduced, the smoothing function should have a compact support. For the second derivative of a function to be exactly reproduced, the first derivative of the smoothing function should also be compactly supported over the same support domain. Therefore, it is clear that the previously discussed requirements on the smoothing function are actually the reproducibility of the SPH kernel approximations or the integral representations for a function and its derivatives.

3.2.3 Consistency of the kernel approximation

The conditions shown in equations (3.33) and (3.34) are derived using the Taylor series expansion. With such conditions satisfied, the SPH approximations for a function and its derivative are be consistent to a given order. This Taylor series expansion based approach is directly related to the consistency concept for the traditional finite difference method in treating partial differential equations (PDE).

Similarly, the consistency concept for the traditional finite element methods (FEM) also applies to the meshfree particle methods. For an FEM approximation to converge, the FEM solution must approach the exact solution when the nodal distance approaches zero. To ensure the convergence, the FEM shape function employed must satisfy a certain degree of consistency. It is well-known that the degree of consistency is often characterized by the order of the polynomial that can be exactly reproduced by the approximation using the shape function (see, e.g., Liu and Quek, 2003). If an approximation can reproduce a constant exactly, the approximation is then said to have zero-th order or C^0 consistency. In general, if an approximation can reproduce a polynomial of up to k-th order exactly, the approximation is said to have k-th order or C^k consistency. In the finite element methods, the concept of consistency is closely related to the concept of completeness of the polynomial basis. In solving any partial differential equations based on a weak form formulation, such as the Galerkin approach, there is a minimum consistency requirement for ensuring the convergence of the discretized equation system. The minimum consistency requirement depends on the order of the PDE. For a PDE of order $2k$, the minimum requirement of the consistency is C^k for the Galerkin formulation. Belytschko et al. (1996; 1998) provided a discussion on the consistency and completeness of a numerical method.

Borrowing the consistency concept from FEM, we argue that for an SPH kernel approximation or an integral representation to exactly reproduce a function, the smoothing function should satisfy some conditions, which can be represented by the *polynomial reproducibility* of the kernel approximation. For a constant (zero order polynomial) field function $f(x) = c$ to be exactly reproduced by the SPH kernel approximation, we should have

$$f(x) = \int_{\Omega} cW(x - x', h)dx' = c \tag{3.36}$$

or

$$\int_{\Omega} W(x - x', h)dx' = 1 \tag{3.37}$$

It is clearly shown, again, that the normalization (unity) condition or the first expression in (3.33) is in fact the condition for the kernel approximation to have the zero-th order consistency.

For a linear function $f(x) = c_0 + c_1 x$ to be reproduced, we should have

$$f(x) = \int_{\Omega} (c_0 + c_1 x')W(x - x', h)dx' = c_0 + c_1 x \tag{3.38}$$

which can be simplified using equation (3.37) as

$$\int_{\Omega} x'W(x - x', h)dx' = x \tag{3.39}$$

Multiplying x to both side of equation (3.37), we have the following identity.

$$\int_{\Omega} xW(x-x',h)dx' = x \tag{3.40}$$

Subtracting equation (3.39) from the above identity leads to

$$\int_{\Omega} (x-x')W(x-x',h)dx' = 0 \tag{3.41}$$

Equation (3.41) is in fact the second expression in equation (3.33). In other words, the second expression in equation (3.33) represents the condition for the linear consistency of the SPH kernel approximations. This expression also provides some insight to the shape of the smoothing function. For equation (3.41) to be satisfied, the smoothing function has to be symmetric so that the first moment can vanish. This is described previously in Section 3.1 as the sixth property of the smoothing function.

For higher order polynomials to be reproduced, similar conclusions can be derived. Let's consider a monomial of k-th order or simply assume $f(x) = c_k x^k$, approximating the function value at the origin $x = 0$ yields

$$f(0) = \int_{\Omega} c_k x'^k W(0-x',h)dx' = 0 \tag{3.42}$$

A more general expression can be obtained by moving the origin to an arbitrary point x to get the approximation of the function under the new coordinate system, which gives

$$\int_{\Omega} (x-x')^k W(x-x',h)dx' = 0 \tag{3.43}$$

It is clear that equation (3.43) is same as the last expression in equation (3.33). Therefore, the conditions shown in equation (3.33) can also be regarded as the conditions for reproducing the k-th order polynomial.

3.2.4 Consistency of the particle approximation

The above discussed consistency concepts are derived from the continuous form of kernel approximation or integral representation. They do not ensure consistency for the discrete form produced after the particle approximation. In meshfree particle methods, the phenomenon in which the discretized equations for the conditions shown in equation (3.33) are not satisfied is called *particle inconsistency* (Morris, 1996; Belytschko et al., 1996). The discrete counterparts of the constant and linear consistency conditions (3.37) and (3.41) are

$$\sum_{j=1}^{N} W(x - x_j, h)\Delta x_j = 1 \qquad (3.44)$$

and

$$\sum_{j=1}^{N} (x - x_j)W(x - x_j, h)\Delta x_j = 0 \qquad (3.45)$$

where N is the total number of particles in the support domain for the given particle located at x.

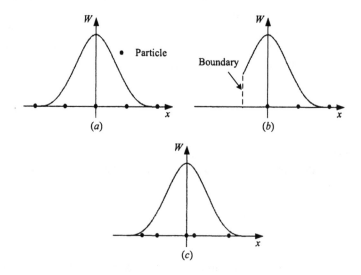

Figure 3.8 SPH particle approximation in one-dimensional case. (a) Particle approximation for an interior particle with regular particle distribution in its support domain. (b) Particle approximation for a particle whose support domain is truncated by the boundary. (c) Particle approximation for a particle with irregular particle distribution in its support domain.

These discretized consistency conditions are not always satisfied. One obvious and simple case is for the particles at or near the boundary of the problem domain so that the support domain intersects with the boundary, as clearly shown in Figure 3.8*b*. Even for regular node (particle) distribution, due to the unbalanced particles contributing to the discretized summation, the LHS of equation (3.44) is less than 1 and the LHS of equation (3.45) will not vanish, due to the truncation of the smoothing function by the boundary, as shown in Figure 3.8*b*.

Another simple case is observed when the particles are irregularly distributed. In such a case, even for the interior particles whose support domains are not truncated, the constant and linear consistency condition in the discretized form will not be exactly satisfied, due to the unbalanced particle contribution, as shown in Figure 3.8c.

Similarly, the discrete counterparts of higher order consistency conditions are also not always exactly satisfied for case 1) particles on and near the boundary, and case 2) irregularly distributed particles.

There are different means to restore the consistency condition for the discrete form. One general approach (Liu et al., 2003b) is given below. For a particle approximation to obtain a *k*-th order consistency in the discrete form, we may write the smoothing function in the following form.

$$W(x - x_j, h) = b_0(x,h) + b_1(x,h)(\frac{x - x_j}{h}) + b_2(x,h)(\frac{x - x_j}{h})^2 + \ldots\ldots$$

$$= \sum_{I=0}^{k} b_I(x,h)(\frac{x - x_j}{h})^I \tag{3.46}$$

After some trivial transformation, the discretized form of equation (3.33) can be written as

$$\left.\begin{array}{l} \sum_{j=1}^{N} [\sum_{I=0}^{k} b_I(x,h)(\frac{x - x_j}{h})^I] \Delta x_j = 1 \\[2em] \sum_{j=1}^{N} [(\frac{x - x_j}{h})][\sum_{I=0}^{k} b_I(x,h)(\frac{x - x_j}{h})^I] \Delta x_j = 0 \\[2em] \vdots \\[2em] \sum_{j=1}^{N} (\frac{x - x_j}{h})^k [\sum_{I=0}^{k} b_I(x,h)(\frac{x - x_j}{h})^I] \Delta x_j = 0 \end{array}\right\} \tag{3.47}$$

or

$$\left.\begin{array}{l}\displaystyle\sum_{I=0}^{k} b_I(x,h)\sum_{j=1}^{N}(\frac{x-x_j}{h})^I \Delta x_j = 1 \\[3mm] \displaystyle\sum_{I=0}^{k} b_I(x,h)\sum_{j=1}^{N}(\frac{x-x_j}{h})^{I+1} \Delta x_j = 0 \\[3mm] \vdots \\[3mm] \displaystyle\sum_{I=0}^{k} b_I(x,h)\sum_{j=1}^{N}(\frac{x-x_j}{h})^{I+k} \Delta x_j = 0 \end{array}\right\}$$
(3.48)

Letting

$$m_k(x,h) = \sum_{j=1}^{N}(\frac{x-x_j}{h})^k \Delta x_j$$
(3.49)

the $k+1$ coefficients $b_I(x,h)$ can then be determined by solving the following matrix equation

$$\underbrace{\begin{bmatrix} m_0(x,h) & m_1(x,h) & \cdots & m_k(x,h) \\ m_1(x,h) & m_2(x,h) & \cdots & m_{1+k}(x,h) \\ \vdots & \vdots & \ddots & \vdots \\ m_k(x,h) & m_{k+1}(x,h) & \cdots & m_{k+k}(x,h) \end{bmatrix}}_{M} \underbrace{\begin{bmatrix} b_0(x,h) \\ b_1(x,h) \\ \vdots \\ b_k(x,h) \end{bmatrix}}_{b} = \underbrace{\begin{Bmatrix} 1 \\ 0 \\ \vdots \\ 0 \end{Bmatrix}}_{I}$$
(3.50)

or

$$Mb = I$$
(3.51)

where M is a moment matrix, b is a vector of coefficients, I is a vector of constants.

After determining the coefficients $b_I(x,h)$, the smoothing function expressed in equation (3.46) can be calculated, which ensures the discretized consistency to k-th order. Therefore, the particle consistency restoring process actually gives an approach to construct some kind of smoothing function for the SPH methods. This approach is very useful. A similar approach called reproducing kernel particle method (RKPM) was proposed earlier in 1995 by Liu and his co-workers (Liu and Chen, 1995; Liu et al., 1995a, b), by correcting the smoothing functions used in the traditional SPH method. In the RKPM, the reproducing kernel function $\overline{W}$ was developed by multiplying the function expressed in equation (3.46) with another window function $W_w(x - x_j, h)$ that is usually a traditional SPH smoothing function such as the cubic spline:

$$\overline{W}(x - x_j, h) = W(x - x_j, h)W_w(x - x_j, h) \qquad (3.52)$$

The $k + 1$ coefficients $b_l(x, h)$ can be determined by solving the same equation

(3.50) except that $m_k(x, h) = \sum_{j=1}^{N} (\dfrac{x - x_j}{h})^k W_w(x - x_j, h)\Delta x_j$.

Comparing with the traditional smoothing function, which is only dependent on the particle distance and applicable for all the particles, the consistency restored smoothing function is particle-wise. It is, therefore, depends on both the distance and position of the interacting particles. The cost-effectiveness for this approach in constructing particle-wise smoothing functions needs to be considered since it will require additional CPU time to solve the particle-wise equation (3.50) for all the particles. Moreover, since all particles are moving, the particle location is changing as well. Hence, the particle-wise smoothing functions need to be computed for every time step. Another problem is that, to solve equation (3.50), the moment matrix M is required to be non-singular. Therefore, the particle distribution must satisfy certain conditions rather than the arbitrary distribution as in the original SPH method.

As far as the approximation is concerned, the restoring of particle consistency is an improvement on the accuracy of the particle approximation, if a method can be used to ensure a non-singular moment matrix M. However, it is noted that restoring the consistency in discrete form leads to some problems in simulating hydrodynamic problems.

- First, the resultant smoothing function is negative in some parts of the region, which can leads to unphysical representation of some field variables, such as negative density, negative energy that can lead to a breakdown of the entire computation.
- Second, the resultant smoothing function may not be monotonically decreasing with the increase of the particle (node) distance.
- Third, the function may not be symmetric. These violate the previously discussed properties of a traditional smoothing function and may result in some serious consequence (see, Chapter 4).

Therefore, although the particle consistency may be restored in this approach, for solving hydrodynamic problems, special cares must be taken.

3.3   Constructing smoothing functions

3.3.1   Constructing smoothing functions in polynomial form

As discussed earlier, the general process of restoring particle inconsistency actually gives a new way to systemically construct the smoothing functions. In this section, we employ the continuous equations (3.33) and (3.34) to construct smoothing functions that can be expressed analytically in explicit forms.

If the smoothing function is assumed to be a polynomial dependent only on the relative distance of the concerned points, it can be assumed to have the following form in the support domain with an influence width of κh.

$$W(x-x',h) = W(R) = a_0 + a_1 R + a_2 R^2 + \ldots + a_n R^n \tag{3.53}$$

It is clear that the smoothing function is a distance function since it depends on the relative distance.

For equation (3.53), if the i-th order derivative of the smoothing function $W^{(i)}(x-x',h)$ at $R=0$ exists, we should have

$$W^{(i)}(0) = \begin{cases} i!a_i & x-x' \to 0_+ \\ (-1)^i i!a_i & x-x' \to 0_- \end{cases} \tag{3.54}$$

where $i=1, 2, \ldots, n$; $(x-x') \to 0_-$ means $(x-x')$ approaches zero from RHS of zero; $(x-x') \to 0_+$ means $(x-x')$ approaches zero from LHS of zero. Hence, $W^{(i+1)}(x-x',h)$ exists at $R=0$ only if the following expressions are satisfied.

$$\begin{cases} a_1, a_2, \cdots, a_i = 0 & i=\text{even} \\ a_1, a_2, \cdots, a_{i-1} = 0 & i=\text{odd} \end{cases} \tag{3.55}$$

In equations (3.33) and (3.34), the first and second derivatives of the smoothing function are involved. Therefore, for the second derivative of the smoothing function to exist, a_1 should vanish in equation (3.53), and possible form of the smoothing function should be

$$W(x - x', h) = W(R) = a_0 + a_2 R^2 + ... + a_n R^n \qquad (3.56)$$

Substituting W into the conditions (3.33) and (3.34), the parameters a_0, a_2, ..., a_n can be calculated from the resultant linear equations, and then the smoothing function can be determined. It can be seen that the resultant expression for the smoothing function can be used for the general purpose of integral representation or the kernel approximation of a function and its derivatives.

3.3.2   Some related issues

Non-negativity vs. high-order reproducibility

As mentioned above, the conditions expressed in equations (3.33) and (3.34) provide a set of general conditions to construct smoothing functions. Similar to the reasons shown in the particle consistency restoring process, a smoothing function derived from this set of conditions will not necessarily be positive in the entire support domain, especially when high order reproduction is required. Such a negative smoothing function may result in unphysical solutions such as negative density (mass) and negative energy. For this reason, the smoothing functions used in the SPH method in the literatures are generally non-negative. On the other hand, according to equation (3.33), for the even moments ($k = 2, 4, 6...$) to be zero, the smoothing function have to be negative in some parts of the region. Therefore, one cannot have both non-negativity and high-order reproducibility at the same time.

Center peak value of the smoothing function

The center peak value of a smoothing function is very important since it determines how much the particle itself will contribute to the approximation. he above discussions show that if a positive smoothing function is used, the highest order of accuracy for the function approximation (expressed in equation (3.13)) is second order. Therefore, the integration $M_2 = \int_\Omega (x - x')^2 W(x - x', h) dx'$ can be used as a rough indicator to measure the accuracy of the function approximation. The smaller the second moment M_2 is, the more accurate the function approximation is. Note that the center peak value of the smoothing function is closely related to M_2, a positive smoothing function with a large center peak value will have a smaller second moment M_2. Therefore, as far as the kernel approximation or the integral representation is concerned, larger center peak value of the smoothing function means better accuracy. This implies that a smoothing function that is close to the Delta function is more accurate in terms of kernel approximations. This is not a surprise, because, as shown in equation (3.11), when W is the Delta function, the integral approximation is exact for all orders.

In summary, in constructing the smoothing function, the center peak value is a factor that needs to be considered.

Piecewise smoothing function

In some circumstances, the piecewise smoothing function is preferred since the shape of the piecewise smoothing function is easier to be controlled by changing the number of the pieces and the locations of the connection points. For example, consider the general form of a smoothing function with two pieces,

$$W(R) = \begin{cases} W_1(R) & 0 \le R < R_1 \\ W_2(R) & R_1 \le R < R_2 \\ 0 & R_2 \le R \end{cases} \tag{3.57}$$

the function itself and the first two derivatives at the connection points should be continuous, i.e.,

$$W_1(R_1) = W_2(R_1) \tag{3.58}$$

$$W_1'(R_1) = W_2'(R_1) \tag{3.59}$$

$$W_1''(R_1) = W_2''(R_1) \tag{3.60}$$

Considering the requirements at these points as well as the compact support property, one possible form of the smoothing function is

$$W(R) = \begin{cases} b_1(R_1 - R)^n + b_2(R_2 - R)^n & 0 \le R < R_1 \\ b_2(R_2 - R)^n & R_1 \le R < R_2 \\ 0 & R_2 \le R \end{cases} \tag{3.61}$$

For more pieces, similar expressions can be used to construct piecewise smoothing functions.

3.3.3   Examples of constructing smoothing functions

Example 3.1   Dome-shaped quadratic smoothing function

If the smoothing function is a quadratic expression of R, and the scaling factor $\kappa = 1$, the smoothing function should be in the following form (see equation (3.56)).

$$W(R,h) = (a_0 + a_2 R^2) = (a_0 + a_2(\frac{r}{h})^2) \tag{3.62}$$

In one-dimensional space, using the first equation in (3.33) (normalization or unity condition) yields

$$1 = 2\int_0^h (a_0 + a_2(\frac{r}{h})^2)dr$$
$$= 2h(a_0 + \frac{a_2}{3}) \tag{3.63}$$

For the smoothing function W to have the compact support property, we have (see, equation (3.34))

$$W(R,h)|_{R=1} = 0 \tag{3.64}$$

which leads to

$$a_0 + a_2 = 0 \tag{3.65}$$

Solving the linear equations (3.63) and (3.65), the coefficients a_0 and a_2 are obtained as $a_0 = 3/4h$ and $a_2 = -3/4h$. Therefore the quadratic kernel in one-dimensional space is found as

$$W(R,h) = \alpha_1(1 - R^2) \tag{3.66}$$

where $\alpha_1 = 3/4\pi$.

The unity condition for the two-dimensional space is

$$1 = \int_0^h (a_0 + a_2 (\frac{r}{h})^2)(2\pi r)dr$$

$$= 2\pi h^2 (\frac{a_0}{2} + \frac{a_2}{4}) \tag{3.67}$$

and that for the three-dimensional space is

$$1 = \int_0^h (a_0 + a_2 (\frac{r}{h})^2)(4\pi r^2)dr$$

$$= 4\pi h^3 (\frac{a_0}{3} + \frac{a_2}{5}) \tag{3.68}$$

Solving the linear equations (3.67) and (3.65), the coefficients a_0 and a_2 for the two-dimensional space can be obtained. The resultant smoothing function in two-dimensional space can be written in the same form as equation (3.66) except that α_d is $2/\pi h^2$.

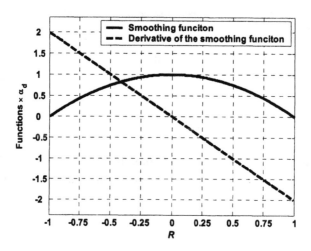

Figure 3.9 The dome-shaped quadratic smoothing function and its first derivative. α_d is $3/4h$, $2/\pi h^2$ and $15/8\pi h^3$ in one-, two- and three-dimensional space, respectively.

Solving the linear equations (3.68) and (3.65), the coefficients a_0 and a_2 for the three-dimensional space can be obtained. The resultant smoothing function in three-dimensional space can be written in the same form as equation (3.66) except that α_d is $15/8\pi h^3$.

This smoothing function was used in the grid free finite integration method (FIM) by Hicks and Liebrock (2000). Note that since the first derivative is not zero on the boundary of the support domain, this smoothing function does not have compact support for its first derivative (as shown in Figure 3.9).

Example 3.2 Quartic smoothing function

The quartic smoothing function used by Lucy (1977) can be constructed as follows.

As the smoothing function is a quartic expression of R, and the scaling factor $\kappa = 1$, the quartic smoothing function should be in the following form.

$$\begin{aligned}
W(R,h) &= (a_0 + a_2 R^2 + a_3 R^3 + a_3 R^4) \\
&= (a_0 + a_2 (\frac{r}{h})^2 + a_3 (\frac{r}{h})^3 + a_4 (\frac{r}{h})^4)
\end{aligned} \tag{3.69}$$

In order to get the four coefficients a_0, a_2, a_3 a_4, four equations are necessary. Using equation (3.35), if the smoothing function and its first two derivatives have compact supports, we have

$$W(R,h)|_{R=1} = a_0 + a_2 + a_3 + a_4 = 0 \tag{3.70}$$

$$W'(R,h)|_{R=1} = 2a_2 + 3a_3 + 4a_4 = 0 \tag{3.71}$$

$$W''(R,h)|_{R=1} = 2a_2 + 6a_3 + 12a_4 = 0 \tag{3.72}$$

Using the first expression (unity condition) in equation (3.33) yields, in one-dimensional space

$$\begin{aligned}
&2\int_0^h (a_0 + a_2 (\frac{r}{h})^2 + a_3 (\frac{r}{h})^3 + a_4 (\frac{r}{h})^4) dr \\
&= 2h(a_0 + \frac{a_2}{3} + \frac{a_3}{4} + \frac{a_4}{5}) = 1
\end{aligned} \tag{3.73}$$

in two-dimensional space

$$\int_0^h (a_0 + a_2(\frac{r}{h})^2 + a_3(\frac{r}{h})^3 + a_4(\frac{r}{h})^4)(2\pi r)dr$$

$$= 2\pi h^2 (\frac{a_0}{2} + \frac{a_2}{4} + \frac{a_3}{5} + \frac{a_4}{6}) = 1 \tag{3.74}$$

and in three-dimensional space,

$$\int_0^h (a_0 + a_2(\frac{r}{h})^2 + a_3(\frac{r}{h})^3 + a_4(\frac{r}{h})^4)(4\pi r^2)dr$$

$$= 4\pi h^3 (\frac{a_0}{3} + \frac{a_2}{5} + \frac{a_3}{6} + \frac{a_4}{7}) = 1 \tag{3.75}$$

Solving equations (3.70), (3.71), (3.72) with any one of equations (3.73), (3.74) and (3.75) simultaneously, the four coefficients can be calculated. The obtained quartic smoothing function is

$$W(R,h) = \alpha_d (1 - 6R^2 + 8R^3 - 3R^4) \tag{3.76}$$

with α_d is $\dfrac{5}{4h}$, $\dfrac{5}{\pi h^2}$ and $\dfrac{105}{16\pi h^3}$ in one-, two- and three-dimensional space, respectively. This smoothing function is exactly the one used by Lucy (1977) in his original paper on SPH. The smoothing function and its first derivative are plotted in Figure 3.1.

Example 3.3 Piecewise cubic smoothing function

If the smoothing function is a piecewise cubic function with the connection point at $R = 1$ and $\kappa = 2$, the smoothing function can have the form expressed in equation (3.61) as follows.

$$W(R,h) = \begin{cases} b_1(1-\frac{r}{h})^3 + b_2(2-\frac{r}{h})^3 & 0 \leq r < h \\ b_2(2-\frac{r}{h})^3 & h \leq r < 2h \\ 0 & r \geq 2h \end{cases} \tag{3.77}$$

Note that the smoothing function shown in equation (3.77) has a compact support. The first two derivatives of this smoothing function also have compact a support, only if these two derivatives exist. For the first two derivatives to exist at the origin, according to equation (3.56), the linear coefficient of the smoothing function should vanish. Therefore, we have

$$-(b_1 + 4b_2) = 0 \tag{3.78}$$

Using the first expression in equation (3.33) (unity condition) in one-dimensional space yields

$$2\left(\int_0^h \left(b_1(1-\frac{r}{h})^3 + b_2(2-\frac{r}{h})^3 \right) dr + \int_h^{2h} b_2(2-\frac{r}{h})^3 dr \right) = 1 \tag{3.79}$$

Equation (3.79) can be simplified as

$$2h(\frac{b_1}{4} + 4b_2) = 1 \tag{3.80}$$

Solving equations (3.78) and (3.80), the resultant parameters in one-dimensional space is found to $b_1 = -\frac{2}{3h}$ and $b_2 = \frac{1}{6h}$. For two- or three-dimensional space, the only difference is the dimension-dependent coefficient α_d which is used to ensure the unity condition in two- or three-dimensional space. The constructed piecewise cubic smoothing function is exactly the most commonly used cubic spline smoothing function (Monaghan and Lattanzio, 1985) in the SPH literatures. The final expression is given in equation (3.6). The smoothing function and its first derivative are plotted in Figure 3.3.

Example 3.4   Piecewise quintic smoothing function

If the smoothing function is a piecewise quintic function with the connection point at $R_1 = 1$ and $R_2 = 2$, the scaling factor $\kappa = 3$, the smoothing function can also take the similar form expressed in equation (3.61). Following the same procedure as in constructing the above piecewise cubic smoothing function, the piecewise quintic smoothing function can be constructed. Using the first expression in equation (3.33), considering the existence of the first two derivatives of the smoothing function, and defining the center peak value, the resultant parameters are $b_1 = 15$, $b_2 = -6$, and $b_3 = 1$. α_d is then found to be

$\frac{1}{120h}$, $\frac{7}{478\pi h^2}$ and $\frac{3}{359\pi h^3}$ in one-, two- and three-dimensional space, respectively. The constructed smoothing function is the one employed by Morris (1997) in simulating low Reynolds number incompressible flow. The final expression is given in equation (3.8). The smoothing function and its first derivative are plotted in Figure 3.5.

Example 3.5   A new quartic smoothing function

The idea of constructing the quartic kernel expressed in equation (3.76) can be changed slightly. Using the following conditions:

- the unity condition,
- compact support of the smoothing function,
- compact support of the first derivative of the smoothing function,
- the centre peak value,

another quartic smoothing function can be constructed. The final expression is given as follows (Liu, Liu and Lam, 2002; Liu et al., 2003b).

$$W(R,h) = \alpha_d \begin{cases} (\frac{2}{3} - \frac{9}{8}R^2 + \frac{19}{24}R^3 - \frac{5}{32}R^4) & 0 \le R \le 2 \\ 0 & R > 2 \end{cases} \tag{3.81}$$

where α_d is $1/h$, $15/7\pi h^2$ and $315/208\pi h^3$ in one-, two- and three-dimensional space, respectively. Note that the centre peak value of this quartic smoothing function is defined as $2/3$. The quartic smoothing function itself and the first two derivatives are shown in Figure 3.10. This quartic function satisfies the normalization condition, the function itself and first derivative have compact support. The presented quartic function is very close to the most commonly used cubic spline (equation (3.6)) with the same center peak value of $2/3$, and monotonically decreases with the increase of the distance. The new quartic function has several advantages over the cubic function.

1) Considering the integration of $M_2 = \int(x-x')^2 W(x-x',h)dx'$, this new quartic function yields a smaller value for the second moment than that using the cubic smoothing function, and therefore as far as the kernel approximation is concerned, this quartic function should, in general, produce more accurate results.
2) Since the stability properties of SPH depend strongly on the second derivative of the smoothing function (Swegle et al., 1994; Morris, 1996; Morris et al., 1997), a smoother smoothing function generally results in a more stable SPH solution. The presented quartic function has a smoother second derivative than the piecewise linear second derivative

of the cubic function, and therefore the stability properties should be superior to those of the cubic function.

3) In addition, this quartic smoothing function has only one piece.

The disadvantage of this quartic smoothing function is that the second derivative is not zero on the boundary of the support domain. This quartic smoothing function and other smoothing function discussed in this chapter are summarized in Table 3.1.

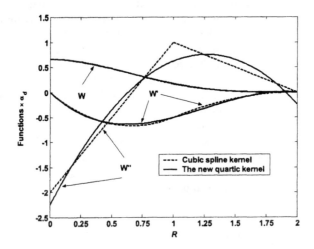

Figure 3.10 The new quartic smoothing function constructed by Liu et al. (2002) using the generalized procedure. α_d is $1/h$, $15/7\pi h^2$ and $315/208\pi h^3$ in one-, two- and three-dimensional space, respectively. This new quartic smoothing function has only one piece, but has very similar shape and first derivative as the popular cubic spline smoothing function. The second derivative of the new quartic smoothing function is continuous.

3.4 Numerical tests

The general approach of constructing smoothing functions has been demonstrated in the examples of the development of the above smoothing functions. The efficiency and accuracy of the constructed smoothing functions have also been shown in various literatures for all these existing smoothing functions.   Presented here are two numerical examples using the newly constructed quartic smoothing function (equation (3.81)). These examples are carried out using the SPH formulation and code to be detailed in the later chapters.   At this stage, readers are advised to pay more attention on the performance of the new quartic smoothing function rather than how the simulation is carried out, as that will be detailed in the later chapters.

Example 3.6   Shock tube problem

The shock tube problem is a good numerical benchmark and was comprehensively investigated by many SPH researchers when studying the SPH method (Monaghan et al., 1983; Hernquist et al., 1989), in which the cubic spline function is employed as the smoothing function.   The shock-tube is a long straight tube filled with gas, which is separated by a membrane into two parts of different pressures and densities.   The gas in each part is initially in an equilibrium state of constant pressure, density and temperature.   When the membrane is taken away suddenly, a shock wave, a rarefaction wave and a contact discontinuity will be produced. The shock wave moves into the region with lower density gas; the rarefaction wave travels into the region with higher density gas; while the contact discontinuity forms near the center and travels into the low-density region behind the shock. Exact solution is available for comparison for this one-dimensional problem.

In this example, the newly constructed quartic function is used as the smoothing function to simulate this shock tube problem. The initial conditions are same as what Hernquist (1989) used, which was introduced by Monaghan (1983) from (Sod, 1978).

$$x \leq 0 \quad \rho = 1 \quad v = 0 \quad e = 2.5 \quad p = 1 \quad \Delta x = 0.001875$$

$$x > 0 \quad \rho = 0.25 \quad v = 0 \quad e = 1.795 \quad p = 0.1795 \quad \Delta x = 0.0075$$

where ρ, p e, and v are the density, pressure, internal energy, and velocity of the gas, respectively. Δx is the particle spacing.

There are 400 particles used in the simulation. All particles have the same mass of $m_i = 0.001875$. 320 particles are evenly distributed in the high-density region [-0.6, 0.0], and 80 particles are evenly distributed in the low density region [0.0, 0.6]. The purpose of this initial particle distribution is to obtain required discontinuous density profile along the tube.

The following equation of state for the ideal gas is used in the simulation.

$$p = (\gamma - 1)\rho e \qquad (3.82)$$

where $\gamma = 1.4$. The time step is set as 0.005 s and the simulation ran for 40 time steps. In resolving the shock, the Monaghan type artificial viscosity (Monaghan,

1992) is used, which also solves to prevent unphysical penetration. The details of the artificial viscosity will be given in Chapter 4.

Figure 3.11, Figure 3.12, Figure 3.13 and Figure 3.14 show the density, pressure, velocity and internal energy profiles, respectively. The solid lines are the exact solution. It can be seen that the obtained results from the new approach agree well with the exact solution in the region of $[-0.4, 0.4]$. The shock is observed around $x = 0.3$; and is resolved within several smoothing lengths. The rarefaction wave is located between $x = -0.3$ and $x = 0$. The contact discontinuity is between $x = 0.1$ and $x = 0.2$. In this simulation, the boundary is not specially treated since for the instant at $t = 0.20$ s, the boundary effect has not propagated to the shock area of our interest, and therefore the shock physics is not yet affected by the boundary.

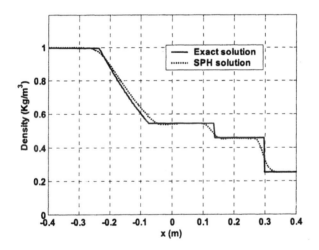

Figure 3.11 Density profiles in the shock tube at $t = 0.20$ s. The shock is observed around $x = 0.3$; and is resolved within several smoothing lengths. The rarefaction wave is located between $x = -0.3$ and $x = 0$. The contact discontinuity is between $x = 0.1$ and $x = 0.2$.

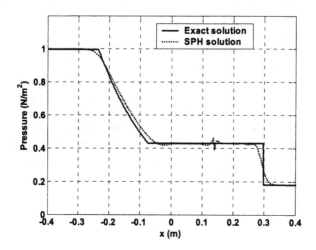

Figure 3.12 Pressure profiles in the shock tube at $t = 0.20$ s. The shock is observed around $x = 0.3$; and is resolved within several smoothing lengths. The rarefaction wave is located between $x = -0.3$ and $x = 0$. The contact discontinuity is between $x = 0.1$ and $x = 0.2$.

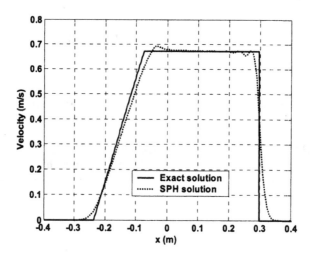

Figure 3.13 Velocity profiles in the shock tube at $t = 0.20$ s. The shock is observed around $x = 0.3$; and is resolved within several smoothing lengths. The rarefaction wave is located between $x = -0.3$ and $x = 0$.

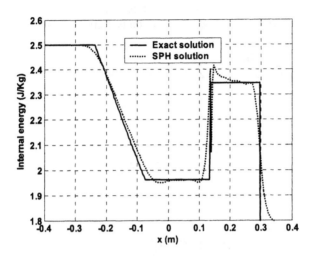

Figure 3.14 Internal energy profiles in the shock tube at $t = 0.20$ s. The shock is observed around $x = 0.3$; and is resolved within several smoothing lengths. The rarefaction wave is located between $x = -0.3$ and $x = 0$. The contact discontinuity is between $x = 0.1$ and $x = 0.2$.

Example 3.7   Two-dimensional heat conduction

The newly constructed quartic smoothing function and the SPH approximation procedure are applied to simulate a simple two-dimensional heat conduction problem. The heat conduction equation is a parabolic PDE given by

$$\rho C \frac{\partial T}{\partial t} = \nabla \cdot (\bar{k} \nabla T) + Q \tag{3.83}$$

where ρ is density, C is the heat capacity, Q is the heat source, and $\bar{k}$ is the heat conduction coefficient. They are all constants in the simulation. T is the field variable of temperature to-be-calculated. In multi-dimensional space, the heat conduction equation for particle i is given as

$$\rho_i C_i (\frac{\partial T}{\partial t})_i = \bar{k} T_{i,\alpha\alpha} + Q_i \tag{3.84}$$

Utilizing equation (3.24) and performing the particle approximation, an SPH equation for multi-dimensional heat conduction is derived as follows,

$$\rho_i C_i (\frac{\partial T}{\partial t})_i = \bar{k} \sum_{j=1}^{N} T_j W_{ij,\alpha\alpha} \frac{m_j}{\rho_j} + Q_i \tag{3.85}$$

where

$$W_{ij,\alpha\beta} = \frac{\partial W_{ij}}{\partial x_j^\alpha \partial x_j^\beta} = \frac{\partial R}{\partial x_j^\alpha}(\frac{d^2 W_{ij}}{dR^2} - \frac{1}{R}\frac{dW_{ij}}{dR})\frac{\partial R}{\partial x_j^\alpha} + \frac{\delta_{\alpha\beta}}{Rh^2}\frac{dW_{ij}}{dR} \tag{3.86}$$

in which

$$\frac{\partial R}{\partial x_j^\alpha} = \frac{1}{h}\frac{x_j^\alpha - x_i^\alpha}{|x_j - x_i|} \tag{3.87}$$

In the above equations, α and β is the number of the dimension of the problem, and

$$\delta_{\alpha\beta} = \begin{cases} 1 & \alpha = \beta \\ 0 & \alpha \neq \beta \end{cases} \tag{3.88}$$

In the numerical example, the geometry of the two-dimensional simulation is set to be a square of 2 m $\times$ 2 m, with the origin of the coordinate located on the center of the square. The material properties used are $\bar{k} = 1$, $\rho = 1$, $C = 1$, and $Q = 10$. The initial temperatures and the temperature on the four boundaries are simply set as $0\,^\circ C$. The time history of the temperature is to be simulated.

The SPH particle distribution used in this simulation is shown in Figure 3.15. A total of 441 particles are used in the simulation, of which 80 particles are located on the boundaries, 361 (19×19) particles are located inside the square domain. The boundary particles are imposed with the boundary value in the temperature evolution process. A simulation is also carried out using FEM with a mesh of 312 elements as shown in Figure 3.16. The simulation is carried out up to 0.1 s with a time step of 0.001 s. Figure 3.17 and Figure 3.18 are the resultant temperature contours obtained by SPH and FEM, respectively. It can be seen that the temperature contours are in a very good agreement. The maximal temperature obtained by SPH is about 2% lower than its counterpart obtained using FEM.

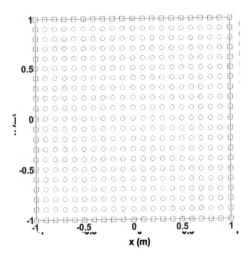

Figure 3.15 The SPH particle distributions for the two-dimensional heat conduction problem. The square markers represent boundary particles, while the circle markers represent the interior particles. The boundary and initial temperatures are set as 0 °C.

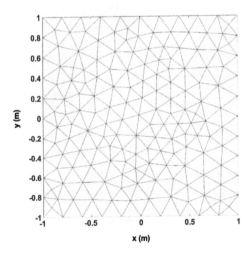

Figure 3.16 The triangular mesh used for the two-dimensional heat conduction analysis using FEM.

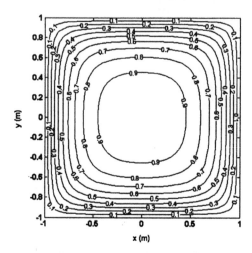

Figure 3.17 The temperature contour obtained by the SPH method for the two-dimensional heat conduction problem.

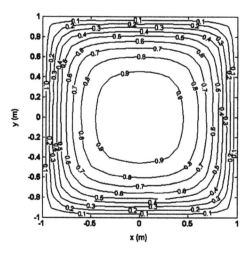

Figure 3.18 The temperature contour obtained by FEM for the two-dimensional heat conduction problem.

3.5   Concluding remarks

The smoothing function is very important in the smoothed particle hydrodynamics (SPH) as well as other meshfree methods since it determines the pattern of the approximation and the effective support domain of a point or particle. In this chapter, based on the Taylor series expansion on the SPH formulations for a function and its derivative approximations, a generalized approach of constructing smoothing functions is introduced. The resultant conditions are systematically derived based on the requirement of the kernel approximations or integral representation for a function and its derivatives.

The particle consistency can be restored through constructing particle-wise smoothing functions for each individual particle. The particle consistency restoration can greatly improve the accuracy of the SPH particle approximation, but may leads problems when these smoothing functions are used in simulating hydrodynamics problems.

We recap the following points of interest.

- The two sets of conditions expressed in equations (3.33) and (3.34) are useful in constructing consistent smoothing functions.
- The discrete and consistent smoothing function has a general form of equation (3.46), where the coefficient vector b can be determined by solving the moment equation (3.50), provided that the moment matrix M is not singular.
- The use of negative (partially) smoothing functions in SPH can lead to unphysical results for hydrodynamic problems or problems where the density or energy is a field variable.
- The center peak value of the smoothing function plays a role in the accuracy of the SPH kernel approximation.
- Piecewise construction of smoothing functions is useful and provides flexibility in smoothing function construction, if proper conditions on the connection points are considered.
- A new quartic smoothing function with some advantages is also constructed and applied to simulate the one-dimensional shock tube problem and a simple two-dimensional heat conduction problem. The results show a good agreement with those obtained using other methods.

Table 3.1 Summary of smoothing functions (Defined for a point or particle at x;
$R = \dfrac{r}{h} = \dfrac{|x - x'|}{h}$; r is the distance between two points or particles at x and x')

Name	Smoothing function $W(R,h)$
Quartic (Lucy, 1977)	$\alpha_d (1+3R)(1-R)^3 \qquad R \leq 1$
Guassian (Gingold and Monaghan, 1977)	$\alpha_d e^{-R^2}$
Piecewise cubic spline (Monaghan and Lattanzio, 1985)	$\alpha_d \begin{cases} \frac{2}{3} - R^2 + \frac{1}{2}R^3 & 0 \leq R < 1 \\ \frac{1}{6}(2-R)^3 & 1 \leq R \leq 2 \end{cases}$
Piecewise quartic (Morris, 1996)	$\alpha_d \begin{cases} (R+2.5)^4 - 5(R+1.5)^4 + 10(R+0.5)^4 & 0 \leq R < 0.5 \\ (2.5-R)^4 - 5(1.5-R)^4 & 0.5 \leq R < 1.5 \\ (2.5-R)^4 & 1.5 \leq R \leq 2.5 \end{cases}$
Piecewise quintic (Morris, 1996)	$\alpha_d \begin{cases} (3-R)^5 - 6(2-R)^5 + 15(1-R)^5 & 0 \leq R < 1 \\ (3-R)^5 - 6(2-R)^5 & 1 \leq R < 2 \\ (3-R)^5 & 2 \leq R \leq 3 \end{cases}$
Quadratic (Johnson et al., 1996b)	$\alpha_d (\frac{3}{16}R^2 - \frac{3}{4}R + \frac{3}{4}) \qquad 0 \leq R \leq 2$
Super Gaussian (Monaghan and Lattanzio, 1985)	$\alpha_d (\frac{3}{2} - R^2)e^{-R^2} \qquad 0 \leq R \leq 2$
Dome-shaped quadratic (Hicks and Liebrock, 2000)	$\alpha_d (1-R^2) \qquad 0 \leq R \leq 1$
New quartic (Liu, Liu and Lam, 2002)	$\alpha_d (\frac{2}{3} - \frac{9}{8}R^2 + \frac{19}{24}R^3 - \frac{5}{32}R^4) \qquad 0 \leq R \leq 2$

Chapter 4

SPH for General Dynamic Fluid Flows

In the previous chapters, the essential formulations and the smoothing functions for the SPH method were provided. It is seen that the SPH kernel and particle approximations can be used for discretization of partial differential equations (PDEs).

This chapter describes the detailed formulation of the SPH method and some interesting applications to general fluid dynamic problems. The SPH formulation is derived by discretizing the Navier-Stokers equations spatially, leading to a set of ordinary differential equations (ODEs) with respect to time. This set of ODEs can then be solved via time integration. The SPH formulation can be employed to simulate general dynamic fluid flows. The following numerical aspects in implementing the SPH formulations will be discussed.

- artificial viscosity
- artificial heat
- physical viscosity
- variable smoothing length
- symmetrization of particle interaction
- zero-energy mode problem
- artificial compressibility
- solid boundary treatment
- choice of time step

Some modifications for these numerical procedures are made to the conventional SPH method to better suit the needs of dynamic fluid flow simulations. Nearest neighboring particle searching algorithms and the pair interaction technique for particle interactions are described. A three-dimensional SPH code is developed and applied to simulate different interesting fluid flow problems, which include incompressible flows with solid boundaries, free surface flows, and compressible flows. The SPH results are validated by comparisons with those obtained using other methods.

103

4.1   Introduction

Problems in computational fluid dynamics (CFD) are generally solved by employing the conventional grid-based numerical methods such as the finite difference method (FDM), finite volume method (FVM) and finite element method (FEM) (Anderson, 1995; Hirsch, 1988). These conventional numerical methods have been have dominated for a long time the subject of computational fluid dynamics. An important feature of these methods is that a corresponding Eulerian (for FDM and FVM) or Lagrangian (for FEM) grid is required as the computational frame to provide spatial discretization for the governing equations. However, when simulating some special problems with large distortions, moving material interfaces, deformable boundaries and free surfaces, these methods can encounter some difficulties. FEM cannot well resolve the problems with large mesh element distortion. The Eulerian methods are inefficient in treating moving material interfaces, deformable boundaries, free surfaces, etc. Though a large amount of work on the numerical schemes for solving the fluid dynamic problems have emerged, special difficulties still exist for problems with the above-mentioned features. Attempts have also been made to combine the best features of the FDM and FEM together by using two-grid systems (Lagrangian and Eulerian) like the methods of Coupled Eulerian Lagrangian (CEL) and Arbitrary Lagrange-Eulerian (ALE). These methods have been implemented in some commercial software packages such as MSC/Dytran (MSC/Dytran, 1997), DYNA2D and DYNA3D (Hallquist, 1988; 1986), LS-DYNA (Hallquist, 1998), and AUTODYN (Century dynamics, 1997). In these coupled methods, the computational information is exchanged either by mapping or by special interface treatment between these two types of grids. These approaches work well for many problems, but are rather complicated and also can cause some inaccuracy in numerical treatments such as the mapping process, as discussed in Chapter 1.

Recently, meshfree methods have been employed to simulate computational fluid dynamic problems (Brackbill et al., 1988; Lam et al., 2000; Liu et al., 1995a; Liu et al., 2001b; Masson and Baliga, 1992; Wu and Liu, 2003). One special approach of them is the smoothed particle hydrodynamics (SPH) (Lucy, 1977; Gingold and Monaghan, 1977). As discussed in the previous chapters, the SPH method is a meshfree particle method of adaptive and pure Lagrangian nature, and is attractive in treating large deformation, tracking moving interfaces or free surfaces, and obtaining the time history of the field variables.

In this chapter, the SPH method is formulated and used for simulating general dynamic fluid flows. The chapter is outlined as follows.

- In Section 2, the Navier-Stokes equations in Lagrangian form are briefly described.

- In Section 3, the SPH (kernel and particle) approximations are applied to the Navier-Stokes equations to derive the corresponding discrete equations of motion in SPH form. The advantages and disadvantages as well as the possible applications of these different forms of SPH equations are discussed.

- In Section 4, major numerical aspects of the SPH method are discussed, with some improvements.

- In Section 5, algorithms for dealing with particle interactions in the SPH method are discussed, which include the nearest neighboring particle searching algorithm and the pairwise interaction technique.

- In Section 6, the SPH method is applied to some problems of dynamic fluid flows including inviscid and viscous, compressible and incompressible flows.

- In Section 7, some remarks and conclusions are given.

4.2   Navier-Stokes equations in Lagrangian form

The basic governing equations of fluid dynamics are based on the following three fundamental physical laws of conservation.

1)   conservation of mass
2)   conservation of momentum
3)   conservation of energy

Different forms of equations can be employed to describe the fluid flows, depending on the specific circumstances (Anderson, 1995; Hirsch, 1988). As discussed in Chapter 1, there are two approaches for describing the physical governing equations, the Eulerian description and Lagrangian description. The Eulerian description is a spatial description, whereas the Lagrangian description is a material description. The fundamental difference of these two descriptions is that the Lagrangian description employs the total time derivative as the combination of local derivative and convective derivative. In conjunction with the Lagrangian nature of the SPH method, the governing equations in Lagrangian form will be discussed and employed in this section. The SPH equations of motion will be derived based on these governing equations in Lagrangian form.

4.2.1 Finite control volume and infinitesimal fluid cell

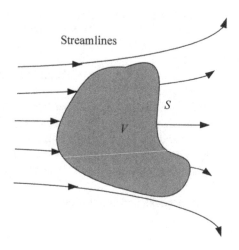

Figure 4.1 Lagrangian finite control volume V with a control surface S.

Consider a closed volume with finite dimensions in a fluid flow system as shown in Figure 4.1. This volume defines a control volume V associated with a closed control surface S which bounds the control volume. In the Lagrangian description, this control volume can move with the fluid flow such that the same material of the fluid is always staying inside the control volume. Therefore, though the fluid flow may result in expansion, compression, and deformation of the Lagrangian control volume, the mass of the fluids contained in the Lagrangian control volume remains unchanged. The Lagrangian control volume is reasonably large with finite dimensions in the flow system and the governing conservation laws can be directly applied to the fluids inside the control volume. Applying the conservation laws to the fluids to Lagrangian finite control volume can result in governing equations in integral form.

Another approach to obtain governing equations is to use the concept of infinitesimal fluid cell. The infinitesimal fluid cell (illustrated in Figure 4.2) can be regarded as a very small clump of fluids associated with a very small control volume δV and a very small control surface δS surrounding the volume δV. δV and δS can be the differential volume dV and the differential surface dS. This infinitesimal fluid cell, on one hand, is large enough so that the assumptions of continuum mechanics are valid, on the other hand, is small enough so that a field property inside it can be regarded as the same throughout the entire cell. Within the Lagrangian description, the infinitesimal fluid cell can move along a streamline with a vector velocity $v = (v_x, v_y, v_z)$ equal to the flow velocity at that point. Applying the conservation laws to the Lagrangian infinitesimal fluid cell, governing equations in the form of partial different equation can be established.

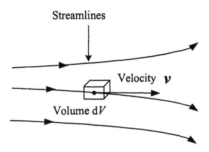

Figure 4.2 Infinitesimal fluid cell in Lagrangian description.

For a Lagrangian control volume, the movement of the fluids inside the control volume V leads to the change of the control surface S. The change of the control surface again results in a volume change of the control volume. As illustrated in Figure 4.3, the volume change of the control volume ΔV due to the movement of dS over a time increment Δt is

$$\Delta V = v\Delta t \cdot ndS \tag{4.1}$$

where n is the unit vector perpendicular to the surface dS.

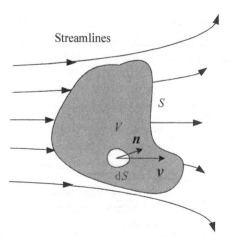

Figure 4.3 Volume change of the Lagrangian control volume.

The total volume change of the entire Lagrangian control volume is therefore the integral over the control surface S

$$\Delta V = \int_S v \Delta t \cdot n dS \tag{4.2}$$

Dividing both sides in equation (4.2) by Δt and applying the divergence theorem yield

$$\frac{\Delta V}{\Delta t} = \int_V \nabla \cdot v dV \tag{4.3}$$

where ∇ is the gradient operator.

If the Lagrangian control volume is downgraded (shrunk) to an infinitesimal fluid cell with volume of δV, so that a field property is equal throughout δV, the following equation can be obtained.

$$\frac{\Delta(\delta V)}{\Delta t} = (\nabla \cdot v) \int_V d(\delta V) = (\nabla \cdot v) \delta V \tag{4.4}$$

Therefore, the time rate of volume change for the infinitesimal fluid cell is

$$\frac{D(\delta V)}{Dt} = (\nabla \cdot v)\delta V \tag{4.5}$$

From equation (4.5), the velocity divergence becomes

$$\nabla \cdot v = \frac{1}{\delta V}\frac{D(\delta V)}{Dt} \tag{4.6}$$

It shows that the velocity divergence can be physically interpreted as the time rate of volume change per unit volume.

4.2.2 The continuity equation

The continuity equation is based on the conservation of mass. For a Lagrangian infinitesimal fluid cell with volume of δV, the mass contained in the control volume is

$$\delta m = \rho \delta V \tag{4.7}$$

where m and ρ are mass and density, respectively.

Since the mass is conserved in the Lagrangian fluid cell, the time rate of mass change is zero. Therefore, we have

$$\frac{D(\delta m)}{Dt} = \frac{D(\rho \delta V)}{Dt} = \delta V\frac{D\rho}{Dt} + \rho\frac{D(\delta V)}{Dt} = 0 \tag{4.8}$$

Equation (4.8) can be rewritten as

$$\frac{D\rho}{Dt} + \rho\frac{1}{\delta V}\frac{D(\delta V)}{Dt} = 0 \tag{4.9}$$

Considering equation (4.6), and replacing the second term in equation (4.9) with the velocity divergence, the continuity equation or the mass conservation equation in Lagrangian form is obtained as

$$\frac{D\rho}{Dt} = -\rho\nabla \cdot v \tag{4.10}$$

4.2.3 The momentum equation

The momentum equation is based on the conservation of momentum, which in the continuum mechanics, is represented by Newton's second law. According to Newton's second law, the net force on a Lagrangian fluid cell equals to its mass multiplying the acceleration of that fluid cell.

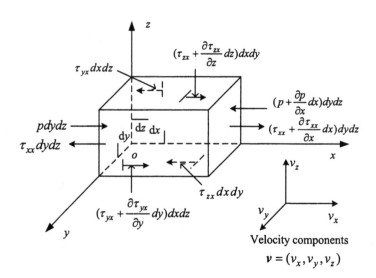

Figure 4.4 Forces in the *x* direction on a Lagrangian infinitesimal fluid cell.

As illustrated in Figure 4.4, the position vector is $x = (x, y, z)$, and the accelerations of the infinitesimal fluid cell in the three directions are $\dfrac{Dv_x}{Dt}$, $\dfrac{Dv_y}{Dt}$ and $\dfrac{Dv_z}{Dt}$, respectively. The net force on the fluid cell consists of body forces and surface forces. The body force may be the gravitational force, magnetic forces and other possible forces acting on the body of the entire fluid cell. The surface force includes

1) the pressure, which is imposed by the outside fluids surrounding the concerned fluid cell,
2) the shear and normal stress, which result in shear deformation and volume change, respectively.

In the x direction, all the forces acting on the Lagrangian infinite fluid cell are

$$-[(p + \frac{\partial p}{\partial x} dx) - p]dydz +$$

$$[(\tau_{xx} + \frac{\partial \tau_{xx}}{\partial x} dx) - \tau_{xx}]dydz +$$

$$[(\tau_{yx} + \frac{\partial \tau_{yx}}{\partial y} dy) - \tau_{yx}]dxdz + \qquad (4.11)$$

$$[(\tau_{zx} + \frac{\partial \tau_{zx}}{\partial z} dz) - \tau_{zx}]dxdy$$

$$= -\frac{\partial p}{\partial x} dxdydz + \frac{\partial \tau_{xx}}{\partial x} dxdydz + \frac{\partial \tau_{yx}}{\partial y} dxdydz + \frac{\partial \tau_{zx}}{\partial z} dxdydz$$

where p the is pressure, τ_{ij} is the stress in the j direction exerted on a plane perpendicular to the i axis. If the body force per unit mass in the x direction is F_x, Newton's second law can be written as

$$m\frac{dv_x}{dt} = \rho dxdydz \frac{dv_x}{dt}$$

$$= -\frac{\partial p}{\partial x} dxdydz \qquad (4.12)$$

$$+ \frac{\partial \tau_{xx}}{\partial x} dxdydz + \frac{\partial \tau_{yx}}{\partial y} dxdydz + \frac{\partial \tau_{zx}}{\partial z} dxdydz$$

$$+ F_x(\rho dxdydz)$$

Therefore the momentum equation in the x direction is

$$\rho\frac{Dv_x}{Dt} = -\frac{\partial p}{\partial x} + \frac{\partial \tau_{xx}}{\partial x} + \frac{\partial \tau_{yx}}{\partial y} + \frac{\partial \tau_{zx}}{\partial z} + \rho F_x \qquad (4.13)$$

Similarly, the momentum equations in the y and z directions are

$$\rho\frac{Dv_y}{Dt} = -\frac{\partial p}{\partial y} + \frac{\partial \tau_{xy}}{\partial x} + \frac{\partial \tau_{yy}}{\partial y} + \frac{\partial \tau_{zy}}{\partial z} + \rho F_y \qquad (4.14)$$

$$\rho \frac{Dv_z}{Dt} = -\frac{\partial p}{\partial z} + \frac{\partial \tau_{xz}}{\partial x} + \frac{\partial \tau_{yz}}{\partial y} + \frac{\partial \tau_{zz}}{\partial z} + \rho F_z \qquad (4.15)$$

For Newtonian fluids, the stress should be proportional to the strain rate denoted by ε through the dynamic viscosity μ

$$\tau_{ij} = \mu \varepsilon_{ij} \qquad (4.16)$$

where

$$\varepsilon_{ij} = \frac{\partial v_j}{\partial x_i} + \frac{\partial v_i}{\partial x_j} - \frac{2}{3}(\nabla \cdot v)\delta_{ij} \qquad (4.17)$$

where δ_{ij} is the Dirac delta function.

4.2.4 The energy equation

The energy equation is based on the conservation of energy, which is a representation of the first law of thermodynamics. The energy equation states that the time rate of energy change inside an infinitesimal fluid cell should equal to the summation of the net heat flux into that fluid cell, and the time rate of work done by the body and surface forces acting on that fluid cell. If neglecting the heat flux, and the body force, the time rate of change of the internal energy e of the infinitesimal fluid cell consists of following two parts.

1) the work done by the isotropic pressure multiplying the volumetric strain
2) the energy dissipation due to the viscous shear forces

Therefore, the energy equation can be written as follows.

$$
\begin{aligned}
\rho \frac{De}{Dt} = & -p\left(\frac{\partial v_x}{\partial x} + \frac{\partial v_y}{\partial y} + \frac{\partial v_z}{\partial z}\right) \\
& + \tau_{xx}\frac{\partial v_x}{\partial x} + \tau_{yx}\frac{\partial v_x}{\partial y} + \tau_{zx}\frac{\partial v_x}{\partial z} \\
& + \tau_{xy}\frac{\partial v_y}{\partial x} + \tau_{yy}\frac{\partial v_y}{\partial y} + \tau_{zy}\frac{\partial v_y}{\partial z} \\
& + \tau_{xz}\frac{\partial v_z}{\partial x} + \tau_{yz}\frac{\partial v_z}{\partial y} + \tau_{zz}\frac{\partial v_z}{\partial z}
\end{aligned}
\qquad (4.18)
$$

4.2.5 Navier-Stokes equations

In summary, the governing equations for dynamic fluid flows can be written as a set of partial differential equations in Lagrangian description. The set of partial differential equations is the well-known Navier-Stokes equations, which state the conservation of mass, momentum and energy. If the Greek superscripts α and β are used to denote the coordinate directions, the summation in the equations is taken over repeated indices, and the total time derivatives are taken in the moving Lagrangian frame, the Navier-Stokes equations consist of the following set of equations.

1) The continuity equation

$$\frac{D\rho}{Dt} = -\rho \frac{\partial v^{\beta}}{\partial x^{\beta}} \qquad (4.19)$$

2) The momentum equation (in the case of free external force)

$$\frac{Dv^{\alpha}}{Dt} = \frac{1}{\rho} \frac{\partial \sigma^{\alpha\beta}}{\partial x^{\beta}} \qquad (4.20)$$

3) The energy equation

$$\frac{De}{Dt} = \frac{\sigma^{\alpha\beta}}{\rho} \frac{\partial v^{\alpha}}{\partial x^{\beta}} \qquad (4.21)$$

In the above equations σ is the total stress tensor. It is made up of two parts, one part of isotropic pressure p and the other part of viscous stress τ.

$$\sigma^{\alpha\beta} = -p\delta^{\alpha\beta} + \tau^{\alpha\beta} \qquad (4.22)$$

For Newtonian fluids, the viscous shear stress should be proportional to the shear strain rate denoted by ε through the dynamic viscosity μ.

$$\tau^{\alpha\beta} = \mu\varepsilon^{\alpha\beta} \qquad (4.23)$$

where

$$\varepsilon^{\alpha\beta} = \frac{\partial v^{\beta}}{\partial x^{\alpha}} + \frac{\partial v^{\alpha}}{\partial x^{\beta}} - \frac{2}{3}(\nabla \cdot v)\delta^{\alpha\beta} \qquad (4.24)$$

If separating the isotropic pressure and the viscous stress, the energy equation can be rewritten as

$$\frac{De}{Dt} = -\frac{p}{\rho}\frac{\partial v^{\beta}}{\partial x^{\beta}} + \frac{\mu}{2\rho}\varepsilon^{\alpha\beta}\varepsilon^{\alpha\beta} \qquad (4.25)$$

4.3   SPH formulations for Navier-Stokes equations

4.3.1   Particle approximation of density

The density approximation is very important in the SPH method since density basically determines the particle distribution and the smoothing length evolution. There are two approaches to evolve density in the conventional SPH method. The first approach is the *summation density*, which directly applies the SPH approximations to the density itself. For a given particle i, the density with summation density approach is written in the form of

$$\rho_i = \sum_{j=1}^{N} m_j W_{ij} \qquad (4.26)$$

where N is the number of particles in the support domain of particle i, and m_j is the mass associated with particle j. W_{ij} is the smoothing function of particle i evaluated at particle j, and is closely related to the smoothing length h.

$$W_{ij} = W(x_i - x_j, h) = W(|x_i - x_j|, h) = W(R_{ij}, h) \qquad (4.27)$$

where $R_{ij} = \dfrac{r_{ij}}{h} = \dfrac{|x_i - x_j|}{h}$ is the relative distance of particle i and j, and r_{ij} is the distance between these two particles. Commonly used smoothing functions are listed in Table 3.1.

Note that W_{ij} has a unit of the inverse of volume. Equation (4.26) simply states that the density of a particle can be approximated by the weighted average of the densities of the particles in the support domain of that particle.

Another approach of particle approximation for density is the *continuity density*, which approximates the density according to the continuity equation using the concepts of SPH approximations plus some transformations. Different transformation or operation on the RHS of the continuity equation (4.19) may lead to different forms of density approximation equations. One possible way is that the SPH approximation is only applied to the velocity divergence part, while the density in the RHS of the continuity equation (4.19) is evaluated on the particle at which the gradient is evaluated, which is

$$\frac{D\rho_i}{Dt} = -\rho_i \sum_{j=1}^{N} \frac{m_j}{\rho_j} v_j^\beta \cdot \frac{\partial W_{ij}}{\partial x_i^\beta} \qquad (4.28)$$

Considering the following expressions of SPH kernel and particle approximations on the gradient of the unity, we have

$$\nabla 1 = \int 1 \cdot \nabla W(x - x', h) dx'$$

$$= \sum_{j=1}^{N} \frac{m_j}{\rho_j} \cdot \frac{\partial W_{ij}}{\partial x_i^\beta} = 0 \qquad (4.29)$$

which can be rewritten as

$$\rho_i \sum_{j=1}^{N} \frac{m_j}{\rho_j} v_i^\beta \cdot \frac{\partial W_{ij}}{\partial x_i^\beta} = \rho_i v_i^\beta \left(\sum_{j=1}^{N} \frac{m_j}{\rho_j} \cdot \frac{\partial W_{ij}}{\partial x_i^\beta} \right) \qquad (4.30)$$

By adding the RHS of equation (4.30), which is zero, into the RHS of equation (4.28), another form of density approximation equation is obtained as

$$\frac{D\rho_i}{Dt} = \rho_i \sum_{j=1}^{N} \frac{m_j}{\rho_j} v_{ij}^\beta \cdot \frac{\partial W_{ij}}{\partial x_i^\beta} \qquad (4.31)$$

where

$$v_{ij}^\beta = (v_i^\beta - v_j^\beta) \tag{4.32}$$

Equation (4.31) introduces velocity difference into the discrete particle approximation, and is usually preferred in the SPH formulations. This is because it accounts for the relative velocities of particle pairs in the support domain. Another benefit is that using the relative velocities in anti-symmetrized form serves to reduce errors arising from the particle inconsistency problem (see, Chapter 3) (Monaghan, 1988; 1982; 1985). Libersky et al. (1991; 1993) in their original work of applying the SPH method to hydrodynamics with material strength employed this density approximation equation.

A more popular form of continuity density is to apply the following identity to place the density inside the gradient operator

$$-\rho \frac{\partial v^\beta}{\partial x^\beta} = -\left(\frac{\partial (\rho v^\beta)}{\partial x^\beta} - v^\beta \cdot \frac{\partial \rho}{\partial x^\beta} \right) \tag{4.33}$$

Similarly, if the SPH approximation is only applied to every gradient, and the velocity at the outside of the second gradient is evaluated at the particle at which the gradients are evaluated, the most frequently used continuity density equation is obtained as

$$\frac{D\rho_i}{Dt} = \sum_{j=1}^{N} m_j v_{ij}^\beta \cdot \frac{\partial W_{ij}}{\partial x_i^\beta} \tag{4.34}$$

This equation shows clearly that the time rate of density change of a particle is closely related to the relative velocities between this particle and all the other particles in the support domain. The gradient of the smoothing function determines the contribution of these relative velocities.

There are advantages and disadvantages for both approaches of summation density and continuity density approximations. The density summation approach conserves the mass exactly since the integration of the density over the entire problem domain is exactly the total mass of all the particles. The continuity density approach does not (Fulk, 1994; Monaghan, 1992). However, the density summation approach has edge effect when being applied to particles at the edge of the fluid domain, and will smooth out the density of the concerned particles and thus lead to spurious results. However, the edge effect can be remedied by using boundary virtual particles or other methods that will be discussed later in this chapter. The edge effect is also called boundary particle deficiency (see, Section 4.4.8). Note that the edge effect not only occurs near the boundaries, but also near the material interfaces if particles from different materials are not allowed to take part in the summation. Another disadvantage for the density summation approach is that it needs more computational effort since the density must be evaluated before other parameters can be evaluated, and it is required to calculate the smoothing function itself. The continuity density approach does not need to calculate density before other parameters, therefore can save considerable computational efforts and is profitable in coding for algorithms of parallel processors.

The summation density approach seems more popular in practical applications of SPH, partly because it well represents the essence of the SPH approximation. Some modifications have been proposed to improve the accuracy of this approach. One possible improvement is to normalize the RHS of equation (4.26) with the SPH summation of the smoothing function itself over the neighboring particles (Randles and Libersky, 1996; Chen et al., 1999a; 2000):

$$\rho_i = \frac{\sum\limits_{j=1}^{N} m_j W_{ij}}{\sum\limits_{j=1}^{N} \left(\dfrac{m_j}{\rho_j}\right) W_{ij}} \tag{4.35}$$

This expression improves the accuracy near both the free boundaries and the material interfaces with a density discontinuity when the summation is taken only on particles from the same material. It is well suited for simulating general fluid flow problems without discontinuities such as shock waves. Derivation and detailed descriptions on the treatments like in equation (4.35) will be given in Chapter 5.

The summation density approach (including its modification in equation (4.35)), and the continuity density approach are all implemented in the SPH code attached in this book. For simulating general fluid phenomena, the modified summation density approach can yield better results. For simulating events with strong discontinuity (e.g. explosion, high velocity impact), the continuity density approach is preferred.

4.3.2 Particle approximation of momentum

The derivation of SPH formulations for particle approximation of momentum evolution is somewhat similar to the continuity density approach, and usually involves some transformations. Again, using different transformations can derive different forms of momentum approximation equations. Directly applying the SPH particle approximation concepts to the gradient on the RHS of the momentum equation (4.20) yields following equation

$$\frac{Dv_i^\alpha}{Dt} = \frac{1}{\rho_i} \sum_{j=1}^{N} m_j \frac{\sigma_j^{\alpha\beta}}{\rho_j} \frac{\partial W_{ij}}{\partial x_i^\beta} \tag{4.36}$$

Adding the following identity

$$\sum_{j=1}^{N} m_j \frac{\sigma_i^{\alpha\beta}}{\rho_i \rho_j} \frac{\partial W_{ij}}{\partial x_i^\beta} = \frac{\sigma_i^{\alpha\beta}}{\rho_i} \left(\sum_{j=1}^{N} \frac{m_j}{\rho_j} \frac{\partial W_{ij}}{\partial x_i^\beta} \right) = 0 \tag{4.37}$$

to equation (4.36) leads to

$$\frac{Dv_i^\alpha}{Dt} = \sum_{j=1}^{N} m_j \frac{\sigma_i^{\alpha\beta} + \sigma_j^{\alpha\beta}}{\rho_i \rho_j} \frac{\partial W_{ij}}{\partial x_i^\beta} \tag{4.38}$$

Equation (4.38) is a frequently used formulation to evolve momentum. One benefit of this symmetrized equation is that it reduces errors arising from the particle inconsistency problem (see, Chapter 3) (Monaghan, 1988; 1982; 1985).
   Considering the following identity

$$\frac{1}{\rho} \frac{\partial \sigma^{\alpha\beta}}{\partial x^\beta} = \frac{\partial}{\partial x^\beta} \left(\frac{\sigma^{\alpha\beta}}{\rho} \right) + \frac{\sigma^{\alpha\beta}}{\rho^2} \frac{\partial \rho}{\partial x^\beta} \tag{4.39}$$

and applying the SPH particle approximation to the gradients lead to

$$\frac{Dv_i^\alpha}{Dt} = \sum_{j=1}^{N} \frac{m_j}{\rho_j} \frac{\sigma_i^{\alpha\beta}}{\rho_j} \frac{\partial W_{ij}}{\partial x_i^\beta} + \frac{\sigma_i^{\alpha\beta}}{\rho_i^2} \sum_{j=1}^{N} \frac{m_j}{\rho_j} \rho_j \frac{\partial W_{ij}}{\partial x_i^\beta} \tag{4.40}$$

After a simple rearrangement, equation (4.40) can be written in the following form

$$\frac{Dv_i^\alpha}{Dt} = \sum_{j=1}^{N} m_j \left(\frac{\sigma_i^{\alpha\beta}}{\rho_i^2} + \frac{\sigma_j^{\alpha\beta}}{\rho_j^2} \right) \frac{\partial W_{ij}}{\partial x_i^\beta} \tag{4.41}$$

This is another very popular formulation for evolving momentum, and is frequently seen in the literatures. Similarly, this symmetrized equation reduces errors arising from the particle inconsistency problem.
   Using equation (4.22), equations (4.38) and (4.41) can be rewritten,

respectively, in more detail as follows.

$$\frac{Dv_i^\alpha}{Dt} = -\sum_{j=1}^{N} m_j \frac{p_i + p_j}{\rho_i \rho_j} \frac{\partial W_{ij}}{\partial x_i^\alpha} + \sum_{j=1}^{N} m_j \frac{\mu_i \varepsilon_i^{\alpha\beta} + \mu_j \varepsilon_j^{\alpha\beta}}{\rho_i \rho_j} \frac{\partial W_{ij}}{\partial x_i^\beta} \qquad (4.42)$$

and

$$\frac{Dv_i^\alpha}{Dt} = -\sum_{j=1}^{N} m_j (\frac{p_i}{\rho_i^2} + \frac{p_j}{\rho_j^2}) \frac{\partial W_{ij}}{\partial x_i^\alpha} + \sum_{j=1}^{N} m_j (\frac{\mu_i \varepsilon_i^{\alpha\beta}}{\rho_i^2} + \frac{\mu_j \varepsilon_j^{\alpha\beta}}{\rho_j^2}) \frac{\partial W_{ij}}{\partial x_i^\alpha} \qquad (4.43)$$

The first parts on the RHS of equations (4.42) and (4.43) are the SPH approximations for the pressure, and the second parts on the RHS of equations (4.42) and (4.43) are the SPH approximations for the viscous force. It is the second part that is concerned with the physical viscosity. Considering equation (4.24), the SPH approximation of $\varepsilon^{\alpha\beta}$ for particle i can be written as

$$\varepsilon_i^{\alpha\beta} = \sum_{j=1}^{N} \frac{m_j}{\rho_j} v_j^\beta \frac{\partial W_{ij}}{\partial x_i^\alpha} + \sum_{j=1}^{N} \frac{m_j}{\rho_j} v_j^\alpha \frac{\partial W_{ij}}{\partial x_i^\beta} - (\frac{2}{3} \sum_{j=1}^{N} \frac{m_j}{\rho_j} v_j \cdot \nabla_i W_{ij}) \delta^{\alpha\beta} \qquad (4.44)$$

Using equation (4.29), we have the following identities

$$\sum_{j=1}^{N} \frac{m_j}{\rho_j} v_i^\beta \frac{\partial W_{ij}}{\partial x_i^\alpha} = v_i^\beta \left(\sum_{j=1}^{N} \frac{m_j}{\rho_j} \frac{\partial W_{ij}}{\partial x_i^\alpha} \right) = 0 \qquad (4.45)$$

$$\sum_{j=1}^{N} \frac{m_j}{\rho_j} v_i^\alpha \frac{\partial W_{ij}}{\partial x_i^\beta} = v_i^\alpha \left(\sum_{j=1}^{N} \frac{m_j}{\rho_j} \frac{\partial W_{ij}}{\partial x_i^\beta} \right) = 0 \qquad (4.46)$$

$$\sum_{j=1}^{N} \frac{m_j}{\rho_j} v_i \cdot \nabla_i W_{ij} = v_i \cdot \left(\sum_{j=1}^{N} \frac{m_j}{\rho_j} \nabla_i W_{ij} \right) = 0 \qquad (4.47)$$

Subtracting these identities from equation (4.44) to include velocity differences into the viscous strain rate, the final SPH approximation of $\varepsilon^{\alpha\beta}$ for particle i is

$$\varepsilon_i^{\alpha\beta} = \sum_{j=1}^{N} \frac{m_j}{\rho_j} v_{ji}^\beta \frac{\partial W_{ij}}{\partial x_i^\alpha} + \sum_{j=1}^{N} \frac{m_j}{\rho_j} v_{ji}^\alpha \frac{\partial W_{ij}}{\partial x_i^\beta} - (\frac{2}{3} \sum_{j=1}^{N} \frac{m_j}{\rho_j} v_{ji} \cdot \nabla_i W_{ij}) \delta^{\alpha\beta} \qquad (4.48)$$

The SPH approximation of $\varepsilon^{\alpha\beta}$ for particle j can be obtained in the same way. After $\varepsilon^{\alpha\beta}$ for particle i and j have been calculated, the acceleration can be calculated by equation (4.42) or (4.43). This approach is straightforward and can model variable viscosity for different fluids.

4.3.3   Particle approximation of energy

For the evolution of the internal energy e in equation (4.25), the part involving strain rate can be approximated using equation (4.48). For the first part involving the pressure work, there are some different ways to derive the corresponding SPH formulations. Considering the following relation

$$-\frac{p}{\rho}\frac{\partial v^{\beta}}{\partial x^{\beta}} = \frac{p}{\rho^2}\left(-\rho\frac{\partial v^{\beta}}{\partial x^{\beta}}\right) = \frac{p}{\rho^2}\frac{D\rho}{Dt} \tag{4.49}$$

the pressure work can be approximated by directly using the continuity density formulation in equation (4.28), (4.31) or (4.34), while evaluating $\frac{p}{\rho^2}$ at the particle concerned. For example, using continuity density in equation (4.34), the pressure work can be approximated as

$$-\frac{p}{\rho}\frac{\partial v_i^{\beta}}{\partial x_i^{\beta}} = \frac{p_i}{\rho_i^2}\sum_{j=1}^{N} m_j v_{ij}^{\beta} \cdot \frac{\partial W_{ij}}{\partial x_i^{\beta}} \tag{4.50}$$

Another way of approximating the pressure work is to use the following identity

$$-\frac{p}{\rho}\frac{\partial v^{\beta}}{\partial x^{\beta}} = -\frac{\partial}{\partial x^{\beta}}\left(\frac{p v^{\beta}}{\rho}\right) + v^{\beta}\frac{\partial}{\partial x^{\beta}}\left(\frac{p}{\rho}\right) \tag{4.51}$$

which leads to

$$-\frac{p}{\rho}\frac{\partial v_i^{\beta}}{\partial x_i^{\beta}} = \sum_{j=1}^{N} m_j \frac{p_j}{\rho_j^2} v_{ij}^{\beta} \cdot \frac{\partial W_{ij}}{\partial x_i^{\beta}} \tag{4.52}$$

Adding equation (4.50) and (4.52) together yields the most popular form of SPH approximation of the pressure work

$$-\frac{p}{\rho}\frac{\partial v_i^\beta}{\partial x_i^\beta} = \frac{1}{2}\sum_{j=1}^{N} m_j (\frac{p_i}{\rho_i^2} + \frac{p_j}{\rho_j^2}) v_{ij}^\beta \cdot \frac{\partial W_{ij}}{\partial x_i^\beta} \qquad (4.53)$$

Note that the variables appear in the above equation symmetrically.

Approximating the time rate of change of density in equation (4.49) using equation (4.31) and evaluating $\dfrac{p}{\rho^2}$ at the particle concerned gives

$$-\frac{p}{\rho}\frac{\partial v_i^\beta}{\partial x_i^\beta} = \frac{p_i}{\rho_i}\sum_{j=1}^{N} \frac{m_j}{\rho_j} v_{ij}^\beta \cdot \frac{\partial W_{ij}}{\partial x_i^\beta} \qquad (4.54)$$

Using the following transformation

$$-\frac{p}{\rho}\frac{\partial v^\beta}{\partial x^\beta} = -\frac{1}{\rho}[\frac{\partial}{\partial x^\beta}(pv^\beta) - v^\beta \frac{\partial p}{\partial x^\beta}] \qquad (4.55)$$

another form of SPH approximation for pressure work is obtained as follows.

$$-\frac{p}{\rho}\frac{\partial v_i^\beta}{\partial x_i^\beta} = \frac{1}{\rho_i}\sum_{j=1}^{N} m_j \frac{p_j}{\rho_j} v_{ij}^\beta \cdot \frac{\partial W_{ij}}{\partial x_i^\beta} \qquad (4.56)$$

Adding equation (4.54) and (4.56) together yields another useful SPH equation for approximating the pressure work

$$-\frac{p}{\rho}\frac{\partial v_i^\beta}{\partial x_i^\beta} = \frac{1}{2}\sum_{j=1}^{N} m_j \frac{p_i + p_j}{\rho_i \rho_j} v_{ij}^\beta \cdot \frac{\partial W_{ij}}{\partial x_i^\beta} \qquad (4.57)$$

Note that the variables appear in the above equation symmetrically.

Since there are many forms of pressure work approximation, the SPH formulations for the internal energy evolution can naturally have many alternatives. The most frequently used are the following two forms

$$\frac{De_i}{Dt} = \frac{1}{2}\sum_{j=1}^{N} m_j (\frac{p_i}{\rho_i^2} + \frac{p_j}{\rho_j^2}) v_{ij}^\beta \frac{\partial W_{ij}}{\partial x_i^\beta} + \frac{\mu_i}{2\rho_i} \varepsilon_i^{\alpha\beta} \varepsilon_i^{\alpha\beta} \qquad (4.58)$$

and

$$\frac{De_i}{Dt} = \frac{1}{2} \sum_{j=1}^{N} m_j \frac{p_i + p_j}{\rho_i \rho_j} v_{ij}^{\beta} \frac{\partial W_{ij}}{\partial x_i^{\beta}} + \frac{\mu_i}{2\rho_i} \varepsilon_i^{\alpha\beta} \varepsilon_i^{\alpha\beta} \qquad (4.59)$$

In the code listed in Chapter 10, these two approaches of evolving momentum (equations (4.42) and (4.43)) and internal energy (equations (4.58) and (4.59)) are all implemented. In our experience, there is not much noticeable difference between the simulation results obtained from these two sets of SPH equations.

The SPH equations for the Navier-Stokes equations for evolving density, momentum and energy are summarized in Table 4.1. It should be noted that neglecting the viscous term in the Navier-Stokes equation yields the Euler equation. Therefore, in equations (4.42) and (4.43) as well as (4.58) and (4.59), neglecting the viscosity approximation and only retaining the pressure approximations result in the SPH equations for the Euler equation. The SPH formulations for the Euler equation will also be used in the next chapters in simulating explosion related large deformation events. The corresponding SPH equations for the Euler equation for evolving density, momentum and energy are listed in Table 4.2.

Table 4.1 SPH equations for the Navier-Stokes equations for evolving density, momentum and energy

Conservation of mass

$$\rho_i = \sum_{j=1}^{N} m_j W_{ij} \tag{4.26}$$

$$\rho_i = \frac{\sum_{j=1}^{N} m_j W_{ij}}{\sum_{j=1}^{N} \left(\frac{m_j}{\rho_j}\right) W_{ij}} \tag{4.35}$$

$$\frac{D\rho_i}{Dt} = -\rho_i \sum_{j=1}^{N} \frac{m_j}{\rho_j} v_j^\beta \cdot \frac{\partial W_{ij}}{\partial x_i^\beta} \tag{4.28}$$

$$\frac{D\rho_i}{Dt} = \rho_i \sum_{j=1}^{N} \frac{m_j}{\rho_j} v_{ij}^\beta \cdot \frac{\partial W_{ij}}{\partial x_i^\beta} \tag{4.31}$$

$$\frac{D\rho_i}{Dt} = \sum_{j=1}^{N} m_j v_{ij}^\beta \cdot \frac{\partial W_{ij}}{\partial x_i^\beta} \tag{4.34}$$

Conservation of momentum

$$\frac{Dv_i^\alpha}{Dt} = \sum_{j=1}^{N} m_j \frac{\sigma_i^{\alpha\beta} + \sigma_j^{\alpha\beta}}{\rho_i \rho_j} \frac{\partial W_{ij}}{\partial x_i^\beta} \tag{4.38}$$

$$\frac{Dv_i^\alpha}{Dt} = \sum_{j=1}^{N} m_j \left(\frac{\sigma_i^{\alpha\beta}}{\rho_i^2} + \frac{\sigma_j^{\alpha\beta}}{\rho_j^2}\right) \frac{\partial W_{ij}}{\partial x_i^\beta} \tag{4.41}$$

Conservation of energy

$$\frac{De_i}{Dt} = \frac{1}{2} \sum_{j=1}^{N} m_j \frac{p_i + p_j}{\rho_i \rho_j} v_{ij}^\beta \frac{\partial W_{ij}}{\partial x_i^\beta} + \frac{\mu_i}{2\rho_i} \varepsilon_i^{\alpha\beta} \varepsilon_i^{\alpha\beta} \tag{4.59}$$

$$\frac{De_i}{Dt} = \frac{1}{2} \sum_{j=1}^{N} m_j \left(\frac{p_i}{\rho_i^2} + \frac{p_j}{\rho_j^2}\right) v_{ij}^\beta \frac{\partial W_{ij}}{\partial x_i^\beta} + \frac{\mu_i}{2\rho_i} \varepsilon_i^{\alpha\beta} \varepsilon_i^{\alpha\beta} \tag{4.58}$$

Table 4.2 SPH equations for the Euler equations for evolving density,
momentum and energy

Conservation of mass

$$\rho_i = \sum_{j=1}^{N} m_j W_{ij} \tag{4.26}$$

$$\rho_i = \frac{\sum_{j=1}^{N} m_j W_{ij}}{\sum_{j=1}^{N} \left(\dfrac{m_j}{\rho_j}\right) W_{ij}} \tag{4.35}$$

$$\frac{D\rho_i}{Dt} = -\rho_i \sum_{j=1}^{N} \frac{m_j}{\rho_j} v_j^{\beta} \cdot \frac{\partial W_{ij}}{\partial x_i^{\beta}} \tag{4.28}$$

$$\frac{D\rho_i}{Dt} = \rho_i \sum_{j=1}^{N} \frac{m_j}{\rho_j} v_{ij}^{\beta} \cdot \frac{\partial W_{ij}}{\partial x_i^{\beta}} \tag{4.31}$$

$$\frac{D\rho_i}{Dt} = \sum_{j=1}^{N} m_j v_{ij}^{\beta} \cdot \frac{\partial W_{ij}}{\partial x_i^{\beta}} \tag{4.34}$$

Conservation of momentum

$$\frac{Dv_i^{\alpha}}{Dt} = -\sum_{j=1}^{N} m_j \frac{p_i + p_j}{\rho_i \rho_j} \frac{\partial W_{ij}}{\partial x_i^{\alpha}} \tag{4.60}$$

$$\frac{Dv_i^{\alpha}}{Dt} = -\sum_{j=1}^{N} m_j \left(\frac{p_i}{\rho_i^2} + \frac{p_j}{\rho_j^2}\right) \frac{\partial W_{ij}}{\partial x_i^{\alpha}} \tag{4.61}$$

Conservation of energy

$$\frac{De_i}{Dt} = \frac{1}{2} \sum_{j=1}^{N} m_j \frac{p_i + p_j}{\rho_i \rho_j} v_{ij}^{\beta} \frac{\partial W_{ij}}{\partial x_i^{\beta}} \tag{4.62}$$

$$\frac{De_i}{Dt} = \frac{1}{2} \sum_{j=1}^{N} m_j \left(\frac{p_i}{\rho_i^2} + \frac{p_j}{\rho_j^2}\right) v_{ij}^{\beta} \frac{\partial W_{ij}}{\partial x_i^{\beta}} \tag{4.63}$$

4.4 Numerical aspects of SPH for dynamic fluid flows

4.4.1 Artificial viscosity

In order to simulate problems of hydrodynamics, special treatments or methods are required to allow the algorithms to be capable of modeling shock waves, or else the simulation will develop unphysical oscillations in the numerical results around the shocked region. A shock wave is not a true physical discontinuity, but a very narrow transition zone whose thickness is usually in the order of a few molecular mean free paths. Application of the conservation of mass, momentum, and energy conditions across a shock wave front requires the simulation of transformation of kinetic energy into heat energy. Physically, this energy transformation can be represented as a form of viscous dissipation. This idea leads to the development of the von Neumann-Richtmyer artificial viscosity (von Neumann and Richtmyer, 1950) that is given by

$$\Pi_1 = \begin{cases} a_1 \Delta x^2 \rho (\nabla \cdot v)^2 & \nabla \cdot v < 0 \\ 0 & \nabla \cdot v \geq 0 \end{cases} \qquad (4.64)$$

where Π_1 is the von Neumann-Richtmyer artificial viscosity, and needs only to be present during material compression. a_1 is an adjustable non-dimensional constant. Note that this von Neumann-Richtmyer artificial viscosity is, in fact, a quadratic expression of velocity divergence.

It is found that adding the following linear artificial viscosity term Π_2 has the advantage of further smoothing the oscillations that are not totally dampened by the quadratic artificial viscosity term

$$\Pi_2 = \begin{cases} a_2 \Delta x c \rho \nabla \cdot v & \nabla \cdot v < 0 \\ 0 & \nabla \cdot v \geq 0 \end{cases} \qquad (4.65)$$

where c is the speed of sound, and a_2 is an adjustable non-dimensional constant.

The quadratic von Neumann-Richtmyer artificial viscosity Π_1 and the linear artificial viscosity Π_2 are widely used today for removing numerical oscillations in hydrodynamic simulations using FDM, FVM, FEM, and etc. The introduced artificial viscosity terms spread the shock wave over several mesh cells and regularize the numerical instability caused by the sharp spatial variation (discontinuity). The artificial viscosity terms are usually added to the physical pressure term, and help to diffuse sharp variations in the flow and to dissipate the energy of high frequency term. Formulation and magnitude needed for an artificial viscosity has undergone many refinements over the last few decades.

The SPH method was first applied to treat problems with low or no dissipation. Later an artificial viscosity was developed (Monaghan and Gingold, 1983; Monaghan and Poinracic, 1985; Monaghan, 1987) to allow shocks to be simulated. This Monaghan type artificial viscosity Π_{ij} is the most widely used artificial viscosity so far in the SPH literatures. It not only provides the necessary dissipation to convert kinetic energy into heat at the shock front, but also prevent unphysical penetration for particles approaching each other (Lattanzio et al., 1986; Monaghan, 1989). The detailed formulation is as follows.

$$
\Pi_{ij} = \begin{cases} \dfrac{-\alpha_\Pi \bar{c}_{ij}\phi_{ij} + \beta_\Pi \phi_{ij}^2}{\bar{\rho}_{ij}} & v_{ij} \cdot x_{ij} < 0 \\[4mm] 0 & v_{ij} \cdot x_{ij} \geq 0 \end{cases} \tag{4.66}
$$

where

$$
\phi_{ij} = \frac{h_{ij} v_{ij} \cdot x_{ij}}{\left|x_{ij}\right|^2 + \varphi^2} \tag{4.67}
$$

$$
\bar{c}_{ij} = \frac{1}{2}\left(c_i + c_j\right) \tag{4.68}
$$

$$
\bar{\rho}_{ij} = \frac{1}{2}\left(\rho_i + \rho_j\right) \tag{4.69}
$$

$$
h_{ij} = \frac{1}{2}\left(h_i + h_j\right) \tag{4.70}
$$

$$
v_{ij} = v_i - v_j, \quad x_{ij} = x_i - x_j \tag{4.71}
$$

In the above equations, α_Π, β_Π are constants that are all typically set around 1.0 (Monaghan, 1988; Evrard, 1988). The factor $\varphi = 0.1h_{ij}$ is inserted to prevent numerical divergences when two particles are approaching each other. c and v represent the speed of sound and the particle velocity vector, respectively. The viscosity associated with α_Π produces a bulk viscosity, while the second term associated with β_Π, which is intended to suppress particle interpenetration at high Mach number, is similar to the von Neumann-Richtmyer artificial viscosity. The artificial viscosity given by equation (4.66) is added into the pressure terms in the SPH equations.

Since the Monaghan type artificial viscosity introduces a shear viscosity into the flows especially in regions away from the shock, an artificial viscosity depending on the divergence of the velocity field was employed by Herquist and Katz (1989):

$$\Pi_{ij} = \frac{q_i}{\rho_i^2} + \frac{q_j}{\rho_j^2} \tag{4.72}$$

where

$$q_i = \begin{cases} \alpha_\Pi h_i \rho_i c_i |\nabla \cdot v_i| + \beta_\Pi h_i^2 \rho_i |\nabla \cdot v_i|^2 & \nabla \cdot v < 0 \\ 0 & \nabla \cdot v \geq 0 \end{cases} \tag{4.73}$$

Other modifications for the Monaghan type artificial viscosity were also proposed (Morris and Monaghan, 1997), and are still under investigation.

4.4.2 Artificial heat

The Monaghan type artificial viscosity described above often provides good results when modeling shocks. However, excessive heating can occur under some severe circumstances such as the wall heating from the classic example of a stream of gas being brought to rest against a rigid wall. Noh (1987) fixed this problem by adding an artificial heat conduction term to the energy equation. An SPH form of artificial heat term was derived by Monaghan (1995b) as follows and is added to the energy equation if necessary (Fulk, 1994).

$$H_i = 2 \sum_{j=1}^{N} \frac{\bar{q}_{ij}}{\bar{\rho}_{ij}} \frac{e_i - e_j}{|x_{ij}|^2 + \varphi^2} x_{ij} \cdot \nabla_i W_{ij} \tag{4.74}$$

where

$$q_i = \alpha_\Pi h_i \rho_i c_i |\nabla \cdot v_i| + \beta_\Pi h_i^2 \rho_i |\nabla \cdot v_i|^2 \tag{4.75}$$

$$q_j = \alpha_\Pi h_j \rho_j c_j \left| \nabla \cdot \boldsymbol{v}_j \right| + \beta_\Pi h_j^2 \rho_j \left| \nabla \cdot \boldsymbol{v}_j \right|^2 \tag{4.76}$$

$$\overline{q}_{ij} = q_i + q_j \tag{4.77}$$

Taking into account the artificial viscosity and artificial heat, a very popular set of SPH formulation for the Navier-Stokes equations is

$$\begin{cases} \dfrac{D\rho_i}{Dt} = \sum_{j=1}^{N} m_j v_{ij}^\beta \dfrac{\partial W_{ij}}{\partial x_i^\beta} \\[3mm] \dfrac{Dv_i^\alpha}{Dt} = -\sum_{j=1}^{N} m_j \left(\dfrac{\sigma_i^{\alpha\beta}}{\rho_i^2} + \dfrac{\sigma_j^{\alpha\beta}}{\rho_j^2} + \Pi_{ij} \right) \dfrac{\partial W_{ij}}{\partial x_i^\beta} \\[3mm] \dfrac{De_i}{Dt} = \dfrac{1}{2} \sum_{j=1}^{N} m_j \left(\dfrac{P_i}{\rho_i^2} + \dfrac{P_j}{\rho_j^2} + \Pi_{ij} \right) v_{ij}^\beta \dfrac{\partial W_{ij}}{\partial x_i^\beta} + \dfrac{\mu_i}{2\rho_i} \varepsilon_i^{\alpha\beta} \varepsilon_i^{\alpha\beta} + H_i \\[3mm] \dfrac{Dx_i^\alpha}{Dt} = v_i^\alpha \end{cases} \tag{4.78}$$

4.4.3 Physical viscosity description

Early implementation of SPH was generally used to solve the Euler equation, which is limited to inviscid flows. This is because it is not easy to obtain the SPH expressions of the second derivatives in the physical viscous term for the more general form of the Navier-Stokes equations. From previous discussion in section 3.2, viscous force can be calculated by so-called nested approximations first on the strain rate, and then on the stress. This nested summation approach in treating physical viscosity, is quite straightforward and is fairly attractive if the whole SPH algorithm is properly arranged to minimize the computational effort arising from the nested summations. Multiple materials with different viscosities can also be treated. Moreover, the viscosity involved can even evolve with time. This nested summation in treating physical viscosity was described by Flebbe and his colleagues (Flebbe et al., 1994; Riffert et al., 1995) when simulating astrophysical problems such as the accretion disk and viscous rings. The idea of nested summation in approximating second order derivatives was also applied by Libersky et al. (1991; 1993) when modeling solids with material strength. In the later presented numerical simulations, if physical viscosity is involved, this nested summation approach will be used.

Monahan (1995b) employed another approach to approximate the viscous term when modeling heat conduction, which seems to be more acceptable in simulating low velocity flows. Morris et al. (1997) also used this approach to model low Reynolds number incompressible flows. In this approach, the momentum equation is evolved using

$$
\frac{Dv_i^\alpha}{Dt} = -\sum_{j=1}^{N} m_j \left(\frac{p_i}{\rho_i^2} + \frac{p_j}{\rho_j^2} \right) \frac{\partial W_{ij}}{\partial x_i^\alpha} + \sum_{j=1}^{N} m_j \frac{\mu_i + \mu_j}{\rho_i \rho_j} v_{ij}^\beta \left(\frac{1}{r_{ij}} \frac{\partial W_{ij}}{\partial r_{ij}} \right)
\tag{4.79}
$$

where r_{ij} is the distance between the two particles.

Takeda (1994), when treating the viscous force, directly used the second order derivative of the smoothing function.

It is noted that the viscosity described here is different from the artificial viscosity that is used mainly for resolving shocks. As discussed previously, in the Monaghan type artificial viscosity, the first term involves shear and bulk viscosity. The second term is similar to the von Neumann-Richtmyer viscosity and it is very important in preventing unphysical particle penetration especially for particles that are approaching each other at high speed and almost head-on. For flows with physical viscosity, since the physical viscosity is resolved either by nested summation or other approaches, it's not necessary to use the first term in Π_{ij} (the term associated with α_Π). Retaining the first term in Π_{ij} sometimes may result in spurious large shear viscosity. However, the second term (the term associated with β_Π) is still retained to prevent unphysical particle penetration. This is different to most of the earlier approaches, where the whole artificial viscosity is added into the pressure term in the corresponding SPH formulations. Therefore, when simulating viscous flows, by removing the α_Π term in Π_{ij}, the arisen spurious shear viscosity can be removed. By retaining the β_Π term in Π_{ij}, unphysical particle penetration can well be prevented.

4.4.4   Variable smoothing length

The smoothing length h is very important in the SPH method, which has direct influence on the efficiency of the computation and the accuracy of the solution. If h is too small, there may be not enough particles in the support domain of dimension κh to exert forces on a given particle, which results in low accuracy. If the smoothing length is too large, all details of the particle or local properties may be smoothed out, and the accuracy suffers, too. The particle approximations used by the SPH method depend on having a sufficient and necessary number of particles within the support domain of κh. The computational effort or speed also depends on this particle number. In one, two and three dimensions, the number of neighboring particles (including the particle itself) should be about 5,

21, 57 respectively if the particles are placed in a lattice with a smoothing length of 1.2 times the particle spacing, and $\kappa = 2$.

In early implementation of SPH, the global particle smoothing length was used which depended on the initial average density of the system. Later, the smoothing length was improved to solve problems where the fluid expands or contracts locally so as to maintain consistent accuracy throughout the space by assigning each particle an individual smoothing length according to the variation of the local number density of each particle (Monaghan, 1988; 1992). For problems that are not isotropic such as shock problems, the smoothing length needs to be adapted both in space and time (Hernquist and Katz, 1989; Nelson et al., 1994; Steinmetz and Muller, 1993). Using a tensor smoothing length that is based on an ellipsoidal kernel rather than the traditional spherical kernel, Shapiro, Owen and their co-workers (Shapiro et al., 1996; Owen et al., 1998) developed an adaptive SPH (ASPH) that will be detailed in Chapter 8.

There are many ways to dynamically evolve h so that the number of the neighboring particles remains relatively constant. The simplest approach is to update the smoothing length according to the averaged density

$$h = h_0 (\frac{\rho_0}{\rho})^{1/d} \qquad (4.80)$$

where h_0 and ρ_0 are the initial smoothing length and the initial density respectively. d is the number of dimensions.

Benz (1989) suggested another method to evolve the smoothing length, which takes the time derivative of the smoothing function in terms of the continuity equation

$$\frac{dh}{dt} = -\frac{1}{d}\frac{h}{\rho}\frac{d\rho}{dt} \qquad (4.81)$$

Equation (4.81) can be discretized using the SPH approximations and calculated with the other differential equations in parallel.

4.4.5   Symmetrization of particle interaction

If the smoothing length is set to vary both in time and space, each particle has its own smoothing length. If h_i is not equal to h_j, the influencing domain of particle i may cover particle j but not necessarily vice versa. Therefore, it is possible for particle i to exert a force on particle j without j exerting the same corresponding reaction on i. This is a violation of the Newton's Third Law. In order to overcome this problem, some measures must be taken to preserve the symmetry of particle interactions.

One approach in preserving the symmetry of particle interaction is to modify the smoothing length. There are different ways to perform the modification to produce a symmetric smoothing length. One way to obtain the symmetric smoothing length is to take the arithmetic mean or the average of the smoothing lengths of the pair of interacting particles (Benz, 1989; 1990)

$$h_{ij} = \frac{h_i + h_j}{2} \qquad (4.82)$$

Other ways can also be used to get the symmetric smoothing length using the geometric mean of the smoothing lengths of the pair of the interacting particles

$$h_{ij} = \frac{2h_i h_j}{h_i + h_j} \qquad (4.83)$$

or the maximal value of the smoothing lengths

$$h_{ij} = \min(h_i, h_j) \qquad (4.84)$$

or the minimal value of the smoothing lengths

$$h_{ij} = \max(h_i, h_j) \qquad (4.85)$$

The smoothing function can then be obtained using the symmetric smoothing length

$$W_{ij} = W(r_{ij}, h_{ij}) \qquad (4.86)$$

There are advantages and disadvantages in these different ways to determine the symmetric smoothing length h_{ij}. Taking the arithmetic mean or the maximal value of the smoothing lengths tends to use more neighboring particles and sometimes may overly smooth out the interactions among surrounding particles. Taking the geometric mean or the minimal value of the smoothing length tends to possess less neighboring particles.

Another approach to preserve symmetry of particle interaction is to use directly the average of the smoothing function values (Hernquist and Katz, 1989) without using a symmetric smoothing length, i.e.

$$W_{ij} = \frac{1}{2}(W(h_i) + W(h_j)) \qquad (4.87)$$

These two approaches in preserving the symmetry of the particle interactions are both widely used in the implementation of SPH. No detailed comparison study on these two approaches has been reported so far.

4.4.6   Zero-energy mode

It is well known that the finite difference methods suffer from a spurious zero-energy mode for which the derivative at certain grid point is zero when evaluated by the function values at the regular grid points on the both sides (e.g., second order central difference scheme for evaluating first derivative). It is clear that except for the normal solution in which the function is constant, a solution of sawtooth pattern exists. One notorious example is the simulation of the incompressible flows, in which the pressure gradient is zero although the pressure itself highly oscillates (as indicated by the numbers in Figure 4.5). To treat this problem, an additional grid is employed to stagger from the original grid so that the pressure and velocity are evaluated at different grid systems (Figure 4.6).

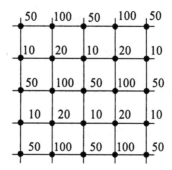

Figure 4.5 A highly oscillating 2D pressure distribution indicated by the numbers. The pressure distribution leads to a zero pressure gradients and hence leads to a spurious velocity contribution.

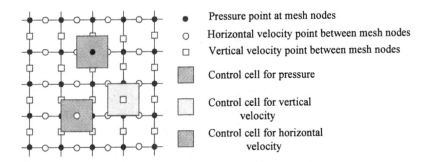

Figure 4.6 A 2D staggered grid in a Cartesian mesh where the pressure is evaluated at the mesh nodes. The velocities are evaluated at points in the middle to two mesh nodes.

In the finite element methods, the zero-energy mode also happens if some kind of elements (e.g. quadrilateral elements) is employed with the use of so-called reduced integration. The velocities of the mesh nodes on the opposite corners of the mesh can be equal in magnitude but opposite in direction (Figure 4.7). This is known as the Hourglass phenomenon. It is seen that this velocity field produces a spurious mode with no strain or volume change in the mesh, because the computation of strain rate tensor involves velocity differencing on the mesh nodes. The resultant stress is also zero, and leads to no resistance to the mesh deformation of Hourglass shape (Figure 4.8). To resolve this problem, the Hourglass viscosity is usually introduced to give artificial rigidity to the system. More discussions on zero-energy mode can be found in Swegle (1978), Belytschko et al; (2000) and many other references.

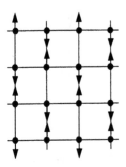

Figure 4.7 A 2D velocity field that leads to spurious strain rate tensor.

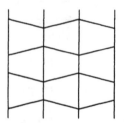

Figure 4.8 Shape of Hourglass deformation.

The same spurious mode problem also occurs in the SPH method when evaluating the derivatives. This can be simply shown for regular particle distribution in one-dimensional case (Figure 4.9). The derivative of the smoothing function at the particle evaluated is zero, and for particles of the same distance on either side of the particle evaluated, the smoothing function derivatives are equal in magnitude and opposite in sign. Considering this anti-symmetric property of the derivative of the smoothing function, the SPH particle approximation of the derivative is actually similar to some kind of central difference scheme in FDM. For example, if there is only one neighboring particle on the either symmetric side of the particle evaluated, the SPH particle approximation of derivative is the same as the second order central difference scheme in the FDM.

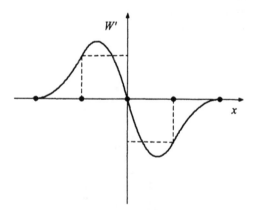

Figure 4.9 An example of regular SPH particle distribution which leads to zero-energy mode.

To avoid the spurious zero-energy mode, an efficient remedy was first proposed by Dyka et al. (1995; 1997) when addressing the tensile stability problem using the so called stress points. The same idea was extended to three dimensions by others (Vignjevic et al., 2000; Randles and Libersky, 2000). The main idea is to use two separate types of particles: velocity particles, which are the points at which the momentum equation is evaluated, and stress particles, which are the points where the stress is evaluated. Figure 4.10 shows a typical setup of the dual particle distribution. It is clear that the idea is also somewhat similar to that associated with the staggered grids in FDM for avoiding zero energy modes.

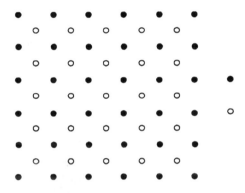

Figure 4.10 A typical setup of dual particle distribution in 2D space.

The zero-energy mode problem in SPH, however, is not as serious as that in the FDM since the particle distribution is usually more irregular rather than strictly regular as in the FDM. Moreover, the initial regular particle distribution may turn to be irregular due to the Lagrangian nature of the particles with the development of the flow. If the particles are not uniformly distributed, summed contributions from these particles generally will not lead to zero derivatives as shown in Figure 4.11. For such circumstances, the contributing particles are actually staggered more or less from the strictly regular particle distribution. In simulating practical hydrodynamic problems with large deformations, particle distributions are usually not uniform.

Note also that even if the particles are uniformly distributed, the zero-energy mode does not usually happen. This is because only very special field variable distributions (such as those shown in Figure 4.5 and Figure 4.7) can trigger the zero-energy mode.

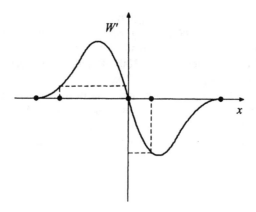

Figure 4.11 An example of irregular SPH particle distribution which does not lead to zero-energy mode.

4.4.7   Artificial compressibility

In the standard SPH method for solving compressible flows, the particle motion is driven by the pressure gradient, while the particle pressure is calculated by the local particle density and internal energy through the equation of state. However, for incompressible flows, the actual equation of state of the fluid will lead to prohibitive time steps that are extremely small. How to effectively calculate the pressure term in the momentum equation is a major task for simulation of incompressible flows. This is also true for other numerical methods like the FDM, not just for the SPH method.   Though it is possible to include the constraint of the constant density into the SPH formulations, the resultant equations are too cumbersome.   In the Moving Particle Semi-implicit (MPS) (Koshizuka et al.; 1998) method for simulating incompressible flows, the pressure term is implicitly calculated using particle interaction with the constant particle number density, while the source term is explicitly calculated.

The fact that a theoretically incompressible flow is practically compressible leads to a concept of artificial compressibility.   The artificial compressibility considers that every theoretically incompressible fluid is actually compressible. Therefore, it is feasible to use a quasi-incompressible equation of state to model the incompressible flow.   The purpose of introducing the artificial compressibility is to produce the time derivative of pressure.   Monaghan (1994) applied the following equation of state for water to model free surface flows:

$$p = B((\frac{\rho}{\rho_0})^\gamma - 1) \tag{4.88}$$

where γ is a constant, and $\gamma = 7$ is used in most circumstances. ρ_0 is the reference density. B is a problem dependent parameter, which sets a limit for the maximum change of the density. In most circumstances, B can be taken as the initial pressure (Morris et al., 1997; Schlatter, 1999). The subtraction of 1 in equation (4.88) can remove the boundary effect for free surface flows. It can be seen that a small oscillation in density may result in a large variation in pressure.

Another possible choice of the artificial equation of state is

$$p = c^2 \rho \tag{4.89}$$

where c is the sound speed. Morris et al. (1997) used this equation of state in modeling low Reynolds number incompressible flows using SPH. Our simulation results using equation (4.89) for the Couette, Poiseuille flows and flow past cylinder show the good agreement with results from other resources. Zhu et al. (1999) applied this expression in a pore-scale numerical model for flow through porous media.

In the artificial compressibility technique, the sound speed is a key factor that deserves careful consideration. If the actual sound speed (e.g. 1480 m/s for water under standard pressure and temperature) is employed, the real fluid is approximated as an artificial fluid, which is ideally incompressible. According to Monaghan (1994), the density variation δ is

$$\delta = \frac{\Delta\rho}{\rho_0} = \frac{|\rho - \rho_0|}{\rho_0} = \frac{V_b^2}{c^2} = M^2 \tag{4.90}$$

where V_b and M are the fluid bulk velocity and Mach number respectively. Since the actually sound speed is very large, and the corresponding Mach number is very small, the density variation δ will be nearly negligible. Therefore, in order to approximate the real fluid as an artificial compressible fluid, a much smaller value than the actual sound speed should be used. This sound speed value, on one hand, should be large enough so that the behavior of the artificial compressible fluid is sufficiently close to the real fluid; on the other hand, should be small enough so that the time step is increased to an acceptable value. Considering the balance of the time step and the incompressible behavior of the artificial compressible fluid, there should be an optimal value for the sound speed. From equation (4.90), it can be seen that in modeling a real fluid as an artificial compressible fluid, the sound speed to be chosen is closely related to the bulk velocity of the flow. Apart from the bulk velocity, the pressure field also needs to be well estimated when choosing the sound speed.

Morris et al. (1997), through considering the balance of pressure, viscous force and body force, proposed an estimate for the sound speed. He argued that the square of the sound speed should be comparable with the largest value of V_b^2/δ, $\upsilon V_b/\delta L$, and FL/δ, i.e.

$$c^2 = \max(\frac{V_b^2}{\delta}, \frac{\upsilon V_b}{\delta L}, \frac{FL}{\delta}) \tag{4.91}$$

where υ ($\upsilon = \mu/\rho$) is the kinetic viscosity, F is the magnitude of the body force, L is the characteristic length scale.

In the applications of the artificial compressibility to incompressible flows, it is useful to use the "XSPH" technique proposed by Monaghan (1989; 1992). In the XSPH, the particle moves in the following way.

$$\frac{dx_i}{dt} = v_i - \varepsilon \sum_j \frac{m_j}{\rho_j} v_{ij} W_{ij} \tag{4.92}$$

where ε is a constant in the range of $0 \le \varepsilon \le 1.0$. It is clear that the XSPH technique includes the contribution from neighboring particles, and thus makes the particle move in a velocity closer to the average velocity of the neighboring particles. The XSPH technique, when applied to incompressible flows, can keep the particles more orderly; when applied to compressible flows, can effectively reduce unphysical penetration between approaching particle. In most circumstances $\varepsilon = 0.3$ seems to be a good choice in simulating incompressible flows. The XSPH technique will also be applied in the simulation of explosions (see chapter 6), in which a larger value ($\varepsilon \ge 0.5$) can be used.

4.4.8 Boundary treatment

Full exploitation of SPH has been hampered by the problem of *particle deficiency* near or on the boundary, which results from the integral that is truncated by the boundary. For particles near or on the boundary, only particles inside the boundary contribute to the summation of the particle interaction, and no contribution comes from outside since there are no particles beyond the boundary (see Figure 4.12). This one-sided contribution does not give correct solutions, because on the solid surface, although the velocity is zero, other field variables such as the density do not necessarily reduce to zero.

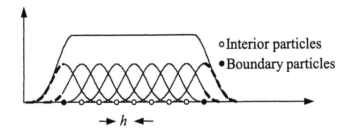

Figure 4.12 SPH kernel and particle approximations for interior and boundary particles.

Recently some improvements have been proposed to treat the boundary condition and solve the particle deficiency problem. Monaghan (1994) used a line of ghost or virtual particles located right on the solid boundary to produce a highly repulsive force to the particles near the boundary, and thus to prevent these particles from unphysically penetrating through the boundary. Campbell (1989) treated the boundary conditions by including the residue boundary terms in the integration by parts when estimating the original kernel integral involving gradients. Libersky and Petschek (1993) first introduced ghost particles to reflect a symmetrical surface boundary condition. Later, Randles and Libersky (1996) proposed a more general treatment of the boundary condition by assigning the same boundary value of a field variable to all the ghost particles, and then interpolating smoothly the specified boundary ghost particle value and the calculated values of the interior particles.

In the studies by Liu et al. (2001b; 2002a), virtual particles are used to treat the solid boundary conditions. They suggested using two types of virtual particles as illustrated in Figure 4.13. The virtual particles of the first type (type I) are located right on the solid boundary and is similar to what Monaghan used (1994). The virtual particles of the second type (type II) fill in the boundary region and are similar to what Libersky used (1993). The virtual particles of the type II are constructed in the following way. For a certain real particle i, if it is located within the distance of κh_i from the boundary, a virtual particle is placed symmetrically on the outside of the boundary. These virtual particles have the same density and pressure as the corresponding real particles but opposite velocity. These virtual particles are not enough to fully prevent the real particles from penetrating outside the boundary. This leads to the application of the virtual particles of the type I, which are used to produce a sufficient repulsive boundary force when a particle approaches the boundary. Therefore, for a boundary particle i , all the neighboring particles $NN(i)$ that are within its influencing area of κh_i can be categorized into three subsets:

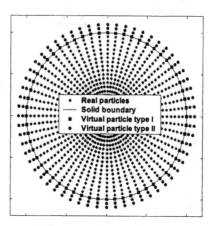

Figure 4.13 Illustration of real particles and the two types of virtual particles used for simulating the solid boundary. (From Liu M. B., Liu G. R. and Lam K. Y., Shock Waves, 12(3):181-195, 2002. With permission.)

a) $I(i)$: all the *interior particles* that are the neighbors of i (real particles);

b) $B(i)$: all the *boundary particles* that are the neighbors of i (virtual particles of type I);

c) $E(i)$: all the *exterior particles* that are the neighbors of i (virtual particles of type II).

These two types of virtual particles are specially marked for contribution in the later summation on the real particles. The virtual particles of the type I take part in the kernel and the particle approximations for the real particles. The position and physical variables for this type of virtual particles, however, do not evolve in the simulation process.

The type I virtual particles are also used to exert a repulsive boundary force to prevent the interior particles from penetrating the boundary. The penalty force is calculated using a similar approach employed for calculating the molecular force of Lennard-Jones form. If a particle of type I is the neighboring particle of a real particle that is approaching the boundary, a force is applied pairwisely along the centerline of these two particles.

$$
PB_{ij} = \begin{cases} D\left[(\dfrac{r_0}{r_{ij}})^{n_1} - (\dfrac{r_0}{r_{ij}})^{n_2} \right] \dfrac{x_{ij}}{r_{ij}^2} & (\dfrac{r_0}{r_{ij}}) \leq 1 \\[4mm] 0 & (\dfrac{r_0}{r_{ij}}) > 1 \end{cases} \tag{4.93}
$$

where the parameters n_1 and n_2 are usually taken as 12 and 4 respectively. D is a problem dependent parameter, and should be chosen to be in the same scale as the square of the largest velocity. The cutoff distance r_0 is important in the simulation. If it is too large, some particles may feel the repulsive force from the virtual particles in the initial distribution, thus leads to initial disturbance and even blowup of particle positions. If it is too small, the real particles may have already penetrated the boundary before feeling the influence of the repulsive force. In most practices, r_0 is usually selected approximately close to the initial particle spacing.

The type II virtual particles do not evolve its parameters. They are produced symmetrically according to the corresponding real particles in each evolution step. Type II virtual particles can both be applied to treat solid boundaries and free surfaces. The numerical tests have shown that this treatment of the boundary is very stable and effective. It not only improves the accuracy of the SPH approximation in the boundary region, but also prevents the unphysical particle penetration outside the solid boundary (Liu et al., 2001b; 2002a).

4.4.9 Time integration

Just as other explicit hydrodynamic methods, the discrete SPH equations can be integrated with standard methods such as the second order accurate Leap-Frog (LF), predictor-corrector and Runge-Kutta (RK) schemes and so on. The advantage of the Leap-Frog (LF) algorithm is its low memory storage required in the computation and the efficiency for one force evaluation per step. In some cases where the smoothing lengths become very small, the time step can become very small to be prohibitive. The Runge-Kutta (RK) integrator with adaptive time-step (Benz, 1990, Hultman and Källander, 1997) has been shown to have an advantage during such conditions. In this case, it is possible to use adaptive time stepping, and the time step is chosen to minimize an estimate of the error in the integration within certain tolerances. In practice, it turns out that this time step may be larger than that estimated using the Courant-Friedrichs-Levy (CFL) condition. However, it needs two force evaluations per step, and thus is computationally expensive. One of the important aspects of SPH time integration is the use of individual time-steps (Kernquist and Katz, 1989). This treatment has advantages for problems where different time-scales naturally develop like cosmological N-body simulations.

The explicit time integration schemes are subject to the CFL condition for stability. The CFL condition states that the computational domain of dependence in a numerical simulation should include the physical domain of dependence, or the maximum speed of numerical propagation must exceed the maximum speed of physical propagation (Anderson, 1995; Hirsch, 1988). This CFL condition requires the time step to be proportional to the smallest spatial particle resolution, which in SPH applications is represented by the smallest smoothing length

$$\Delta t = \min(\frac{h_i}{c}) \tag{4.94}$$

The time step to be employed in an SPH application is closely related to the physical nature of the process. Monaghan (1989; 1992) gave two expressions when taking into account the viscous dissipation and the external force

$$\Delta t_{cv} = \min\left(\frac{h_i}{c_i + 0.6\left(\alpha_\Pi c_i + \beta_\Pi \max(\phi_{ij})\right)}\right) \tag{4.95}$$

$$\Delta t_f = \min(\frac{h_i}{f_i})^{\frac{1}{2}} \tag{4.96}$$

where f is the magnitude of force per unit mass (acceleration). Equation (4.95) is obtained from equation (4.94) by adding the viscous force term. Combining equations (4.95) and (4.96) together with two corresponding safety coefficients λ_1 and λ_2, the typical time step is calculated using

$$\Delta t = \min(\lambda_1 \Delta t_{cv}, \lambda_2 \Delta t_f) \tag{4.97}$$

Monaghan (1992) suggested that $\lambda_1 = 0.4$ and $\lambda_2 = 0.25$.

Morris et al. (1997) gave another expression for estimating time step when considering viscous diffusion

$$\Delta t = 0.125 \frac{h^2}{\upsilon} \tag{4.98}$$

where υ ($\upsilon = \mu/\rho$) is the kinetic viscosity.

4.5   Particle interactions

4.5.1   Nearest neighboring particle searching (NNPS)

In the SPH method, since the smoothing function has a compact support domain, only a finite number of particles are within the support domain of dimension κh of the concerned particle, and are used in the particle approximations. These particles are generally referred to as nearest neighboring particles (NNP) for the concerned particle. The process of finding the near nearest particles is commonly referred to as nearest neighboring particle searching (NNPS). Unlike a grid-based numerical method, where the position of neighbor grid-cells are well defined once the grids are given, the nearest neighboring particles in the SPH for a given particle can vary with time. Three NNPS approaches, all-pair search, linked-list search algorithm, and tree search algorithm, are popular in SPH applications.

All-pair search

A direct and simple NNPS algorithm is the *all-pair search* approach (as shown in Figure 4.14). For a given particle i, the all-pair approach calculates the distance r_{ij} from i to each and every particle j ($= 1, 2, \cdots, N$), where N is the total number of particles in the problem domain. If the distance r_{ij} is smaller than the dimension of the support domain for particle i, κh, particle j is found belonging to the support domain of particle i. If the symmetric smoothing length is employed, particle i is also within the support domain of particle j. Therefore, particle i and particle j are a pair of neighboring particles. This searching is performed for all the particles. The all-pair search approach is carried out for particles $i = 1, 2, \cdots, N$, and the searching is performed for all the particles $j = 1, 2, \cdots, N$. It is clear that the complexity of the all-pair search approach is of order $O(N^2)$. Note that the NNPS process is necessary at all the time steps, and thus the computational time in this all-pair search approach is simply too long, and is intolerable especially for cases with large number of particles. It is, therefore, used only for problems of very small scale.

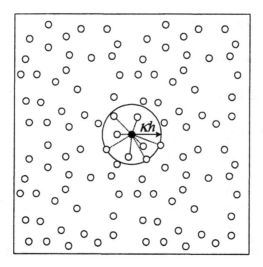

Figure 4.14 All-pair search algorithm for searching the nearest neighboring particles in two-dimensional space. For each particle, the distance from anther particle is compared with the dimension of the support domain of the particle to check if these two particles are neighboring particles.

Linked-list algorithm

The *linked-list* search algorithm works well for cases with spatially constant smoothing length. As pointed by Monaghan and Gingold (1983), substantial savings in computational time can be achieved by using cells as a bookkeeping device. If all the particles are assigned to cells and identified through linked-lists, the computational time can be greatly reduced since the NNPS process is only necessary for a certain group of particles. Monaghan (1985) described the procedure for carrying out the nearest neighboring particle searching using linked-list. More details of the method were also given by Hockney and Eastwood (1988) in their discussion of short-range forces in particle simulation methods. Rhoades (1992) has also given such an algorithm to search the nearest neighboring particles, which is reported to be efficient, particularly on vector computers. Simpson (1995) also described the linked-list algorithm in detail when addressing the numerical techniques for three-dimensional SPH simulations applied to accretion disks.

In the implementation of the linked-list algorithm, a temporary mesh is overlaid on the problem domain (Figure 4.15). The mesh spacing is selected to match the dimension of the support domain. For smoothing kernels having compact support of κh, the mesh spacing should be set to κh. Then for a given particle i, its nearest neighboring particles can only be in the same grid cell or the immediately adjoining cells. Therefore, the search is confined only over 3, 9 or 27 cells for one-, two- or three-dimensional space, respectively, if $\kappa = 2$. The linked-list algorithm allows each particle to be assigned to a cell and for all the particles in a cell to be chained together for easy access. If the average number of particles per cell is sufficiently small, the complexity of the linked-list algorithm is of order $O(N)$. The problem with the linked-list methods is that when variable smoothing length is used, the mesh spacing may not be optimal for every particle. Therefore, it can be less efficient.

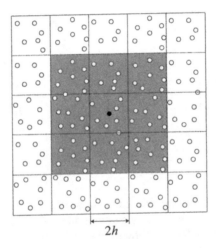

Figure 4.15 Linked-list algorithm for searching the nearest neighboring particles in two-dimensional space. The smoothing length is spatially constant. In the figure, the dimension of the support domain is $2h$ for the particles.

Tree search algorithm

Tree search algorithm works well for problems with variable smoothing lengths. It involves creating ordered trees according to the particle positions. Once the tree structure is created, it can be used efficiently to find the nearest neighboring particles. In this book, an adaptive hierarchy tree search method is adopted to suit the needs of adaptive smoothing lengths. This tree method recursively splits the maximal problem domain into octants that contain particles, until the leaves on the tree are individual particles (Figure 4.16). After the tree structure is constructed, the search process can be performed.

For a given particle i, a cube with the side of $2\kappa h_i$ is used to enclose the particle, which is located at the center of the cube. At each level, check to see if the volume of the search cube overlaps the volume represented by the current node in the tree structure. If not, discontinue the descent down on that particular path. If yes, continue the tree descent and proceed down to the next level repeatedly until the current node represents a particle. Then check to see if the particle is within the support domain of the given particle i. If yes, it's recorded as a neighboring particle.

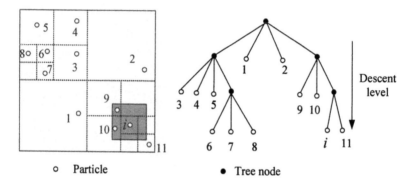

Figure 4.16 Tree structure and tree search algorithm in two-dimensional space. The tree is constructed by recursively splitting the maximal problem domain into octants that contain particles, until the leaves on the tree are individual particles. The tree search algorithm is performed by checking if the volume of the search cube (shaded area) for a given particle overlaps with the volume represented by the current node.

The complexity of this tree search algorithm is of order $O(N \log N)$ (Hernquist and Katz, 1989). Numerical tests show the SPH method combining with tree search method is very efficient and robust especially for large number of particles of variable smoothing lengths.

Another efficient algorithm of NNPS is the so-called Bucket algorithm that was proposed by Liu and Tu (2001; 2002), and used in the MFree2D$^{\copyright}$ code (see, Liu, 2002).

4.5.2   Pairwise interaction

A very time consuming part for the SPH simulations is to identify all the interacting particles. After determining the interacting particles, the particle approximations can be carried out in the summation process over all the interacting particles. One simple way for carrying out the particle interaction is shown in the example on the summation density approach (equation (4.26)) according to the following pseudo-code.

```
DO I=1, NTOTAL
    RHO(I)=0                    | Initialization
END DO


DO I=1, NTOTAL
    DO K=1, NNP(I)      | NNP(I): total number of neighboring particles for i.
    J=LIST(I, K)        | LIST(I, K): the list of NNP for particle i.
    Calculate the smoothing function W of particle i and j.
    Calculate the smoothing function derivative DXDW(D)
        RHO(I) = RHO(I)+ MASS(J)*W    |MASS: particle mass
                                      |RHO: particle density

    END DO
END DO
```

LIST is an array that is allocated to store the neighboring particles for all the particles in the problem domain. Assuming the maximal number of neighboring particles is MAXNPP, the size of LIST is NTOTAL times MAXNPP. Therefore LIST is a very large array. The smoothing function and its derivatives are either calculated at each summation process, or obtained from other large arrays (the same size as that of LIST) that store the previously calculated smoothing function and its derivatives. It is clear that this simple way to calculate the particle interaction is not computationally efficient and needs a large storage.

Since the interactions between particles are based on particle-particle pair, a pairwise technique can be used to reduce the computational effort and save storage in the SPH simulations. The pairwise interaction technique (Hockney and Eastwood, 1988; Hernquist and Katz , 1989; Riffert et al., 1995) is carried out with the process of nearest neighboring particle searching, and stores necessary data for the later SPH summation process. The following pseudocode shows the loop over all interacting particles

```
NIAC=0
DO I=1, NTOTAL                          | Initialization
      COUNTIAC(I)=0
END DO

DO I=1,NTOTAL-1                         | NTOTAL: total number of particles
   DO J=I+1, NTOTAL
```

$$\text{IF } (\,|x(i) - x(j)| \le \kappa(\frac{h_i + h_j}{2})\,) \text{ then } \mid \textit{NNPS algorithms}$$

```
      NIAC=NIAC+1      | NIAC: total number of the interacting pairs
      PAIR_I(NIAC)=I   | PAIR_I: array storing the 1ˢᵗ particles in the  pair
      PAIR_J(NIAC)=J   | PAIR_J: array storing the 2ⁿᵈ particles in the pair
      COUNTIAC (I)= COUNTIAC (I)+1 |COUNTIAC: number of neighbors
      COUNTIAC (J)= COUNTIAC (J)+1 |
      Calculate the smoothing function: W(NIAC) for the NIAC-th pair
      Calculate the derivatives: DWDX(D,NIAC) for the NIAC-th pair
                                        | D: dimension index
      END IF
   END DO
END DO
```

In the above loop, NTOTAL is the total number of particles; the total number of the interacting pairs is recorded in NIAC; the arrays PAIR_I, and PAIR_J contain the indices of the particles in the interacting pair; the number of neighboring particles for each particle are saved in the array COUNTIAC, which can be used in the evolution of the smoothing length to keep a roughly constant number of NNP. After the storage of the pairwise arrays of PAIR_I, PAIR_J, W, and DWDX, the SPH summations can be carried out in a single loop rather than two loops where the pairwise interaction technique is not used. For example the summation of density according to equation (4.26) is coded as follows.

```
DO I =1, NTOTAL
      RHO(I)=MASS(I)*W(0)               | Considering self effect
END DO

DO K=1,NIAC
      I=PAIR_I(K)
      J=PAIR_J(K)
      RHO(I) = RHO(I) + MASS(J)*W(K)    |RHO: particle density
      RHO(J) = RHO(J) + MASS(I)*W(K)    |MASS: particle mass
END DO
```

Other particle summation processes can also be carried out in a similar way. As can be seen, by using the pairwise interaction technique, only one loop is needed in the particle summation, the interacting data such as the indices of interacting particles as well as the corresponding smoothing function and its derivatives are stored for only one half of all the interactions. It is obvious that this pairwise interaction technique saves storage and improves computational efficiency.

4.6 Numerical examples

A series of numerical tests have been carried out to examine the ability and efficiency of the presented SPH formulations and the SPH code implemented in simulating fluid dynamic problems (Lam et al., 2000). As can be seen from the previous discussions, the presented SPH method with various formulations is able to simulate different dynamic fluid flow problems, such as inviscid or viscous flows, compressible or incompressible flows. The followings are some applications to incompressible flows, free surface flows and compressible flows. These applications and numerical examples not only examine the effectiveness of the SPH method, but also provide foundations for the complex explosion problems to be simulated in the next chapters. In these applications, the cubic spline function (see, Section 3.1) is used. The SPH code used to generate these examples is attached in Chapter 10 with detailed descriptions.

4.6.1 Applications to incompressible flows

In the finite difference methods, the incompressible flows have been widely studied and will be studied in this section using the SPH method. In the SPH simulation of incompressible flows, the artificial compressibility technique (see, Section 4.4.7) is employed to model the incompressible fluid as slightly compressible by selecting a proper equation of state. Three standard benchmarking cases, the Poiseuille flow, Couette flow, and the shear driven cavity problem are studied. For the Poiseuille flow, Couette flow and the shear driven cavity problem, equation (4.89) is employed. For the later simulation of water discharge with free surface, equation (4.88)) is used to model the compressibility of water.

Example 4.1 Poiseuille flow

The Poiseuille flow involves flow between two parallel stationary infinite plates placed at $y = 0$ and $y = l$. The initially stationary fluid is driven by a body force F (e.g., pressure difference or external force), gradually flows between the two plates, and finally arrives at a steady state. In the classical hydrodynamics, the flow velocity at a point in the Poiseuille flow can be obtained by solving the Navier-Stokes momentum equation. For the planar Poiseuille flow at low Reynolds number confined between two parallel stationary infinite plates, the steady state continuity equation for the fluid confined in the z direction flowing in the x direction is

$$-\frac{dp}{dx} = \frac{d\tau_{zx}}{dz} \tag{4.99}$$

For isotropic fluids,

$$\tau_{xz} = -\mu \frac{dv_x}{dz} \tag{4.100}$$

Substituting equation (4.100)) into equation (4.99), the one-dimensional Navier-Stokes momentum equation is then obtained as

$$\rho \frac{dv_x}{dt} = -\frac{dp}{dx} = -\mu \frac{d^2 v_x}{dz^2} \tag{4.101}$$

Solving equation (4.101) yields the following solution for the velocity

$$v_x(z) = -\frac{1}{2\upsilon} \frac{dp}{dx}(a^2 - z^2) \tag{4.102}$$

where a is a constant determined by the boundary conditions. If the non-slip boundary conditions are assumed, the tangential component of the velocity vanishes on the boundary. In this case, a is equal to one half of the dimension in the z direction, i.e. $a = l/2$. If the driven force is an external force, then the pressure difference in equation (4.102) can be replaced by that external force.

Morris et al. (1997) provided a series solution for the time dependent behavior of the Poiseuille flow

$$v_x(z,t) = \frac{F}{2\upsilon} z(z-l)$$

$$+ \sum_{n=0}^{\infty} \frac{4Fl^2}{\upsilon\pi^3(2n+1)^3} \sin(\frac{\pi z}{l}(2n+1)) \exp(-\frac{(2n+1)^2 \pi^2 \upsilon}{l^2} t) \tag{4.103}$$

In this example, $l = 10^{-3}$ m, $\rho = 10^{3}$ kg/m^3, the kinetic viscosity $\upsilon = 10^{-6}$ m^2/s, and the driven body force $F = 2 \times 10^{-4}$ m/s^2. According to equation (4.102), the peak fluid velocity is $v_0 = 2.5 \times 10^{-5}$ m/s, which corresponds to a Reynolds number of Re $= 2.5 \times 10^{-2}$ according to the following expression

$$ \mathrm{Re} = \frac{v_0 l}{\upsilon} \qquad (4.104) $$

When simulating the Poiseuille using the SPH method, several implementation models can be used to approximate the two infinite parallel plates. The first one is to simulate the flow between two parallel plates that are sufficiently long compared to the dimension perpendicular to the flow direction. The second one is to simulate the flow confined in an annulus. These two models are time consuming. The third way is to apply the *periodic boundary* in the flow direction both in the particle movement and particle interaction. In the particle movement, a particle that leaves the specified region through a particular boundary face immediately reenters the region through the opposite face. In the particle interaction, a particle located within the dimension of the support domain from a boundary interacts with particles in an adjacent copy of the system, or equivalently with particles near the opposite boundary. Therefore, this wraparound effect of the periodic boundary condition is taken into consideration in both the integration of the equations of motion when moving the particles and the interaction computations between neighboring particles.

In the following simulation, the problem domain is a rectangular of 0.0005 m × 0.001 m, and is modeled with 20 × 40 particles. 40 type I virtual particles are employed (Figure 4.17). The smoothing length used is slightly more (say, 1.1 times) than the initial particle spacing. The time step is set to 10^{-4} s. After about 5000 steps, the simulation results reach a steady state. The particle distribution and velocity quiver plotted at the 6000-th step are shown in Figure 4.18 and Figure 4.19, respectively. It can be seen that some particles that are originally located in the downstream are wraparounded to the upstream. This is the representation of the use of the periodic boundary effect. Figure 4.20 shows the comparison between the velocity profiles obtained using the SPH method and those by the series solution at $t = 0.01$ s, 0.1 s, and the final steady state $t = \infty$. It is found that the SPH results are in a very good agreement with less than 0.5% discrepancy with the analytical series solution.

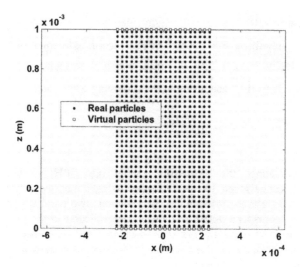

Figure 4.17 Initial geometry and particle distribution for the Poiseuille flow. The interior real and the type I virtual particles are shown in the figure.

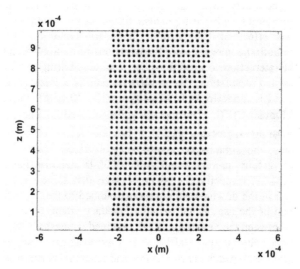

Figure 4.18 Particle distribution for the Poiseuille flow obtained using the SPH method at the 6000-th step. Only the interior real particles are shown. It is seen that some particles that are originally located in the downstream are wraparounded to the upstream after imposing the periodic boundary condition.

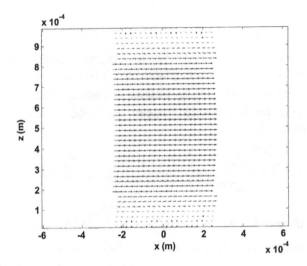

Figure 4.19 Velocity quiver for the Poiseuille flow obtained using the SPH method at the 6000-th step. It is seen that some particles that are originally located in the downstream are wraparounded to the upstream after imposing the periodic boundary condition.

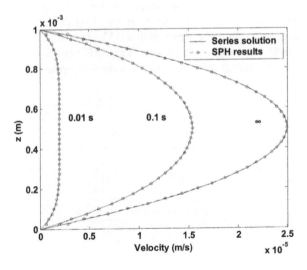

Figure 4.20 Velocity profiles for the Poiseuille flow.

Example 4.2   Couette flow

The Couette flow involves a fluid flow between two infinite plates that are initially stationary and horizontally placed at $y = 0$ and $y = l$. The flow is generated after the upper plate suddenly moves at a certain constant velocity v_0. The series solution for the horizontal velocity field in the Couette flow is given as (Morris at al., 1997)

$$v_x(z,t) = \frac{v_0}{l}z + \sum_{n=1}^{\infty}\frac{2v_0}{n\pi}(-1)^n \sin(\frac{n\pi}{l}z)\exp(-\upsilon\frac{n^2\pi^2}{l^2}t) \qquad (4.105)$$

In this example, $l = 10^{-3}$ m, $\upsilon = 10^{-6}$ m^2/s, $\rho = 10^3$ kg/m^3, $v_0 = 2.5 \times 10^{-5}$ m/s. The corresponding Reynolds number is Re $= 2.5 \times 10^{-2}$. The geometry and the initial particle distribution are the same as the case of Poiseuille flow (Example 4.1). The smoothing length and time step are exactly the same. Similarly the simulation results reach a steady state after about 5000 steps. The particle distribution and velocity quiver at the 6000-th step are obtained and shown in Figure 4.21 and Figure 4.22, respectively. It is seen that some particles that are originally located in the downstream are wraparounded to the upstream due to the imposing of the periodic boundary condition. Figure 4.23 shows the comparison between the velocity profiles obtained using the SPH method at the 6000-th steps and the analytical series solution. The results at $t = 0.01$ s, 0.1 s, and the final steady state $t = \infty$ are plotted. The SPH results are also in good agreement with the analytical solution with less than 0.5% errors.

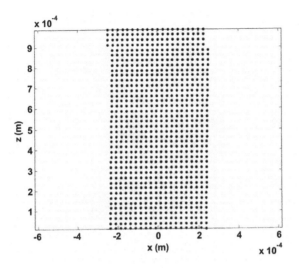

Figure 4.21 Particle distribution for the Couette flow obtained using the SPH method at the 6000-th step. Only the interior real particles are shown. It is seen that some particles that are originally located in the downstream are wraparounded to the upstream after imposing the periodic boundary condition.

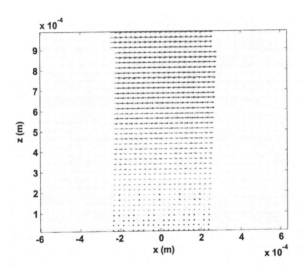

Figure 4.22 Velocity quiver for the Couette flow obtained using the SPH method at the 6000-th step. It is seen that some particles that are originally located in the downstream are wraparounded to the upstream after imposing the periodic boundary condition.

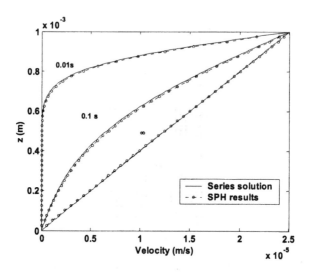

Figure 4.23 Velocity profiles for the Couette flow.

Example 4.3   Shear driven cavity problem

The classic shear driven cavity problem is the fluid flow within a closed square generated by moving the top side of the square at a constant velocity V_{Top} while the other three sides remain stationary. The flow will reach a steady state, and form a recirculation pattern. In the simulation, the dimension of the side of the square domain is $l = 10^{-3}$ m; the kinetic viscosity and density are $\upsilon = 10^{-6}$ m²/s and $\rho = 10^3$ kg/m³ respectively. The top side of the square moves at a velocity of $V_{Top} = 10^{-3}$ m/s, thus the Reynolds number for this case is 1. As shown in Figure 4.24, a total of 1600 (40×40) real particles (circular dots) are placed in the square region, while 320 (81 on each side of the square) virtual particles (unfilled circles) are placed right on the four edges. A constant time step of 5×10^{-5} s is used. It takes approximately 3000 steps to reach a steady state.

Figure 4.25 and Figure 4.26 show the particle and velocity distribution at the steady state, in which the recirculation pattern of the flow can be observed clearly. Results are compared with those by the finite difference method with 41×41 grid. Figure 4.27 shows the non-dimensional vertical velocity profile along the horizontal centerline, while Figure 4.28 shows the non-dimensional horizontal velocity profile along the vertical centerline. It can be seen from

Figure 4.27 and Figure 4.28, that the results from the present method and those from the finite difference method agree well, while the SPH method slightly underpredicts the values compared to the finite difference method. More close solution can be obtained by fine tuning of the parameters in the SPH algorithms, as well be discussed in Chapter 10.

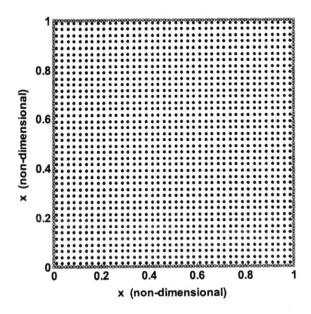

Figure 4.24 Initial particle distributions. The interior real particles (dots) and type I virtual boundary particles (circles) are shown.

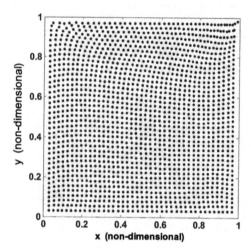

Figure 4.25 Particle distributions at the steady state for the shear driven cavity problem. Only the real interior particles (dots) are shown.

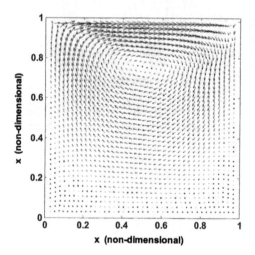

Figure 4.26 Steady state velocity distributions for the shear driven cavity flow. The length of the arrows represents the magnitude of the velocities.

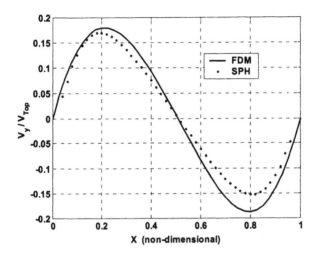

Figure 4.27 Non-dimensional vertical velocities along the horizontal centerline.

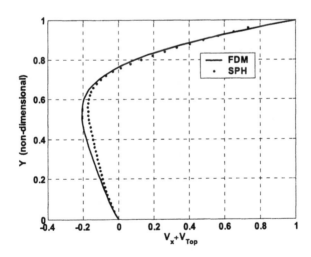

Figure 4.28 Non-dimensional horizontal velocities along the vertical centerline.

4.6.2   Applications to free surface flows

The study of free surface flows is very important in applications in environmental and many other industries. Special treatment needs to be made to properly model the free surface. This section presents applications of the SPH method to free surface flows. Three cases are studied. The first case is a clump of water splashing onto a solid step, the second case is a water discharge problem with partial opening of the water gate at the bottom of the water container, and the third case is the water flow after the sudden collapse of the dam. These problems were investigated by Monaghan (1994) and Lam et al. (2000). The viscous effect of the water is neglected here. In the simulation, the equation (4.88) is used to model the water as a slightly compressible fluid. Both the type I and type II virtual particles are used to simulate the solid walls and the ground.

Example 4.4   Water splash

In this example, a clump of water in circular shape with a diameter of 0.05 m splashes onto the solid step. The water clump is modeled using particles and the SPH method is used for simulating the movement of the water particles.

Figure 4.29 shows the particle distribution of the water particles at six representative instants when the water clump impacts on the tip of the solid step. The incidence angle of the water clump is 30° and the incidence velocity of the water clump is 10 m/s. It is seen that the water particles gradually approach the tip of the step, collide with the solid step, and then splash away from the step.

Figure 4.30 shows the particle distribution of the water particles at six representative instants when the water clump impacts on the corner of the solid step. The incidence angle of the water clump is 45° and an incidence velocity of the water clump is 10 m/s. It can be observed that the water particles gradually approach the corner of the step, collide with the solid step, push together, and then reflect from the corner of the step. Note that for this example, there is no data from other sources readily available for comparison. However, from the particle evolutions shown in Figure 4.29 and Figure 4.30, it is demonstrated that the SPH method is capable of capturing the major features of such free surface flow problems.

Example 4.5   Water discharge

Figure 4.31 shows the particle distribution of the water particles being discharged with 12% of the gate opening at four representative instants. It can be seen that the particles distribute orderly with the flow of water before the gate. The streamline can be clearly seen. Water particles eject out of the gate bottom due to the pressure force, splash high outside and finally fall to the ground due to the gravity. Near the region of the gate bottom, the water flows rather evenly with potential energy transformed into kinetic energy. Since the water splashes high outside the gate and then falls to the ground, a cavity forms during the course of the flow.

Example 4.6 Dam collapse

If the gate opens fully in a sudden, the water discharge problem becomes a dam collapsing problem. In this case, the initial water lever is denoted by HT_0, the initial surge front is SR_0, and they are all set at 25 meters, meaning that the water block is of square shape.

The SPH code is used to simulation for the fall the water. Figure 4.32 shows the particle distribution at six representative instants obtained from the SPH code. The water particles flow orderly forward with increasing surge front SR and decreasing water lever HT. The numerical results from the present work (denoted by the subscript p), the experimental data (denoted by the subscript exp), and the results by Monaghan (1994) (denoted by the subscript m) are compared and shown in Table 4.3. The surge front SR and the water lever HT at different instants are normalized by the initial water lever HT_0, while the time is normalized by $\sqrt{HT_0 / g}$, where g is the gravity constant. The results, especially the surge fronts, are more accurate than what obtained by Monaghan (1994). This is due to the contributions of the type II boundary virtual particles used in the present summation process, which increase the drag force for the particles near the bottom.

Table 4.3 Water level and water surging front for the dam collapsing problem

Time	HT_{exp}	HT_m	HT_p	SR_{exp}	SR_m	SR_p
0.71	0.90	0.90	0.90	1.33	1.56	1.45
1.39	0.76	0.75	0.75	2.25	2.50	2.38
2.10	0.57	0.56	0.56	3.22	3.75	3.50
3.20	0.32	0.37	0.35	3.80	3.00	3.88

* Variables Subscripts

HT—Height of water; p—Present solution;
SR—Surge front of water. m—Results from Monaghan (1994)
  exp—Experimental data (Monaghan 1994).

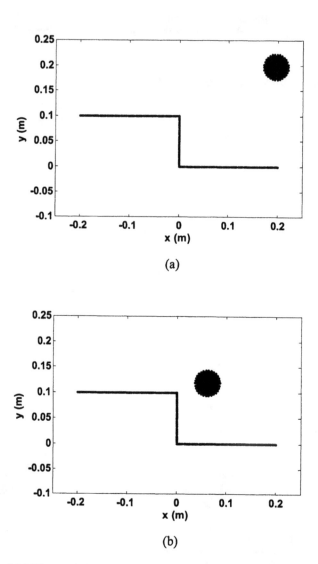

Figure 4.29 The particle distribution of the water particles at six representative instants when the water clump impacts on the tip of the solid step. The incidence angle is 30° and the incidence velocity is 10 m/s. The type II virtual particles are used but not shown. The water particles gradually approach the tip of the step. (To be continued.)

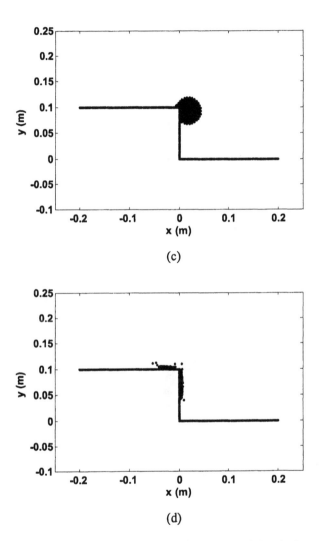

(c)

(d)

Figure 4.29 The particle distribution of the water particles at six representative instants when the water clump impacts on the tip of the solid step. The incidence angle is 30° and the incidence velocity is 10 m/s. The type II virtual particles are used but not shown. The water particles collide with the solid step, and then splash away from the step. (To be continued.)

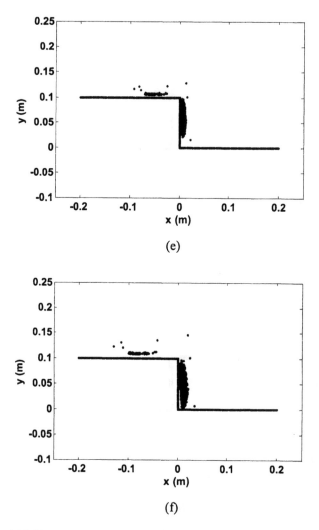

(e)

(f)

Figure 4.29 The particle distribution of the water particles at six representative instants when the water clump impacts on the tip of the solid step. The incidence angle is 30° and the incidence velocity is 10 m/s. The type II virtual particles are used but not shown. The water particles splash away from the step.

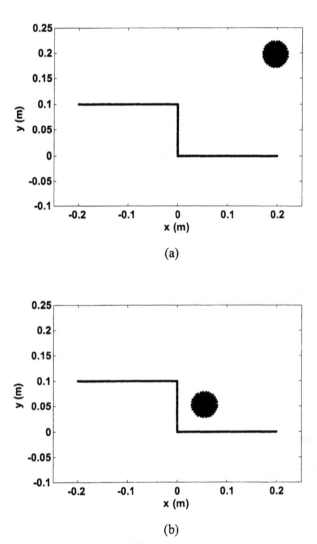

(a)

(b)

Figure 4.30 The particle distribution of the water particles at six representative instants when the water clump impacts on the corner of the solid step. The incidence angle is 45° and the incidence velocity is 10 m/s. The type II virtual particles are used but not shown. The water particles gradually approach the corner of the step. (To be continued.)

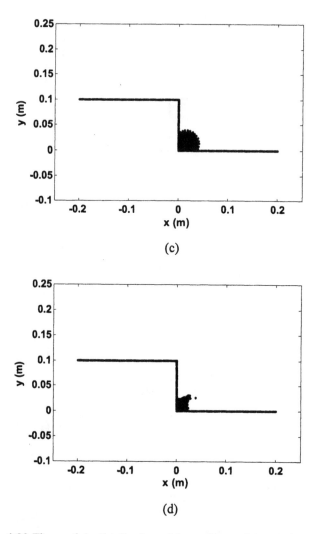

(c)

(d)

Figure 4.30 The particle distribution of the water particles at six representative instants when the water clump impacts on the corner of the solid step. The incidence angle is 45° and the incidence velocity is 10 m/s. The type II virtual particles are used but not shown. The water particles collide with the solid step, and push together. (To be continued.)

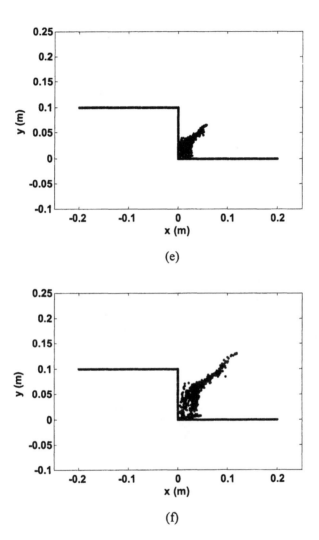

Figure 4.30 The particle distribution of the water particles at six representative instants when the water clump impacts on the corner of the solid step. The incidence angle is 45° and the incidence velocity is 10 m/s. The type II virtual particles are used but not shown. The water particles reflect from the corner of the step.

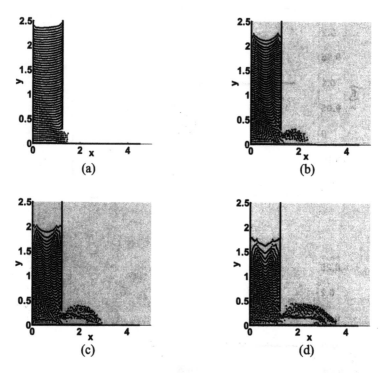

Figure 4.31 Particle distributions at four representative instants t = 0.5, 1.0, 1.5 and 2.0 s for the water discharge problem. The gate is 12% open at the bottom of the water container. The type II virtual particles are used but not shown.

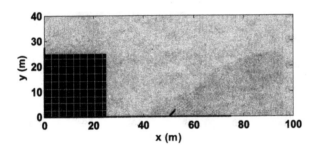

(a)

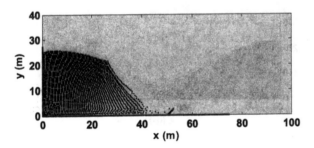

(b)

Figure 4.32 The particle distribution of the falling water at six representative instants for the dam collapse problem. The type II virtual particles are used but not shown. (To be continued.)

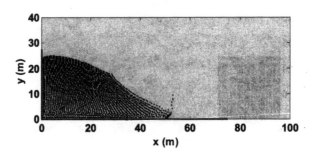

(c)

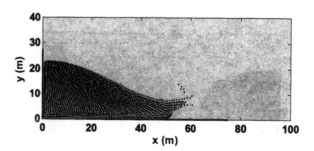

(d)

Figure 4.32 The particle distribution of the falling water at six representative instants for the dam collapse problem. The type II virtual particles are used but not shown. (To be continued.)

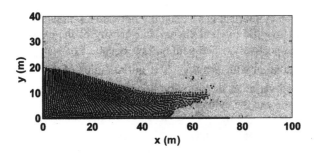

(e)

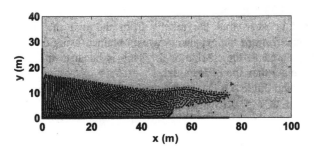

(f)

Figure 4.32 The particle distribution of the falling water at six representative instants for the dam collapse problem. The type II virtual particles are used but not shown.

4.6.3   Applications to compressible flows

Example 4.7   Gas expansion

In this example, a cylindrical gas expanding outwards is simulated using the SPH method.   Initial specific internal energy and density of the explosive are 1 J/kg and 4 kg/m^3, respectively.   The gamma law equation of state $p = (\gamma - 1)\rho u$ is used in the simulation with $\gamma = 1.4$.   The initial radius of the cylinder is 0.1 m. Because the filed variables do not vary along the axial direction of the cylinder, only a cross-section of the cylinder is modeled using particles.

In the simulation, 20 particles are placed along the radial direction while 60 particles are placed along the tangential direction.   This exactly same problem is also simulated using the commercial software package MSC/Dytran (a hydrocode using the finite volume method) for comparison.   In the Dytran model, the symmetry of the problem is made use of, and only a section of the cross-section of the cylinder is modeled with 400 volume cells along the radial direction.   It is, in fact, treated as a quasi-one-dimensional problem.   Therefore, the resolution of the Dytran model is about 7 times higher that the SPH model along the radial direction.

The results from the SPH method at the instants of 0.05, 0.1 and 0.15 s are shown in Figure 4.33, together with those obtained using the fine MSC/Dytran model.   It can be seen that the results from the two methods are in good agreement.   The front of the expansion wave obtained using the Dytran model is smoother compared to the SPH model.   This is because the Dytran model is about 7 times finer than the SPH model.

Table 4.4 and Table 4.5 list, respectively, the density and pressure along the radial direction at $t = 0.15$ s.   It is seen that except for the last location (corresponding to the last particle in the SPH simulation), the density and pressure profiles obtained using the SPH method are very close to those obtained using MSC/Dytran.   The difference of the results at the last location is caused by the different resolution of these two different models.   Note that the MSC/Dytran is a grid-based hydrocodes, in which the results are interpolated with volumes of cells.   In contrast, the SPH results are obtained from the particle and Lagrangian simulation, where no real particles exist beyond the last particle.   Hence, it may be justifiable that the density and pressure obtained using the SPH method are bigger than those obtained using MSC/Dytran.

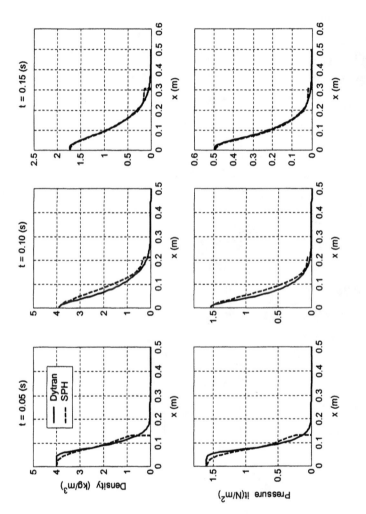

Figure 4.33 Density and pressure profiles for the two-dimensional gas expansion problem at t=0.05, 0.10, and 0.15 s.

Table 4.4 Results for the simulation of 2D gas expansion. Density along the
radial direction at $t = 0.15$ s

Location (m)	SPH results (Kg/m³)	MSC/Dytran results (Kg/m³)
0.0070729	1.719	1.719
0.014232	1.710	1.728
0.021425	1.690	1.711
0.028626	1.658	1.682
0.0359	1.611	1.651
0.043427	1.548	1.587
0.051312	1.471	1.486
0.059628	1.380	1.405
0.068489	1.281	1.320
0.077905	1.178	1.209
0.087981	1.0721	1.102
0.098778	0.96623	1.001
0.11053	0.86046	0.884
0.12354	0.75473	0.777
0.13771	0.65214	0.661
0.15358	0.54242	0.558
0.17245	0.4201	0.439
0.19764	0.27104	0.316
0.23889	0.17114	0.174
0.30684	0.13268	0.050

Table 4.5 Results for the simulation of 2D gas expansion. Pressure along the radial direction at $t = 0.15$ s.

Location (m)	SPH results (N/m^2)	MSC/Dytran results (N/m^2)
0.0070729	1.719	1.719
0.014232	1.710	1.728
0.021425	1.690	1.711
0.028626	1.658	1.682
0.0359	1.611	1.651
0.043427	1.548	1.587
0.051312	1.471	1.486
0.059628	1.380	1.405
0.068489	1.281	1.320
0.077905	1.178	1.209
0.087981	1.0721	1.102
0.098778	0.96623	1.001
0.11053	0.86046	0.884
0.12354	0.75473	0.777
0.13771	0.65214	0.661
0.15358	0.54242	0.558
0.17245	0.4201	0.439
0.19764	0.27104	0.316
0.23889	0.17114	0.174
0.30684	0.13268	0.050

4.7   Concluding remarks

Smoothed particle hydrodynamics has a number of advantages over the conventional grid-based numerical methods, and is an effective technique for solving fluid dynamic problems.   This chapter presented the SPH implementations based on the Navier-Stokes equations for general dynamic fluid flows.   The resultant SPH equations of motion can simulate different flow problems such as inviscid or viscous flows, compressible or incompressible flows.   The standard artificial viscosity is used to model the inviscid flow problems in resolving shock waves.   A SPH expression of physical viscosity constructed from the viscous shear force is employed for viscous flow problems. For incompressible flow problems, the concept of artificial compressibility is applied by selecting a special equation of state to model the incompressible fluid as quasi-incompressible.   Extra virtual particles (type I and/or type II) are used to treat the solid boundary condition.   This treatment can prevent real particles from penetrating outside the boundary and can also improve the accuracy of numerical results.

A series of numerical tests have been carried out for different dynamic fluid flow problems.   For incompressible flows such as the Poiseuille flow, Couette flow and the shear driven cavity problem, the presented SPH method can obtain satisfactory results.   The advantages of the SPH method in treating free surface flows can be clearly seen in the simulation of the water discharge problems with the gate partially opened and the dam collapsing, as well as the problems of water clump impacting on a solid step.   The SPH method has also simulated the gas expansion process very well.

Compared to the grid-based method, the SPH method can successfully simulate such problems at reasonable accuracy with less computational effort.   It shows that the SPH method, with proper modifications, is an effective alternative to conventional numerical methods for dynamic fluid flow simulations.   The numerical examples not only verify the validity of the SPH method and the code, but also provide a good foundation for extending the code to simulate large deformation hydrodynamic phenomena such as high explosive explosions, which will be discussed in detail in later chapters.

Chapter 5

Discontinuous SPH (DSPH)

Chapter 4 has formulated the SPH method for general CFD problems that have very large deformations. The capability of the SPH method has also been demonstrated via a number of example problems of fluid flows.

In this chapter, a SPH formulation is presented for simulating physical phenomena with discontinuities. As the formulation allows discontinuity of field functions in the support domain, it is, therefore, termed in this book as discontinuous SPH or DSPH for the convenience of description.

The DSPH formulation was initially presented by Liu et al. (2003e) for handling shock waves using the SPH method. The formulation is based on the Taylor series expansion in the piecewise continuous regions on both sides of the discontinuity. The formulation is originated from an existing improved version of the SPH method called corrective smoothed particle method (CSPM). The DSPH not only improves the boundary deficiency problem in the traditional SPH method, but also restores the kernel consistency of the corrective smoothed particle method (CSPM) in the discontinuous regions. The resultant kernel and particle approximations consist of a primary part similar to that in the CSPM, and an additional part derived from the discontinuity treatment.

A numerical study is carried out to examine the performance of the DSPH formulation. The results show that the DSPH not only remedies the boundary deficiency problem but also well simulates the discontinuity that may exist in the field variable functions. The DSPH formulation is also applied to simulate the one-dimensional shock tube problem with fairly good results. These numerical tests suggest that the DSPH is attractive in simulating hydrodynamic problems with discontinuities such as shock waves.

5.1 Introduction

The SPH method has been successfully applied to a vast range of CFD problems as shown in Chapter 4. There are, however, some inherent difficulties associated with the SPH method. One of the difficulties is the boundary deficiency problem (Monaghan, 1992; Randles and Libersky, 1996), which is in the kernel approximation, caused by the truncated integral by the boundary, and in the particle approximation, caused by the insufficient particles in the summation process since no particle exists beyond the boundary. In the hydrodynamic simulations, the boundary deficiency problem results in lower density near or on the boundary, and finally yields spurious pressure gradients on the surface (Morris et al., 1997).

Different remedies have been proposed to treat the boundary deficiency problem. Campbell (1989) suggested treating the boundary condition by including the residue boundary terms in the integration by parts when estimating the original kernel integral involving gradients. Libersky and Petschek (1991) first introduced ghost particles to treat the free surface condition with the ghost particles possessing the same density but opposite velocity of the reflected real particles. Later, Randles and Libersky (1996) proposed a more general treatment of the boundary condition by assigning the boundary values of a field variable to all ghost particles, and then interpolating smoothly the values specified on the boundary ghost particles and the calculated values of the interior particles.

Johnson and Beissel (1996) proposed a normalized smoothing function (NSF) for axisymmetric problems based on the condition of uniform strain rate and then extended it to three-dimensional geometry (Johnson et al., 1996b). In their implementation, corrective factors were employed and adjusted in the particle approximation of the strain rates so that, for velocity fields that yield constant values of normal velocity strains, the normal velocity strains will be exactly reproduced.

To remedy the boundary deficiency problem, Randles and Libersky (1996) derived a normalization formulation for the density approximation

$$\rho_i = \frac{\displaystyle\sum_{j=1}^{N} m_j W_{ij}}{\displaystyle\sum_{j=1}^{N} \left(\frac{m_j}{\rho_j}\right) W_{ij}} \tag{5.1}$$

and a normalization for the divergence of the stress tensor σ

$$(\nabla \cdot \sigma)_i = \frac{\left(-\sum_{j=1}^{N}\left(\frac{m_j}{\rho_j}\right)(\sigma_j - \sigma_i) \otimes \nabla_i W_{ij}\right)}{\left(-\sum_{j=1}^{N}\left(\frac{m_j}{\rho_j}\right)(x_j - x_i) \otimes \nabla_i W_{ij}\right)} \qquad (5.2)$$

where m and x are the mass and position vector, W is the smoothing function, and $\otimes$ is the tensor product. Equation (5.1) not only takes care of the density deficiency at free boundaries, it also solves the contact boundary problem with a density discontinuity if the neighbors of a certain particle are taken only from the same material, and not from different materials. With equation (5.2), linear field of stresses can be reproduced (Randles and Libersky; 1996).

Recently, in a series of papers (Chen et al., 1999a; 1999b; 1999c; 2000), Chen and his co-workers have developed a corrective smoothing particle method (CSPM). The corrective smoothing particle method provides an approach to normalize the kernel and particle approximations in the traditional SPH method so as to solve the boundary deficiency problem. It was reported that the CSPM can also reduce the so-called tensile instability inherent in the traditional SPH method. The method is straightforward and flexible compared with other improvement techniques for the traditional SPH. The advantages of the CSPM over traditional SPH were shown on a series of applications to transient heat conduction, structure dynamics and nonlinear dynamic problems. The corrective smoothed particle method does not change the traditional smoothing function, and therefore does not have the problems related to negative smoothing functions and asymmetric particle interactions. However, since the method is still based on the Taylor series expansion for a function that should be sufficiently smooth in the entire support domain, it is not applicable to problems with discontinuities such as hydrodynamic problems that generate shock waves.

In this chapter, the corrective smoothed particle method (CSPM) is first introduced. Based on the CSPM, a DSPH formulation is then developed to approximate discontinuous field variables. The DSPH is based on the Taylor series expansion performed separately in both sides of a discontinuity, rather than in the entire domain as in the CSPM. The resultant kernel and particle approximations for a discontinuous function and its derivatives consist of two parts, one part similar to that in the CSPM, and an additional correction part from the discontinuity.

This chapter is outlined as follows.

- In Section 2, the corrective smoothed particle method is discussed.
- In Section 3, the DSPH formulation is introduced.

- In Section 4, a study is conducted to examine the numerical performance of the DSPH formulation.
- In Section 5, the DSPH formulation is applied to simulate the one-dimensional shock tube problem. The results are compared with those from other versions of the SPH method and the analytical solution.
- In Section 6, some remarks and conclusion are given.

5.2   Corrective smoothed particle method (CSPM)

5.2.1   One-dimensional case

Based on the Taylor series expansion, the corrective smoothed particle method (CSPM) provides an approach to normalize the kernel and particle approximation in the SPH method. Therefore, it reduces the possible errors inherent in the traditional SPH kernel and particle approximations. The CSPM procedure may be summarized as follows.

Expanding the Taylor series for a function at a point or particle, multiplying both sides with the smoothing function, and then integrating over the support domain yield the kernel approximation and then the later particle approximation of the function at that particle. Similarly, if replacing the smoothing function with its first or higher order derivatives in the multiplication process, the kernel and the particle approximations of the first and higher order derivatives for the function can be obtained.

In a one-dimensional space, if a function $f(x)$ is assumed to be sufficiently smooth in a domain that contains x, performing the Taylor series expansion for $f(x)$ in the vicinity of x_i yields

$$f(x) = f_i + (x - x_i)f_{xi} + \frac{(x - x_i)^2}{2} f_{xxi} + \dots \tag{5.3}$$

Multiplying both sides of equation (5.3) by a smoothing function W defined in the local support domain Ω of x_i and integrating over the support domain yield

$$\int_{\Omega} f(x)W_i(x)dx = f_i \int_{\Omega} W_i(x)dx + f_{xi} \int_{\Omega} (x - x_i)W_i(x)dx$$

$$+ \frac{f_{xxi}}{2!} \int_{\Omega} (x - x_i)^2 W_i(x)dx + ... \tag{5.4}$$

In equations (5.3) and (5.4),

$$f_i = f(x_i) \tag{5.5}$$

$$f_{xi} = f_x(x_i) = (\partial f / \partial x)_i \tag{5.6}$$

$$f_{xxi} = f_{xx}(x_i) = (\partial^2 f / \partial x^2)_i \tag{5.7}$$

$$W_i(x) = W(x - x_i, h) \tag{5.8}$$

If the terms involving derivatives in equation (5.4) are neglected, a corrective kernel approximation for function $f(x)$ at x_i is obtained as

$$f_i = \frac{\int_{\Omega} f(x)W_i(x)dx}{\int_{\Omega} W_i(x)dx} \tag{5.9}$$

Similarly, if replacing $W_i(x)$ in equation (5.4) with $W_{ix}(x) (= \partial W_i(x)/\partial x)$, and neglecting the terms related to the second and higher derivatives, a corrective kernel approximation for the first derivative of $f(x)$ at x_i is derived as

$$f_{xi} = \frac{\int_{\Omega} (f(x) - f(x_i))W_{ix}(x)dx}{\int_{\Omega} (x - x_i)W_{ix}(x)dx} \tag{5.10}$$

In equations (5.9) and (5.10), the numerators are actually the kernel approximations for a function and its first derivative in the traditional SPH method. The denominators (or the normalization factors) are the representation of the normalization property of the smoothing function (the denominator in equation (5.10) equals to the denominator in equation (5.9) when applying the integration by parts and neglecting the surface term). Therefore, it can be said that, normalizing the kernel approximations in the traditional SPH method results in the corrective kernel approximations in the corrective smoothed particle method.

Note that the integration of the smoothing function is unity for particles whose support domains does not intersect with the boundary, and is non-unity for particles whose support domains intersects with the boundary. It is clear that the truncation error for f_i and f_{xi} is of order $(x - x_i)^2$ for the interior particles and $(x - x_i)$ for the particles near or on the boundary. Obviously, ignoring the normalization factors or assuming the integration of the smoothing function W to always be unity for all the particles either in interior region or near a boundary is the essential reason for the boundary deficiency in the traditional SPH method.

The particle approximations for equations (5.9) and (5.10) at particle i become

$$f_i = \frac{\displaystyle\sum_{j=1}^{N}\left(\frac{m_j}{\rho_j}\right)f_j W_{ij}}{\displaystyle\sum_{j=1}^{N}\left(\frac{m_j}{\rho_j}\right)W_{ij}} \tag{5.11}$$

and

$$f_{xi} = \frac{\displaystyle\sum_{j=1}^{N}\left(\frac{m_j}{\rho_j}\right)(f_j - f_i)\nabla_i W_{ij}}{\displaystyle\sum_{j=1}^{N}\left(\frac{m_j}{\rho_j}\right)(x_j - x_i)\nabla_i W_{ij}} \tag{5.12}$$

where N is the total number of particles in the support domain.

The same idea can be extended to approximate higher derivatives of the function if multiplying equation (5.3) with the same order derivatives of the smoothing function and again neglecting the derivative terms whose order is higher than that of the one being approximated.

From the above equations, it can be seen that the difference between the CSPM and the traditional SPH method is the construction of the kernel approximations in integral representation. The kernel approximations as well as the particle approximations in the CSPM and those in the traditional SPH are connected together by the denominators that actually act as a kind of corrective normalization.

5.2.2 Multi-dimensional case

Similar results can be obtained following the same procedure using the Taylor series expansion in a multi-dimensional space.

$$f(x) = f_i + (x^\alpha - x_i^\alpha)f_{i,\alpha} + \frac{(x^\alpha - x_i^\alpha)(x^\beta - x_i^\beta)}{2}f_{i,\alpha\beta} + \dots \tag{5.13}$$

where α, β are the dimension indices from 1 to 3.

$$f_{i,\alpha} = f_\alpha(x_i) = (\partial f/\partial x^\alpha)_i \tag{5.14}$$

$$f_{i,\alpha\beta} = f_{\alpha\beta}(x_i) = (\partial^2 f/\partial x^\alpha \partial x^\beta)_i \tag{5.15}$$

Multiplying both sides of equation (5.13) by the smoothing function W, integrating over the support domain Ω and neglecting the derivative terms yield the corrective kernel approximation for the function $f(x)$.

$$f_i = \frac{\int_\Omega f(x)W_i(x)dx}{\int_\Omega W_i(x)dx} \tag{5.16}$$

The three first derivatives $f_{i,\alpha}$ at particle i can be obtained in the form of the following three coupled equations

$$f_{i,\alpha}\int_\Omega \left(x^\alpha - x_i^\alpha\right)W_{i,\beta}(x)dx = \int_\Omega \left(f(x) - f(x_i)\right)W_{i,\beta}(x)dx \tag{5.17}$$

where

$$W_{i,\beta}(x) = \partial W(x_i)/\partial x^\beta \tag{5.18}$$

It is noted that the form of kernel approximation for a function in a multi-dimensional space is the same as the one in one-dimensional space.

For the kernel approximation of the three first derivatives, the expression is a little complicated since the three equations representing the kernel approximations of the three derivatives are actually coupled together, and a matrix inversion must be performed to obtain the result. Approximating the integrals in equation (5.17) by summing over the discretized particles yields the following set of particle approximation equations

$$X_{i,\alpha\beta} f_{i,\alpha} = Y_{i,\beta} \qquad (5.19)$$

where

$$X_{i,\alpha\beta} = \sum_{j=1}^{N} \frac{m_j}{\rho_j} (x_j^\alpha - x_i^\alpha) W_{ij,\beta} \qquad (5.20)$$

$$Y_{i,\beta} = \sum_{j=1}^{N} \frac{m_j}{\rho_j} (f(x_j) - f(x_i)) W_{ij,\beta} \qquad (5.21)$$

Equations (5.11), (5.12), (5.20) and (5.21) are the basic formulae in the CSPM that are used to generate the discrete system equations for partial differential equations of the conservation laws of physics. The CSPM was shown to successfully improve the boundary deficiency and reduce the tensile instability in the traditional SPH method (Chen et al., 1999b). The correction is straightforward and works well, and hence it will be the starting point for the DSPH formulation.

5.3   DSPH formulation for simulating discontinuous phenomena

This section introduces a DSPH formulation for simulating phenomena with discontinuities (Liu et al., 2003e). As shown in Section 5.2, the CSPM is straightforward and works well. It is extended in this section for problem with discontinuity. Since the following discussions are based on one-dimensional space, the position vector x is, in fact, a scalar. The detailed formulation for the DSPH is as follows.

5.3.1   DSPH formulation

Let's consider a one-dimensional problem. Examine the kernel approximation for any function $f(x)$ in the support domain of x_i. The support domain is bounded by a and b with a dimension of $2\kappa h$. shown in Figure 5.1. Assume that function $f(x)$ has an *integrable* discontinuity at d in the support domain and it is located in the right half of the support domain, i.e., $x_i < d \le b$, the integration of the multiplication of $f(x)$ and the smoothing function W over the entire support domain can be divided into two parts

$$\int_a^b f(x)W_i(x)dx = \int_a^d f(x)W_i(x)dx + \int_d^b f(x)W_i(x)dx \qquad (5.22)$$

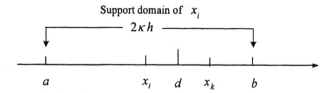

Figure 5.1 Kernel approximations for a one-dimensional function with a discontinuity at point d. The support domain of x_i is bounded by a and b with a dimension of $2\kappa h$.

Expanding $f(x)$ in the first integral on the right hand side around point x_i, and around another arbitrary point x_k in the second integral, where $d \le x_k \le b$ gives

$$
\begin{aligned}
&\int_a^b f(x)W_i(x)dx\\
&= \int_a^d W_i(x)\Big[f(x_i)+(x-x_i)f'(x_i)+r(h^2)\Big]dx\\
&\quad + \int_d^b W_i(x)\Big[f(x_k)+(x-x_k)f'(x_k)+r(h^2)\Big]dx\\
&= f(x_i)\int_a^d W_i(x)dx + f(x_k)\int_d^b W_i(x)dx\\
&\quad + f'(x_i)\int_a^d (x-x_i)W_i(x)dx + f'(x_k)\int_d^b (x-x_k)W_i(x)dx\\
&\quad + r(h^2)
\end{aligned}
\qquad (5.23)
$$

where r stands for the residual.
Rearranging by combining some similar terms with some transformations yields

$$\int_a^b f(x)W_i(x)dx$$

$$= f(x_i)\int_a^b W_i(x)dx + [f(x_k)-f(x_i)]\int_d^b W_i(x)dx$$

$$+f'(x_i)\int_a^b (x-x_i)W_i(x)dx$$

$$+\int_d^b [(x-x_k)f'(x_k)-(x-x_i)f'(x_i)]W_i(x)dx + r(h^2)$$

(5.24)

Note that the kernel W is assumed to be even, normalized, and has a compact support. By the compact support property of W, we have $|x-x_i|\leq \kappa h$, $|x-x_k|\leq \kappa h$. In the above equation, we assume that $f'(x)$ must exist and be bounded in $[a,d)\cup(d,b]$. In other words, the derivative of $f(x)$ exists, and when x approaches d from both sides, it is bounded within a finite limit. Hence, the last two terms (excluding the residual term) on the RHS of the above equation can be bounded by terms of order of h respectively. Therefore, the above equation can be rewritten as:

$$\int_a^b f(x)W_i(x)dx$$

$$= f(x_i)\int_a^b W_i(x)dx + [f(x_k)-f(x_i)]\int_d^b W_i(x)dx + r(h)$$

(5.25)

Rewriting the above equation yields

$$f(x_i) = \frac{\int_a^b f(x)W_i(x)dx}{\int_a^b W_i(x)dx} - \left\{\frac{[f(x_k)-f(x_i)]\int_d^b W_i(x)dx}{\int_a^b W_i(x)dx}\right\} + r(h) \qquad (5.26)$$

Similar results can be obtained for $f'(x_i)$,

$$\int_a^b f(x)W_{ix}(x)dx$$

$$= \int_a^d W_{ix}(x)\Big[f(x_i)+(x-x_i)f'(x_i)+r(h^2)\Big]dx$$

$$+ \int_d^b W_{ix}(x)\Big[f(x_k)+(x-x_k)f'(x_k)+r(h^2)\Big]dx$$

$$= f(x_i)\int_a^b W_{ix}(x)dx+[f(x_k)-f(x_i)]\int_d^b W_{ix}(x)dx \qquad (5.27)$$

$$+f'(x_i)\int_a^b (x-x_i)W_{ix}(x)dx$$

$$+ \int_d^b [(x-x_k)f'(x_k)-(x-x_i)f'(x_i)]W_{ix}(x)dx$$

$$+r(h^2)$$

which is simplified as

$$f'(x_i) = \frac{\int_a^b [f(x)-f(x_i)]W_{ix}(x)dx}{\int_a^b (x-x_i)W_{ix}(x)dx}$$

$$-\{\frac{[f(x_k)-f(x_i)]\int_d^b W_{ix}(x)dx}{\int_a^b (x-x_i)W_{ix}(x)dx} \qquad (5.28)$$

$$+\frac{\int_d^b [(x-x_k)f'(x_k)-(x-x_i)f'(x_i)]W_{ix}(x)dx}{\int_a^b (x-x_i)W_{ix}(x)dx}\}+r(h)$$

Compared with equations (5.9) and (5.10), the first expressions on the right hand side of equations (5.26) and (5.28) are actually the corrective kernel approximations in the CSPM. If the discontinuous function $f(x)$ and its first derivative are approximated in this way as a continuous function, and the second terms on the right hand side of equations (5.26) and (5.28) are neglected, the resultant kernel approximations in the presence of a discontinuity are inconsistent. If the second terms are retained, the resultant kernel approximations are consistent up to the first order.

If the domain is discretized by particles, since two particles cannot be located at the same position, the discontinuity should be always located between two particles. In deriving the above equations (5.26) and (5.28), since the point x_k is arbitrarily selected, it can be taken as the particle that is nearest to and on the right hand side of the discontinuity.

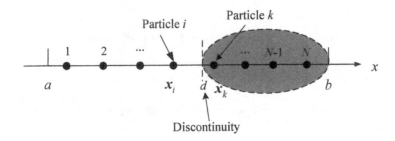

Figure 5.2 Particle approximations for a function with a discontinuity at point d. In the process from the kernel approximation to particle approximation, the arbitrary point x_k is at particle k that is the nearest particle on the right hand side of the discontinuity. The total number of particles in the support domain of $[a, b]$ is N.

Then the particle approximation of the discontinuous function is

$$f_i = P_i + A_i \tag{5.29}$$

where

$$P_i = \frac{\displaystyle\sum_{j=1}^{N}\left(\frac{m_j}{\rho_j}\right)f_j W_{ij}}{\displaystyle\sum_{j=1}^{N}\left(\frac{m_j}{\rho_j}\right)W_{ij}} \tag{5.30}$$

and

$$A_i = -\frac{[f(x_k)-f(x_i)]\displaystyle\sum_{j=k}^{N}\left(\frac{m_j}{\rho_j}\right)W_{ij}}{\displaystyle\sum_{j=1}^{N}\left(\frac{m_j}{\rho_j}\right)W_{ij}} \tag{5.31}$$

The particle approximation of the first derivative of the discontinuous function is

$$f'(x_i) = P_i' + A_i' \tag{5.32}$$

where

$$P_i' = \frac{\displaystyle\sum_{j=1}^{N}\left(\frac{m_j}{\rho_j}\right)(f(x_j) - f(x_i))\nabla_i W_{ij}}{\displaystyle\sum_{j=1}^{N}\left(\frac{m_j}{\rho_j}\right)(x_j - x_i)\nabla_i W_{ij}} \tag{5.33}$$

and

$$A_i' = -\frac{[f(x_k) - f(x_i)]\displaystyle\sum_{j=k}^{N}\left(\frac{m_j}{\rho_j}\right)\nabla_i W_{ij}}{\displaystyle\sum_{j=1}^{N}\left(\frac{m_j}{\rho_j}\right)(x_j - x_i)\nabla_i W_{ij}}$$

$$-\frac{\displaystyle\sum_{j=k}^{N}\left(\frac{m_j}{\rho_j}\right)\left[(x_j - x_k)f'(x_k) - (x_j - x_i)f'(x_i)\right]\nabla_i W_{ij}}{\displaystyle\sum_{j=1}^{N}\left(\frac{m_j}{\rho_j}\right)(x_j - x_i)\nabla_i W_{ij}} \tag{5.34}$$

The particle approximations consist of two parts, the primary part similar to those in the CSPM approximation and the additional part developed for the discontinuity treatment. It should be noted that the summations of the numerators of the additional parts are only carried out for particles in the right portion of the support domain $[d, b]$ shaded in Figure 5.2. The coordinates of these particles are $x_j(j = k, k+1, \cdots, N)$ that satisfy $(x_k - x_j)(x_k - x_i) \leq 0$.

Particles at the same side as i do not contribute to the summation in the additional parts. Otherwise, the additional part would become zero, and the modification term would disappear.

Equations (5.26), (5.28), (5.29) and (5.32) consist of the DSPH formulation for approximating (kernel and particle) a discontinuous function and its first order derivative. Higher order derivatives at the both sides of the discontinuity can also be approximated following the same procedure and can also be divided into two primary and additional parts. With some modifications, the DSPH formulation can be extended to multi-dimensional space.

5.3.2   Discontinuity detection

After the problem domain is represented with particles, the function is then in discretized form. Since the particles are free to move and the physical phenomena evolve, an efficient algorithm is needed to detect the location of the discontinuities based on the discrete function values at the current time step. If the discontinuities can be well located, the function and its derivative can be approximated using the formulae derived in Section 5.3.1.

An algorithm has been developed by Liu et al. (2003e) to locate the discontinuities. In their algorithm, the discontinuity detection is carried out by calculating the ratio of the local difference of the function values to the maximal difference of the function values in the entire problem domain. If the largest ratio for a particle is larger than a prescribed value defined as a criterion for the discontinuity, a discontinuity is captured. The discontinuity detection process is shown as follows.

$Given\ \Psi$ | Ψ: *a prescribed criterion for the discontinuities*
$Find\ f_{max}\ and\ f_{min},\ calculate\ \Delta = \left| f_{\max} - f_{\min} \right|$
For i from 1 to N
  $k = 1$
  $\delta = 0$
  For j from 2 to N
    $If\ \left| x_j - x_i \right| \leq \kappa h$, *then*
      $If\ \left| (f_j - f_i)/\Delta \right| \geq \Psi\ and\ \left| (f_j - f_i)/\Delta \right| \geq \delta\ then$
        $\delta = \left| (f_j - f_i)/\Delta \right|$
        $k = j$
      End if
    End if
  End for
End for

The located k is the particle that is closest to the discontinuity and is on the opposite side of the particle being evaluated. It should be noted that in the above algorithm, we aimed for clarity and the computational efficiency is not a major concern. However, it can be optimized using the previously described nearest particle searching algorithms to find the neighboring particles for a certain particle, and at the same time to determine if the discontinuity particle is one of the neighbors. Since the smoothing function is compactly supported, only particles near to the discontinuity can feel the influence of the discontinuity at the stage of particle approximation. The maximal function difference can also be taken within the neighboring particles in the support domain, but this perhaps would detect more discontinuities than the actual ones. Taking the maximal function difference in the entire problem domain can capture relatively large discontinuities. A problem is the value of Ψ that is pre-defined to capture a discontinuity. A very small Ψ may result in too many discontinuities, while a very large Ψ may lose sight of some discontinuities or even miss the discontinuities. Therefore, an engineering judgment and experience are usually necessary. Ψ can also be updated in the loops.

The discontinuity detection can also be carried out by checking if the global or local largest gradient of the field variables is bigger than a pre-defined criterion for continuity. Development of more efficient algorithms for discontinuity detection is required, especially for problems in multi-dimensional spaces.

5.4   Numerical performance study

Example 5.1   Discontinuous function simulation

In order to verify the effectiveness and accuracy of the DSPH formulation in approximating discontinuous functions, a piecewise continuous polynomial is used for the examination. Numerical results from the original CSPM and the traditional SPH method are also given for comparison.

The piecewise continuous function to-be-examined is

$$f(x) = \begin{cases} x^2 & -1 \le x \le 0.5 \\ -0.5 - x^2 & 0.5 < x \le 1 \end{cases} \tag{5.35}$$

The first derivative of this piecewise continuous function can be easily obtained as

$$f'(x) = \begin{cases} 2x & -1 \le x \le 0.5 \\ -2x & 0.5 < x \le 1 \end{cases} \tag{5.36}$$

Note that the function and its first derivative are continuous except at the point of $x = 0.5$.

Forty particles are equally distributed at the center of 40 equal sections along the problem domain of $[-1, 1]$ with a uniform particle volume of 0.05 ($m_i / \rho_i = \Delta x_i = 2/40$). In practical applications, the SPH method should have boundary conditions implemented, either by using virtual (ghost) particles or other methods. In this numerical performance study, computations using the CSPM, the DSPH and the traditional SPH are all without the implementation of boundary conditions. This is to prevent the features in the numerical results near the boundary from being concealed, and to get more distinguishable results for comparison. In this study, the cubic spline smoothing function is used (see, Section 3.1).

The approximation is carried out by obtaining numerical values at current step from the existing numerical values at the particles at the previous step. There is no physical dynamics involved in this numerical performance study. Only the accuracy of the numerical approximations of the given function and its first derivative are studied using the different versions of the SPH approximations. Therefore, in the approximation evolution, the errors present in the current step can propagate from one step to the next step, and may be magnified.

Note that when the SPH method is used to simulate a well-posed practical problem, the physics is governed by the conservation equations. If a stable numerical scheme is used, the errors present in the current step will always be controlled by the use of the governing equations. Therefore, the errors should be more or less within some level, and will not be magnified from one step to the next step.

Figure 5.3 and Figure 5.4 show the approximated results for the function and its first derivative after one step of approximation. The boundary deficiency for the traditional SPH method can be clearly seen either from the approximation result of the function or that of the first derivative of the function. The discontinuities in the function and in its first derivative around $x = 0.5$ are not correctly captured. The CSPM successfully remedied the boundary deficiency problem. However, the discontinuities in the function and in the derivative are not well resolved using the CSPM. The DSPH not only improves the results near the boundary, but also successfully models the discontinuity in the function and the derivation. The incorrect results by the traditional SPH and the CSPM around the discontinuities are originated from the inconsistency in the traditional SPH or in the CSPM. In other words, if a discontinuous function and its first derivative are approximated as a continuous function, and the second terms on the right hand side of equations (5.26) and (5.28) are neglected, the resultant kernel approximations for modeling the discontinuity are inconsistent. The obtained numerical value at the discontinuity is therefore incorrect. This incorrect numerical value at the discontinuity influences the approximation of the neighboring particles and therefore the inaccuracy propagates away from the discontinuity evolution. In the DSPH, since the second corrective terms are employed, the resultant kernel approximations are consistent up to the first order. The obtained numerical values at the discontinuity are much more accurate. The DSPH obviously performs very well in approximating the discontinuity, and is a significant improvement from the original CSPM.

If the approximation is performed for more steps of evolution, the mis-approximation of the discontinuities will further propagate to other neighboring particles in the following steps. Therefore, it is a good examination approach to approximate the function and its derivatives for more steps using these different versions of SPH formation, so as to show their difference in performance more clearly. Figure 5.5 and Figure 5.6 show the approximation results obtained after five steps for the approximation of functions and derivatives, respectively. It is clear that the DSPH produces much more accurate results in reproducing the discontinuity of the original discontinuous function. The approximated function and its derivative by the DSPH are much more accurate than those by the SPH and the CSPH.

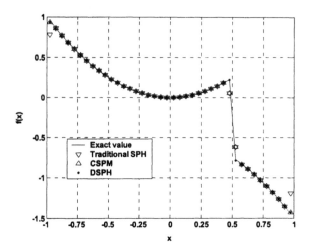

Figure 5.3 Approximated results for the discontinuous function defined by equation (5.35). The results are obtained after one step of approximation using different versions of SPH formulation.

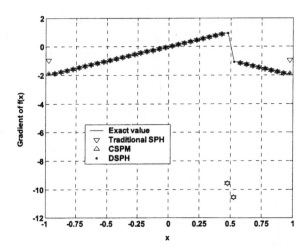

Figure 5.4 Approximated results for the first derivative of the discontinuous function defined by equation (5.36). The results are obtained after one step of approximation using different versions of SPH formulation.

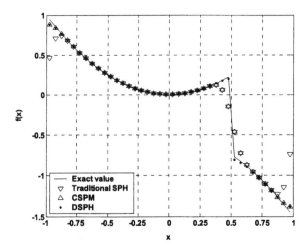

Figure 5.5 Approximated results for the discontinuous function defined by equation (5.35) after five evolution steps of approximation using different versions of SPH formulation.

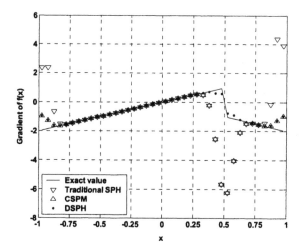

Figure 5.6 Approximated results for the first derivative of the discontinuous function defined by evolution equation (5.36) after five steps of approximation using different versions of SPH formulation.

5.5   Simulation of shock waves

The good performance in the function approximation of the DSPH provides confidence to apply it to physical problems with discontinuity such as shock problems.

Example 5.2   Shock discontinuity simulation

The DSPH is applied to simulate the one-dimensional shock tube problem that has been previously simulated in Example 3.1. The results are compared with those obtained using other methods. In this example, the initial conditions and material parameters are

$$x \leq 0 \quad \rho = 1 \quad v = 0 \quad e = 2.5 \quad p = 1 \quad \Delta x = 0.001875$$
$$x > 0 \quad \rho = 0.25 \quad v = 0 \quad e = 1.795 \quad p = 0.1795 \quad \Delta x = 0.0075$$

where ρ, p e, and v are the density, pressure, internal energy, and velocity of the gas, respectively. Δx is the particle spacing.

A total of 400 particles are deployed in the one-dimensional problem domain. All particles have the same mass of $m_i = 0.001875$. 320 particles are evenly distributed in the high-density region $[-0.6, 0.0]$, and 80 particles are evenly distributed in the low-density region $[0, 0.6]$. The initial particle distribution is to obtain the required discontinuous density profile along the tube. The equation of state for the ideal gas $p = (\gamma - 1)\rho e$ is employed in the simulation with $\gamma = 1.4$.

In solving the Euler equation using the CSPM, the density can be approximated using equation (5.11) by replacing $f(x)$ with density ρ, and is given by

$$\rho_i = \frac{\displaystyle\sum_{j=1}^{N} m_j W_{ij}}{\displaystyle\sum_{j=1}^{N} \left(\frac{m_j}{\rho_j}\right) W_{ij}} \tag{5.37}$$

As discussed earlier, the density evolution in the traditional SPH is the numerator on the RHS of equation (5.37), i.e.,

$$\rho_i = \sum_{j=1}^{N} m_j W_{ij} \tag{5.38}$$

The density approximation in the DSPH can be conducted by replacing $f(x)$ with density ρ in equation (5.29) as follows

$$\rho_i = \frac{\displaystyle\sum_{j=1}^{N} m_j W_{ij}}{\displaystyle\sum_{j=1}^{N} \left(\frac{m_j}{\rho_j}\right) W_{ij}} - \frac{[\rho_k - \rho_i]\displaystyle\sum_{j=k}^{N} \left(\frac{m_j}{\rho_j}\right) W_{ij}}{\displaystyle\sum_{j=1}^{N} \left(\frac{m_j}{\rho_j}\right) W_{ij}} \tag{5.39}$$

The momentum equation and the energy equation can be solved in the following way for particle i.

$$< \frac{Dv}{Dt} >_i = -\frac{1}{\rho_i} < \nabla p >_i \qquad (5.40)$$

$$< \frac{De}{Dt} >_i = -\frac{p_i}{\rho_i} < \nabla \cdot v >_i \qquad (5.41)$$

It is seen that approximation of the change rate of momentum and the energy can be obtained from the approximation of the pressure gradient and the velocity divergence. The approximation of the pressure gradient and the velocity divergence can be obtained simply substituting the pressure and velocity into equation (5.12) in the CSPM and (5.32) in the DSPH, respectively. The approximations of the momentum and energy in the traditional SPH method are the numerators of the corresponding CSPH approximation equations.

In the simulation, the time step is 0.005 s and the simulation is carried out for 40 steps. In resolving the shock, the Monaghan type artificial viscosity (see, Section 4.4) is used, which also serves to prevent unphysical penetration. Figure 5.7, Figure 5.8, Figure 5.9 and Figure 5.10 show, respectively, the density, pressure, velocity and internal energy profiles.

It can be seen that the obtained results from the DSPH agree well with the exact solution in the region $[-0.4, 0.4]$. The shock is observed at around $x = 0.3$; and is resolved within several smoothing lengths. The rarefaction wave is located between $x = -0.3$ and $x = 0$. The contact discontinuity is between $x = 0.1$ and $x = 0.2$. Just as reported by Hernquist and Katz (1989) and Monaghan (1983), the traditional SPH with artificial viscosity yields comparable results in capturing shock physics. The DSPH gives slightly better results than the traditional SPH method though only 40 steps passed. If more dynamic steps are involved, the better performance of the DSPH is more evident (Liu et al., 2003e).

The CSPM failed to captures the major shock physics. In this example, the CSPM performs poorer even than the traditional SPH method. The poor performance of the CSPM in simulating shock waves should arise from the denominator that in fact acts as a normalization factor. Observing equation (5.37), it is found that the summation of the smoothing function around the discontinuity region is far from the unity. When it is used to normalize the summation of the density, the local features of shock wave can be smoothed out, and hence the shock physics can be concealed. In contrast, the traditional SPH does not suffer from this smoothing effect, and the DSPH benefits from the addition part in resolving the shock physics. Their performance, therefore, should be better than that of the CSPM. It is clear that the difference between the results from the CSPM and the DSPH is originated from the additional corrective part that approximates the discontinuity.

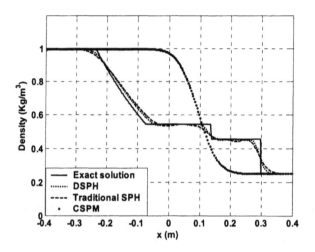

Figure 5.7 Density profiles for the shock tube problem obtained using different versions of SPH formulation. The DSPH gives the best result. The CSPM failed to capture the major shock physics. ($t = 0.2$ s)

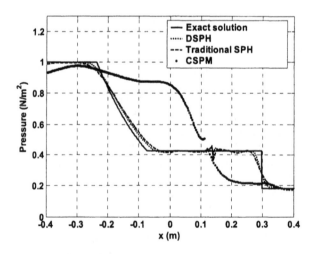

Figure 5.8 Pressure profiles for the shock tube problem obtained using different versions of SPH formulation. The DSPH gives the best result. The CSPM failed to capture the major shock physics. ($t = 0.2$ s)

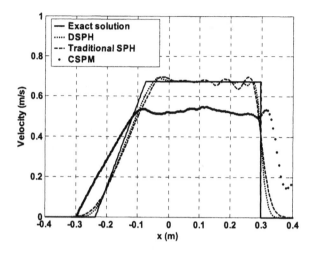

Figure 5.9 Velocity profiles for the shock tube problem obtained using different versions of SPH formulation. The DSPH gives the best result. The CSPM failed to capture the major shock physics. ($t = 0.2$ s)

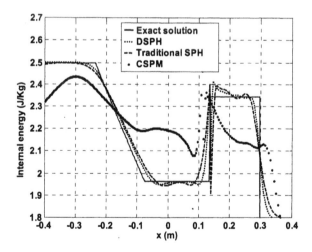

Figure 5.10 Energy profiles for the shock tube problem obtained using different versions of SPH formulation. The DSPH gives the best result. The CSPM failed to capture the major shock physics. ($t = 0.2$ s)

5.6   Concluding remarks

The traditional smoothed particle hydrodynamics (SPH) suffers from the boundary deficiency problem that leads to inaccurate results. Based on the Taylor series expansion, the corrective smoothed particle method (CSPM) provides a simple way of normalization to the traditional SPH approximation, and thus improves the boundary deficiency.

In this chapter, a DSPH formulation has been developed to simulate physical phenomena with discontinuities. The formulation employs Taylor series expansion separately in both sides of the discontinuities, rather than in the entire support domain. The resultant kernel and particle approximations for either the function or its derivatives consist of two parts. One is the same as that of the CSPM; another additional part is derived from the discontinuity treatment. The presence of the additional part ensures the consistency of the kernel approximations for field functions with discontinuity.

The numerical studies show that the DSPH not only remedies the boundary deficiency problem, but also well simulates the discontinuity. The DSPH is also applied to simulate the shock tube problem, and gives very good results. These numerical examples suggest that the DSPH, if accompanied with an effective discontinuity detecting algorithm, is attractive in simulating problems with discontinuities such as shock physics. Extension to multi-dimensions is possible if an effective algorithm for detecting discontinuities in multi-dimensional space can be developed. Detecting discontinuity lines or surfaces for two- or three-dimensional space is a rather challenging task, and is still under development.

Chapter 6

SPH for Simulating Explosions

In the previous chapters, the SPH method is introduced for solving CFD problems including problems with shock waves. This chapter extends and applies the SPH method to simulate the complicated explosion process of high explosives (HE). The combination of adaptive, meshfree and Lagrangian nature of the SPH method plus an explicit algorithm make the SPH method very attractive in treating highly dynamic phenomena with large deformations and large inhomogeneities that occur in the extremely transient HE explosion process.

A variable smoothing length model is employed in this chapter to ensure a minimal and sufficient number of neighboring particles to contribute to the discrete particle approximations at the current time step, so as to ensure a proper adaptability of the SPH to the drastic movement of particles.

The SPH formulation is performed based on the Euler equations, as the explosion is an extremely fast phenomenon. The JWL equation of state for high explosives is incorporated into the SPH equations.

Two numerical examples, a one-dimensional TNT slab detonation and a two-dimensional TNT explosive gas expansion, are investigated with comparisons to the results from other sources. The simulation results show that the SPH method can give a very good prediction for both the magnitude and the form of the detonation waves as well as the pressure distributions in the explosion process. The major physics of the HE explosion can be well captured in the simulation.

The SPH method is also applied to investigate the shaped charge explosion. The effects of different detonation ignition models, cavity shape, and charge length are investigated in detail. The key features in the shaped charge explosions are also captured by the SPH simulation. Discussions on the numerical results are presented.

6.1   Introduction

The explosion of a high explosive (HE) charge can rapidly convert the original explosive charge into gaseous products with extremely high pressure through a chemical reaction process. The high pressure can lead to damages to nearby personnel and structures. A typical HE explosion consists of the detonation process in the HE and the later expansion process of the gaseous products towards the surrounding medium. During the detonation process, the detonation wave advances outwards through the HE at a constant velocity. In the expansion process, rarefaction wave propagates inwards in the gas behind the detonation wave if the HE ideally explodes in a vacuum or in a medium such as air, water, soil, etc.

Theoretical solutions to the explosion of high explosive are only limited to some simple cases. Experimental studies need to resort to dangerous and expensive firing trials, and sometimes certain physical phenomena related to the explosions cannot be scaled in a practical experimental setup. Recently, more and more analyses of high explosive explosions are based on numerical simulations with the advancement of the computer hardware and computational techniques (Fickett and Wood, 1966; Zhang, 1976; Mader, 1979; 1998; Lam and Liu et al. 1996; Liu and Lam et al, 1998). However, numerical simulations of the HE explosions are generally very difficult for the conventional grid-based numerical methods. First, during the detonation process in the explosion, a very thin reaction zone divides the domain into two inhomogeneous parts and produces large deformations. Second, in the expansion process, there are free surfaces and moving interfaces involved. Traditional Lagrangian techniques such as the finite element methods are capable of capturing the history of the detonation events associated with each material. It is, however, difficult to apply practically, since the severely distorted mesh may result in very inefficient small time step, and may even lead to the breakdown of the computation. Traditional Eulerian techniques, such as the finite difference methods or finite volume methods, can well resolve the problem due to the large deformations in the global motions, but it's very difficult to analyze the details of the flow because of the lack of history and the smearing of information as the mass moves through the fixed-in-space Eulerian mesh (see, Chapter 1).

In this chapter, the novel application of the SPH method to the simulation of HE explosion is presented. As demonstrated in the previous chapters, smoothed particle hydrodynamics (SPH), as a meshfree, Lagrangian and particle method, has distinct advantages over the conventional grid-based numerical methods for CFD problems. In the SPH method, particles are used to represent materials at discrete locations. The history of the fluid particles can be naturally obtained, and thus it is easy to trace material interfaces, free surfaces and moving boundaries. The meshless feature of the SPH method overcomes the difficulties related to large deformations, and makes it fairly attractive in simulating the high explosive explosion.

This chapter is outlined as follows.

- In Section 2, the HE explosion physics and the governing equations are described.
- In Section 3, the basic SPH equations for simulating HE explosions are discussed.
- In Section 4, an adaptive smoothing length evolution algorithm suitable for the HE explosion simulation is detailed.
- In Section 5, two numerical examples are presented with comparisons to results from other sources.
- In Section 6, the SPH method is applied to investigate complicated cases of shaped charge explosion.
- In Section 7, some remarks and conclusions are given.

6.2   HE explosions and governing equations

6.2.1   Explosion process

The HE explosion process involves a violent chemical reaction which converts the original explosive charge of high energy density into gaseous products with very high temperature and pressure. The HE explosion occurs with extreme rapidity, releasing a great deal of heat. A typical HE explosion consists of the detonation process through the HE at a constant detonation velocity and the later expansion process of the gaseous products to the surrounding medium (Figure 6.1). The detonation process is accompanied by the propagation of the reactive wave that advances through the explosive with a constant velocity related to the particular type of explosive concerned. In a steady state detonation process, the reaction rate is essentially infinite and the chemical equilibrium is attained. After the completion of the detonation, the detonation-produced explosive gas expands outwards. This gas expansion generally involves moving material interfaces if the explosive is surrounded by outside medium or free surfaces if the explosion occurs in the vacuum.

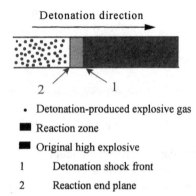

Detonation direction

- Detonation-produced explosive gas
- Reaction zone
- Original high explosive

1   Detonation shock front

2   Reaction end plane

Figure 6.1 Detonation in a 1D high explosive. The reaction end plane is an interface of the pressurized high explosive charge and the explosive gas produced in the detonation process.

6.2.2   HE steady state detonation

If the reaction rate of the explosive is essentially infinite and if a chemical equilibrium is attained, the detonation is in a steady state, and the propagation of the detonation front is governed solely by the thermodynamics and the hydrodynamics of the HE. In the steady state, the speed of the detonation wave is constant. Environmental factors such as the temperature and pressure in the surroundings of the HE do not have much influence on the propagation of the detonation shocks once the detonation is ignited and attains the steady state.

As shown in Figure 6.1, advancing through the high explosive with a detonation velocity D is a shock front that compresses the HE from an original state point (p_0, ρ_0) to another specific state point with a raised pressure p_1, as shown in Figure 6.2 along the Hugoniot curve for the HE. Completion of the reaction leaves the pressure and volume at point (p, ρ) on the Hugoniot for the reaction produced gas. The gaseous products then expand (Taylor wave) along the isentrope through (p, ρ).

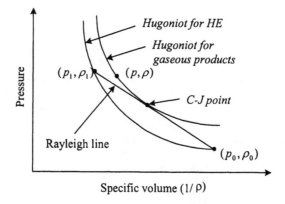

Figure 6.2 The Hugoniots and Raleigh line for explosives. (From Liu et al., Computers & Fluids, 32(3): 305-322, 2003. With permission.)

The parameters before and after the detonation shock can be related by applying the laws of conservation of mass, momentum and energy across the detonation shock propagation (Zhang, 1976).

$$D - U_0 = V_0 \sqrt{\frac{p - p_0}{V_0 - V}} \tag{6.1}$$

$$U - U_0 = (V_0 - V) \sqrt{\frac{p - p_0}{V_0 - V}} \tag{6.2}$$

$$e - e_0 = \frac{1}{2}(p + p_o)(V_0 - V) \tag{6.3}$$

where D is the detonation (propagation) velocity. p, V, U, and e are, respectively, the pressure, specific volume, velocity and specific internal energy of the reaction produced explosive gas right after the detonation shock. The parameters with subscript "0" are the corresponding variables for the original high explosive ahead of the shock.

Equations (6.1) and (6.2) are derived from the conservation of mass and momentum. Equation (6.1) describes the relationship of the detonation velocity with the pressure and specific volume behind the shock, and is called Rayleigh equation. It determines the Rayleigh line in Figure 6.2. Equation (6.3) is derived from the conservation of energy, and is called Rankine-Hugoniot equation. It determines the Rankine-Hugoniot (or simplified as Hugoniot) curve in Figure 6.2. Calculating the Hugoniot curve from equation (6.3) requires the knowledge of the equation of state of the explosive gas (commonly, but somewhat misleadingly called an equation of state for the high explosive), which can be expressed in terms of the pressure and the specific volume of the explosive gas. Chapman and Jouguet's (C-J) hypothesis states that for a plane detonation wave to be propagated steadily, the Rayleigh line, must be tangent to the Hugoniot curve of the gaseous detonation products at the C-J point (Mader, 1979; 1998). Compared to the pressure at the C-J point (C-J pressure), the pressure in the original HE, p_0, is very small and can be neglected. Detailed derivation on equations (6.1)~ (6.3) and more descriptions on HE detonation can be found in (Zhang, 1976; Mader, 1998).

6.2.3   Governing equations

Since the detonation and expansion speed are extremely high, the gaseous products can be assumed to be inviscid and the explosion process is adiabatic. Therefore the Euler equation can be used to model the explosion process together with a suitable equation of state.

$$\begin{cases} \dfrac{D\rho}{Dt} = -\rho\nabla \cdot v \\[2mm] \dfrac{Dv}{Dt} = -\dfrac{1}{\rho}\nabla p \\[2mm] \dfrac{De}{Dt} = -\dfrac{p}{\rho}\nabla \cdot v \\[2mm] p = p(\rho,e) \end{cases} \qquad (6.4)$$

where, ρ, e, p, v and t are density, internal energy, pressure, velocity vector and time, respectively. The first three equations in equation (6.4) state the conservation of mass, momentum, and energy while the fourth equation is the equation of state (EOS). In this work, TNT is used in the simulation as an example of HEs. The detonation velocity of TNT is known from experiments as 6930 m/s.

For the explosive gas, the standard Jones-Wilkins-Lee (JWL) (Dobratz, 1981) equation of state can be employed. The pressure of the explosive gas is given by

$$p = A(1-\frac{\omega\eta}{R_1})e^{-\frac{R_1}{\eta}} + B(1-\frac{\omega\eta}{R_2})e^{-\frac{R_2}{\eta}} + \omega\eta\rho_0 e \qquad (6.5)$$

where η is the ratio of the density of the explosive gas to the initial density of the original explosive. e is the internal energy of the high explosive per unit mass. A, B, R_1, R_2 and ω are coefficients obtained by fitting the experimental data. The values of the corresponding coefficients (Shin et al., 1998) are listed in Table 6.1.

Table 6.1 Material parameters and coefficients in the equation of state for TNT obtained via experiments

Symbol	Meaning	Value
ρ_0	Initial density	1630 Kg/m^3
D	Detonation velocity	6930 m/s
P_{CJ}	CJ pressure	2.1×10^{10} N/m^2
A	Fitting coefficient	3.712×10^{11} N/m^2
B	Fitting coefficient	3.21×10^9 N/m^2
R_1	Fitting coefficient	4.15
R_2	Fitting coefficient	0.95
ω	Fitting coefficient	0.30
E_0	Detonation energy per unit mass	4.29×10^6 J/Kg

Another possible choice of the equation of state for the explosive gas is to simply use that for the real gas that takes the form of

$$p = (\gamma - 1)\rho e \qquad (6.6)$$

where γ is generally taken as 3 for most of the high explosives, and in some circumstances taken as 1.4 (Zhang, 1976). Equation (6.6) is also known as the gamma law for ideal gas. It is also employed in the later numerical examples when the artificial model of explosion is used.

6.3   SPH formulations

Since the explosive gas is assumed to be inviscid, and the explosion process is
regarded as adiabatic, the Euler equation governs the HE explosion. According
to the SPH kernel and particle approximations, some possible forms of the SPH
equations are given in Table 4.2. In this chapter, the following set of discretized
form of the Euler equation with artificial viscosity is applied to simulate the
explosion process numerically.

$$
\begin{cases}
\dfrac{D\rho_i}{Dt} = \sum_{j=1}^{N} m_j (v_i - v_j) \cdot \nabla_i W_{ij} \\[2ex]
\dfrac{Dv_i}{Dt} = -\sum_{j=1}^{N} m_j \left(\dfrac{p_i}{\rho_i^2} + \dfrac{p_j}{\rho_j^2} + \Pi_{ij} \right) \nabla_i W_{ij} \\[2ex]
\dfrac{De_i}{Dt} = \dfrac{1}{2} \sum_{j=1}^{N} m_j \left(\dfrac{p_i}{\rho_i^2} + \dfrac{p_j}{\rho_j^2} + \Pi_{ij} \right) \left(v_i - v_j \right) \cdot \nabla_i W_{ij} \\[2ex]
\dfrac{Dx_i}{Dt} = v_i
\end{cases}
\tag{6.7}
$$

The above SPH equations are the equations of motion in the form of a set of
simultaneous ordinary differential equations with respect to time, and can be
integrated numerically.   Using one of the standard techniques such as the
leapfrog (LF), predictor-corrector and Runge-Kutta (RK) schemes, the numerical
integration can be carried out easily for the field variables at every particle.

In the practice, the leapfrog method is very popular for its low requirement
on memory storage and its computational efficiency, and hence it is used here.
In the leapfrog method, the particle velocities and positions are offset by a half
time step when integrating the equations of motion. At the end of the first time
step (t_0), the change in density, energy and velocity are used to advance the
density, energy and velocity at half a time step, while the particle positions are
advanced in a full time step.

$$\left\{ \begin{array}{l} t = t_0 + \Delta t \\[2mm] \rho_i(t_0 + {}^{\Delta t}\!/_2) = \rho_i(t_0) + \dfrac{\Delta t}{2} D\rho_i(t_0) \\[3mm] e_i(t_0 + {}^{\Delta t}\!/_2) = e_i(t_0) + \dfrac{\Delta t}{2} De_i(t_0) \\[3mm] v_i(t_0 + {}^{\Delta t}\!/_2) = v_i(t_0) + \dfrac{\Delta t}{2} Dv_i(t_0) \\[3mm] x_i(t_0 + \Delta t) = x_i(t_0) + \Delta t \cdot v_i(t_0 + {}^{\Delta t}\!/_2) \end{array} \right. \tag{6.8}$$

In order to keep the calculations consistent at each subsequent time step, at the start of each subsequent time step, the density, energy and velocity of each particle need to be predicted at half a time step to coincide the position.

$$\left\{ \begin{array}{l} \rho_i(t) = \rho_i(t - {}^{\Delta t}\!/_2) + \dfrac{\Delta t}{2} D\rho_i(t - \Delta t) \\[3mm] e_i(t) = e_i(t - {}^{\Delta t}\!/_2) + \dfrac{\Delta t}{2} De_i(t - \Delta t) \\[3mm] v_i(t) = v_i(t - {}^{\Delta t}\!/_2) + \dfrac{\Delta t}{2} Dv_i(t - \Delta t) \end{array} \right. \tag{6.9}$$

At the end of the subsequent time step, the particle density, internal energy, velocity and position are advanced in the standard leapfrog scheme

$$\left\{ \begin{array}{l} t = t + \Delta t \\[2mm] \rho_i(t + {}^{\Delta t}\!/_2) = \rho_i(t - {}^{\Delta t}\!/_2) + \Delta t \cdot D\rho_i(t) \\[3mm] e_i(t + {}^{\Delta t}\!/_2) = e_i(t - {}^{\Delta t}\!/_2) + \Delta t \cdot De_i(t) \\[3mm] v_i(t + {}^{\Delta t}\!/_2) = v_i(t - {}^{\Delta t}\!/_2) + \Delta t \cdot Dv_i(t) \\[3mm] x_i(t + \Delta t) = x_i(t) + \Delta t \cdot v_i(t + {}^{\Delta t}\!/_2) \end{array} \right. \tag{6.10}$$

Note that the LF scheme is conditionally stable. The stability condition is the so-called CFL (Courant-Friedrichs-Levy) condition, which typically results in a time step proportional to the smoothing lengths. In this work, the time step is taken as (Hernquist and Katz, 1989).

$$\Delta t = \min \left(\frac{\xi h_i}{h_i \nabla \cdot v_i + c_i + 1.2(\alpha_\Pi c_i + \beta_\Pi h_i |\nabla \cdot v_i|)} \right) \qquad (6.11)$$

where ξ is the Courant number, taken around 0.3, α_Π and β_Π are the two coefficients used in the Monaghan type artificial viscosity (see, Section 4.4.1). Similar to equation (4.91), this time step also considers the sound speed of the media and the viscous dissipation. In our experience, the time step given by equation (6.11) seems more stable for explosion simulation.

The standard Monaghan type artificial viscosity Π_{ij} is used in the simulation to capture the shock wave, to stabilize the numerical scheme, and to prevent particle penetration. It should be noted that the parameters α_Π and β_Π used in the standard Monaghan type artificial viscosity are usually taken as $\alpha_\Pi = 1$ and $\beta_\Pi = 1$ or 2. In most circumstances, these values are effective to prevent unphysical penetration and can produce reasonably good results. However, in the numerical simulation of high explosive explosion, since the pressure is extremely high and the detonation velocity are extremely large, these values are not sufficient to prevent particles from unphysical penetration. In our experience, a significantly large β_Π, say 10, is found to be effective to prevent unphysical penetration in simulating HE explosions.

In the later numerical examples to be presented, the cubic spline smoothing function (see, Section 3.1) is used. In treating the boundary, the virtual particles are used (see, Section 4.4.8). As discussed in the previous chapters that the reflection virtual particles (type II) improve the boundary deficiency and contribute to the summation together with interior the real particles. If a solid boundary is considered, besides the type II virtual particles, the type I virtual particles are also used on the solid boundary to impose repulsive force to the interior real particles approaching the boundary and to prevent unphysical penetration.

6.4   Smoothing length

The smoothing length h is very important in the SPH method to ensure an efficient, robust and stable adaptability in the computational process. It has direct influence on both the efficiency of the computation and accuracy of the solution. If h is too small, there may be not enough particles in the support domain of dimension κh to exert forces on the particle, which results in low accuracy. If the smoothing length is too large, all details of the particle or local properties may be smoothed out, and the accuracy will also be affected. There are diversified ways of determining the smoothing length (Nelson and Papaloizou, 1994; Hernquist and Katz, 1989). Section 4.4.4 provides a discussion on this issue. Nearly all these forms of smoothing length are problem-dependent, which is limited for general use.

In the HE explosion simulations, there are large density inhomogeneities in the problem domain, and the particles can move drastically in the explosion process. In order to make the adaptability of the SPH method more robust, accurate and more applicable in modeling problems with large density inhomogeneities, an effective, adaptive and robust smoothing length model is necessary. The smoothing length h should vary both in space and time (see, Section 4.4.4).

Liu et al. (2002a) has developed such an adaptive model to evolve the smoothing length both in space and time. In their model, the smoothing length is adapted in three sequential steps:

1. initialization of the smoothing length for each particle;
2. calculating the time rate of change of the smoothing length of each particle;
3. smoothing length optimization and relaxation.

When the smoothing length is calculated in the above combined procedure, only a minimal and necessary number of neighboring particles will contribute to the discrete summations.

6.4.1 Initial distribution of particles

Suppose h_i^0 is the initial smoothing length of particle i, while N_i^0 is the number of the initial neighboring particles within the support domain of κh_i^0 for particle i. For example, in one-, two- or three-dimensional space, N_i^0 may be taken as 5, 21 and 57, respectively. In the three-dimensional space, N_i^0 corresponds to the number of neighboring particles on a cubic lattice with a smoothing length slightly bigger than the particle spacing, e.g., 1.2 times the particle spacing, and the cubic spline-smoothing kernel that extends to $2h$. The particle has a mass of m_i and remains unchanged during the computation. The initial smoothing length of h_i^0 can be calculated based upon initial density of the particle. If the initial density of particle i is ρ_i^0, then for the three-dimensional case (assume that the support domains of all the particles are of spherical shape), we have

$$\sum_{j=1}^{N_i^0} m_j = \frac{4}{3}\pi\left(2h_i^0\right)^3 \rho_i^0 \qquad (6.12)$$

Hence h_i^0 can be obtained for three-dimensional cases as

$$h_i^0 = c_3 \times \left(\frac{\sum_{j=1}^{N_i^0} m_j}{\rho_i^0}\right)^{1/3} , c_3 = \left(\frac{3}{32\pi}\right)^{1/3} \qquad (6.13)$$

Similar results can also be derived for two-dimensional cases

$$h_i^0 = c_2 \times \left(\frac{\sum_{j=1}^{N_i^0} m_j}{\rho_i^0}\right)^{1/2} , c_2 = \left(\frac{1}{4\pi}\right)^{1/2} \qquad (6.14)$$

and for one-dimensional cases

$$h_i^0 = c_1 \times \left(\frac{\sum_{j=1}^{N_i^0} m_j}{\rho_i^0}\right) , c_1 = \frac{1}{4} \qquad (6.15)$$

In a concise form, h_i^0 can be written as

$$h_i^0 = c_d \times \left(\frac{\sum\limits_{j=1}^{N_i^0} m_j}{\rho_i^0} \right)^{1/d} \tag{6.16}$$

where d is the number of dimension. Equation (6.16) was derived by Liu et al. (2002a) when investigating the water mitigation problems.

6.4.2   Updating of smoothing length

The smoothing length of each particle is updated using the following most commonly used expression by Benz (1989).

$$\frac{Dh_i^n}{Dt} = -\frac{h_i^n}{\rho_i^n d} \frac{D\rho_i^n}{Dt} \tag{6.17}$$

where h_i^n, ρ_i^n, v_i^n and $\nabla_i^n W_{ij}$ are the smoothing length, density, velocity, and kernel gradient of particle i at time step n.

Referring to equation (6.7) and using the concepts of kernel and particle approximations of SPH, the above equation can be written in the following form of particle approximation.

$$\frac{Dh_i^n}{Dt} = -\frac{h_i^n}{\rho_i^n d} \sum_{j=1}^{N} m_j (v_i^n - v_j^n) \cdot \nabla_i^n W_{ij} \tag{6.18}$$

The smoothing length at the next step becomes

$$h_i^{n+1} = h_i^n + \frac{Dh_i^n}{Dt} \Delta t \tag{6.19}$$

Equations (6.18) and (6.19) are commonly used to update the smoothing length. It works well for problems of homogenous, slowly expanding or contracting fluids. However, for problems with extremely large deformation, or problems with significant density inhomogeneities, it may not always be very accurate and stable. Hence a procedure of optimization and relaxation is required to improve the updating of the smoothing length for explosion simulations.

6.4.3   Optimization and relaxation procedure

In order to model the problems with large density inhomogeneities, Liu et al. (2002a) proposed the following procedure for the smoothing length optimization and relaxation to update h_i^{n+1}. The procedure aims to have each particle interact with a roughly constant number of nearby neighboring particles.

This procedure consists of two steps: the prediction step and the correction step.

Prediction step: update h_i^{n+1} with the following expression

$$h_i^{n+1} = h_i^n + \theta\left(\frac{Dh_i^n}{Dt}\Delta t\right) \tag{6.20}$$

where θ is a relaxation factor that is taken as 1.0 initially and then adjusted slightly around 1.0 in the later optimization and relaxation steps.

Correction step: once h_i^{n+1} is obtained, the current number of the neighboring particles N_i^{n+1} can be determined. If N_i^{n+1} is found to be roughly the same as N_i^0, it is regarded desirable. If it's different from N_i^0 by a prescribed tolerance of a few percent, the relaxation factor θ is then adjusted around 1.0 to get a new h_i^{n+1} that leads to a new N_i^{n+1}. This process is repeated until N_i^{n+1} is roughly the same as N_i^0.

6.5   Numerical examples

A three-dimensional SPH code for simulating HE explosion is developed based on the SPH code listed in Chapter 10. A series of numerical tests have been carried out using this code. Two numerical examples are presented here. The first case is a one-dimensional TNT slab detonation. The second case is a two-dimensional TNT explosive gas expansion. The simulation results of the one-dimensional case show that the SPH can give good predictions for both the magnitude and the form of the detonation wave. For the two-dimensional case, the numerical results from SPH show good agreement with the results obtained using the software MSC/Dytran (MSC, 1997). It suggests that the SPH method and the code can also successfully simulate the transient explosive gas expansion process.

Example 6.1 One-dimensional TNT slab detonation

The simulation of one-dimensional TNT slab detonation is useful since early analyses based on the assumption of spherical charge detonating from the charge center can also be simplified as a one-dimensional problem. The one-dimensional detonation process is, therefore, often regarded as a good benchmark problem for testing numerical methods in simulating explosions. Due to its particle nature, the SPH method can be easily extended to other detonation problems and can simulate various detonation problems, e.g. arbitrary charge shape, different detonation orientation, multiple charges and so on.

In this simulation, a 0.1 m long TNT slab is detonated at one end. Shin et al. (1997) have simulated the same case using a coupled Lagrangian-Eulerian analysis with the commercial software package MSC/Dytran. The same dimensions and parameters as those used by Shin et al. (1997) are used in this simulation for the sake of comparison.

In Shin's simulation, the wall boundary conditions were used to forbid material transport from outside. In this simulation, the symmetric condition is used. Therefore the detonation of the 0.1 m long slab from one end to the other is equivalent to the detonation of a 0.2 m long slab that is ignited at the middle point and advances to both ends. Before the detonation, particles are evenly distributed along the slab. The initial smoothing length is one and a half times the particle spacing. After the detonation, a plane detonation wave is produced. According to the detonation velocity, it takes around 14.4 µs to complete the detonation throughout the TNT slab. The equation of state given by equation (6.5) is used.

In order to investigate the effects of different particle resolutions, analyses are carried out using 250, 500, 1000, 2000 and 4000 particles along the slab. Figure 6.3, Figure 6.4, Figure 6.5 and Figure 6.6 show, respectively, the pressure, density, velocity and the internal energy profiles along the slab at 1 µs interval from 1 to 14 µs obtained using 4000 particles. Figure 6.7 shows the peak pressures at 1 µs interval with the complete pressure profiles at 7 and 14 µs for different particle resolutions. The dashed lines in Figure 6.3 and Figure 6.7 represent the experimentally determined C-J detonation pressure, which is, according to the Chapman and Jouguet's hypothesis, the pressure at the tangential point of the Hugoniot curve and the Rayleigh line (see, Figure 6.2), and represents the pressure at the equilibrium plane at the trailing edge of the very thin chemical reaction zone (Mader, 1979; 1998). For this one-dimensional TNT slab detonation problem, the experimental C-J pressure is 2.1×10^{10} N/m^2 (Shin et al., 1997; Zhang, 1976). It can be seen from Figure 6.3 and Figure 6.7, that with the process of the detonation, the detonation pressure converges to the C-J pressure. It is shown on Figure 6.7 that more particles along the slab result in sharper pressure profiles with bigger peak pressures. The results are quite accurate and are comparable to the results obtained by Shin. The detonation shock is resolved within several smoothing lengths. The number of particles is more than the number of elements (2000) that Shin used, and the resulted detonation shock fronts are sharper.

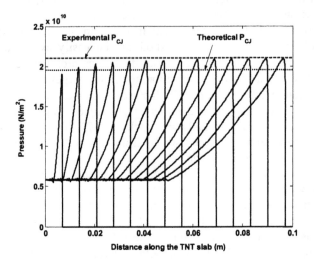

Figure 6.3 Pressure profiles along the one-dimensional TNT slab during the detonation process. The pressure profiles are plotted at 1 μs interval from 1 μs to 14 μs. (From Liu et al., Computers & Fluids, 32(3): 305-322, 2003. With permission.)

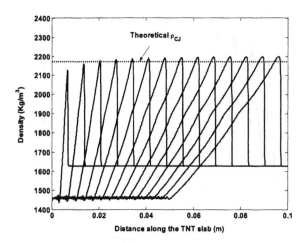

Figure 6.4 Density profiles along the one-dimensional TNT slab during the detonation process. The density profiles are plotted at 1 μs interval from 1 μs to 14 μs. (From Liu et al., Computers & Fluids, 32(3): 305-322, 2003. With permission.)

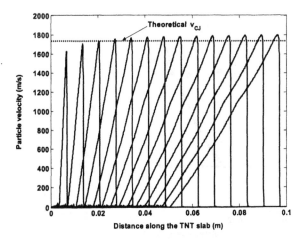

Figure 6.5 Velocity profiles along the one-dimensional TNT slab during the detonation process. The velocity profiles are plotted at 1 μs interval from 1 μs to 14 μs. (From Liu et al., Computers & Fluids, 32(3): 305-322, 2003. With permission.)

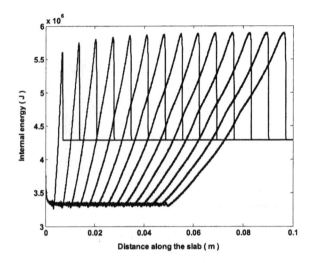

Figure 6.6 Internal energy profiles along the one-dimensional TNT slab during the detonation process. The profiles are plotted at 1 μs interval from 1 μs to 14 μs.

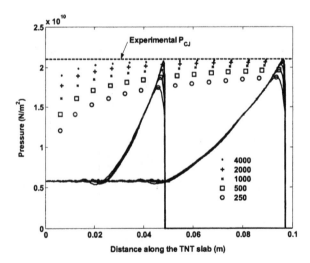

Figure 6.7 Peak pressures in the one-dimensional TNT slab at 1 μs interval with the complete pressure profiles at 7 and 14 μs for different particle resolutions. (From Liu et al., Computers & Fluids, 32(3): 305-322, 2003. With permission.)

If the detonation-produced high explosive gas is assumed to satisfy the gamma law equation of state, approximate theoretical values for pressure, density and particle velocity at the C-J point can be calculated according to the following expressions with the thermal expansion coefficient, $\gamma = 3$ (Mader, 1979; 1998; Zhang, 1976),

$$P_{CJ} = \frac{1}{\gamma+1} \rho_o D^2 \tag{6.21}$$

$$\rho_{CJ} = \frac{\gamma+1}{\gamma} \rho_o \tag{6.22}$$

$$V_{CJ} = \frac{1}{\gamma+1} D \tag{6.23}$$

The approximate theoretical values for pressure, density and particle velocity at the C-J point are found to be 1.957×10^{10} N/m^2, 2173 Kg/m^3 and 1733 m/s, respectively. These theoretical values are shown in Figure 6.3, Figure 6.4 and Figure 6.5 by the dotted lines. The theoretical results are obtained based on idealized assumptions, and are constant. These theoretical results are larger than the SPH solution at earlier time and smaller than the SPH solution at later time, as shown in Figure 6.3, Figure 6.4 and Figure 6.5. Since the C-J pressure calculated using the SPH method is much closer to the experimental value rather than the approximate theoretical value, it suggests that the numerical results are more reasonable. For TNT, the thermal expansion coefficient, γ, should be slightly smaller than 3, which is generally accepted for most high explosives (Zhang, 1976). More suitable values of γ for TNT can result in closer agreement between the theoretical values and the presented numerical results.

Table 6.2 shows the time history of the internal, kinetic and total energy produced by the SPH code. With the advancement of the detonation process, the total kinetic energy gradually increases, while the total internal energy reduces. The total energy remains approximately constant, and conserves the original total energy of 6.9927×10^8 J very well. This suggests that the SPH method is numerically stable.

For this one-dimensional TNT detonation case, if the free boundary applies at the left end (ignition point), after the ignition, the produced gas will expand leftwards with a rarefaction wave propagating rightwards through the gas behind the detonation wave. Analytical solution exists for this problem (Zhang, 1976).

The analytical formulae for pressure, density, and velocity transients in the explosive gas behind the detonation wave are

Table 6.2 Time history of energy balance

Time	Internal energy	Kinetic energy	Total energy	Error
(μs)	(J)	(J)	(J)	(%)
1	6.9603e+008	3.2418e+006	6.9927e+008	0.
2	6.9255e+008	6.7183e+006	6.9927e+008	0.
3	6.8907e+008	1.0206e+007	6.9927e+008	0.
4	6.8558e+008	1.3696e+007	6.9927e+008	0.
5	6.8209e+008	1.7184e+007	6.9928e+008	0.0014
6	6.7860e+008	2.0676e+007	6.9928e+008	0.0014
7	6.7511e+008	2.4163e+007	6.9928e+008	0.0014
8	6.7163e+008	2.7653e+007	6.9928e+008	0.0014
9	6.6813e+008	3.1146e+007	6.9928e+008	0.0014
10	6.6465e+008	3.4635e+007	6.9928e+008	0.0014
11	6.6115e+008	3.8127e+007	6.9928e+008	0.0014
12	6.5767e+008	4.1614e+007	6.9928e+008	0.0014
13	6.5418e+008	4.5105e+007	6.9928e+008	0.0014
14	6.5069e+008	4.8597e+007	6.9928e+008	0.0014

$$P = \frac{16}{27}\frac{\rho_0}{D}(\frac{x}{2t}+\frac{D}{4})^3 \tag{6.24}$$

$$\rho = \frac{16}{9}\frac{\rho_0}{D}(\frac{x}{2t}+\frac{D}{4}) \tag{6.25}$$

$$v = \frac{x}{2t}-\frac{D}{4} \tag{6.26}$$

Figure 6.8, Figure 6.9 and Figure 6.10 show the distribution of the pressure, density and velocity at 1 and 2 μs after ignition. The results obtained using the SPH method and the analytical solutions are plotted together for comparison. It can be seen that the pressure transients obtained by the SPH method nearly coincide with those from the analytical solution. The calculated C-J pressure using the SPH method is very close to the analytical solution of 1.957×10^{10} N/m^2.

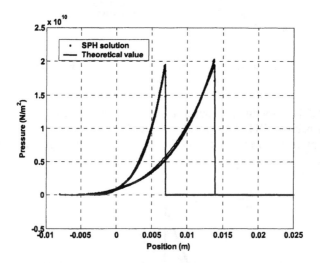

Figure 6.8 Pressure distributions at 1 and 2 μs after the detonation of a one-dimensional TNT slab ignited at the left free end. (From Liu et al., Computational Mechanics, 30(2):106-118, 2003. With permission.)

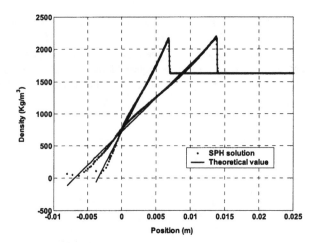

Figure 6.9 Density distributions at 1 and 2 μs after the detonation of a one-dimensional TNT slab ignited at the left free end. (From Liu et al., Computational Mechanics, 30(2):106-118, 2003. With permission.)

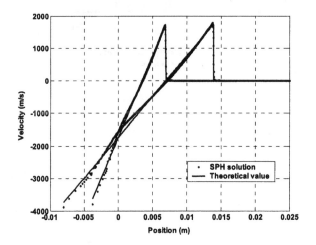

Figure 6.10 Velocity distributions at 1 and 2 μs after the detonation of a one-dimensional TNT slab ignited at the left free end. (From Liu et al., Computational Mechanics, 30(2):106-118, 2003. With permission.)

Example 6.2   Two-dimensional explosive gas expansion

The previous numerical test shows the ability of the SPH method in simulating detonation shocks. In this 2D test, a cross-section of a cylindrical TNT charge is considered. It is assumed that the ignition occurs on the axial-symmetrical axes at the same time. Therefore, the field variables are independent of the axial direction, and the problem can be treated as a two-dimensional case in the x-y plane. The expansion process of this two-dimensional explosive gas is simulated to model the expansion of the explosive gas behind the detonation shock. The initial radius of the TNT explosive cylinder is 0.1 m. In the simulation, 20 particles are placed along the radial direction, while 60 particles are placed along the tangential direction (Figure 6.11). This numerical is also simulated using the commercial software package MSC/Dytran for comparison, with 400 volume cells along the radial direction.

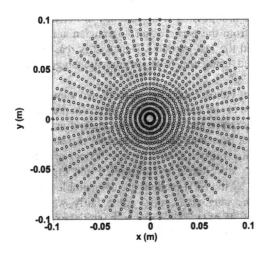

Figure 6.11 Initial geometry and particle distributions for the 2D explosive (TNT) gas expansion. 20 particles are placed along the radial direction, while 60 particles are placed along the tangential direction.

For this example, only the expansion process is simulated with the assumption that the detonation speed is infinite, the HE converts into gas in an instant, and the gas occupies the original space of the HE without energy loss. This artificial model is idealized to allow the simulation emphasize only on the expansion process of the already produced explosive gas with extremely high pressure. It is very commonly used in solving some practical HE explosion problems when the major concern is the later effects of the explosion after the detonation of the explosive charge. The equation of state takes the form given in equation (6.6) with $\gamma = 1.4$. The surrounding outside space of the explosive is assumed to be a vacuum.

Figure 6.12 shows a good agreement between the distribution of density and pressure transients along the radial direction obtained using the SPH method and those using MSC/Dytran. The results are plotted at four arbitrarily selected instants $t = 0.02, 0.4, 0.08$ and 0.1 ms. In the figure, the solid lines are the results by MSC/Dytran that is a grid-based hydro-code based on the finite volume method (FVM). The dashed lines are the results from the SPH method.

Table 6.3 and Table 6.4 show the density and pressure profiles along the radial direction at $t = 0.08$ ms. It is seen that except for the last location (corresponding to the last particle in the SPH simulation), the density and pressure profiles obtained using the SPH method are very close to those obtained using MSC/Dytran. Note that the MSC/Dytran is a grid-based hydrocodes, in which the results are interpolated with volumes of cells. In contrast, the SPH results are obtained from the particle and Lagrangian simulation, where no real particles exist beyond the last particle. Hence, it is reasonable that the density and pressure obtained using the SPH method are bigger than those obtained using MSC/Dytran.

In order to further validate the SPH method, two viewpoints at 0.1125 m and 0.05 m away from the center along the radial direction are used to investigate the time history of the pressure and density. The viewpoints are selected in such a way that the first point is located outside the initial explosive area and the second point inside the initial explosive area. The time history of pressure and density at these two viewpoints are plotted in Figure 6.13 and Figure 6.14, respectively. The results obtained by the SPH method are comparable with those obtained by MSC/Dytran. At the later stage, they agree fairly well. Note that for the viewpoint located at 0.05 m (which is inside the initial explosive gas), the results from MSC/Dytran and those from SPH agree well even at early time stages. For the viewpoint located at 0.1125 m (which is outside the initial explosive gas), the disagreement of the initial results of SPH and MSC/Dytran at $t = 0.01$ ms is the right representation of the nature of these two different methods used. The SPH method is a Lagrangian particle method. At the instant of 0.01 ms, the particles have not yet reached this position (0.1125 m). Therefore, the pressure and density is practically zero at this instant. However, in the simulation by MSC/Dytran, the grid-based methods are employed. The results are interpolated within the volumes of cells, so the pressure and density is nonzero at this instant. From the physical nature of this problem, before the gas reaches a certain point, the local density and pressure should be zero. Hence the initial results from SPH at 0.1125 m are more reasonable.

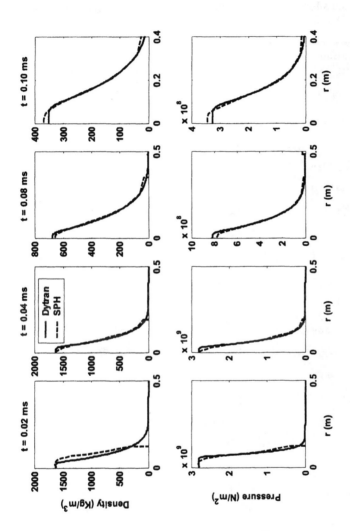

Figure 6.12 Results for the simulation of 2D explosive (TNT) expansion. Density and pressure profiles in the radial direction at four instants, $t = 0.02$ ms, 0.04 ms, 0.08 ms and 0.10 ms. (From Liu et al. (2003), Computers & Fluids, 32(3): 305-322. With permission.)

Smoothed Particle Hydrodynamics

Table 6.3 Results for the simulation of 2D explosive (TNT) gas expansion.
Density along the radial direction at $t = 0.08$ ms

Location (m)	SPH results (Kg/m^3)	MSC/Dytran results (Kg/m^3)
0.006615	659.45	678.10
0.014385	657.35	677.95
0.023498	651.30	676.17
0.032959	637.62	665.64
0.041174	617.43	643.88
0.048286	594.17	607.37
0.055887	565.93	572.85
0.065411	529.96	532.59
0.076852	488.97	482.71
0.088221	449.71	430.85
0.097884	414.87	396.99
0.1079	377.36	360.76
0.11967	333.75	319.13
0.13581	279.68	274.26
0.15209	233.76	231.48
0.16992	192.93	188.95
0.19125	153.13	148.57
0.2193	109.89	105.24
0.25599	62.84	65.58
0.34893	30.30	14.95

Table 6.4 Results for the simulation of 2D explosive (TNT) gas expansion. Pressure along the radial direction at t = 0.08 ms

Location (m)	SPH results (N/m^2)	MSC/Dytran results (N/m^2)
0.006615	7.7652e+008	8.1959e+008
0.014385	7.7482e+008	8.195e+008
0.023498	7.6792e+008	8.1664e+008
0.032959	7.4703e+008	7.9903e+008
0.041174	7.1132e+008	7.5514e+008
0.048286	6.6868e+008	7.0323e+008
0.055887	6.1937e+008	6.4812e+008
0.065411	5.6429e+008	5.855e+008
0.076852	5.0795e+008	5.1051e+008
0.088221	4.5372e+008	4.3577e+008
0.097884	4.0278e+008	3.8885e+008
0.1079	3.4803e+008	3.4041e+008
0.11967	2.9078e+008	2.8709e+008
0.13581	2.3113 e+008	2.3268e+008
0.15209	1.8160e+008	1.8401e+008
0.16992	1.3966e+008	1.3905e+008
0.19125	1.0220e+008	9.9941e+007
0.2193	6.7387e+007	6.2466e+007
0.25599	3.2765e+007	3.3134e+007
0.34893	1.8010+007	5.3327e+006

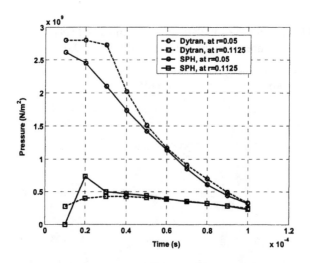

Figure 6.13 Results for the simulation of 2D explosive (TNT) expansion. Comparison of pressure at two locations, $x = 0.05$ m and $x = 0.1125$ m, along the radial direction. (From Liu et al. (2003), Computers & Fluids, 32(3): 305-322. With permission.)

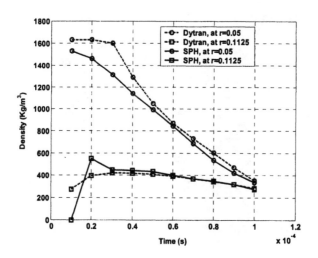

Figure 6.14 Results for the simulation of 2D explosive (TNT) expansion. Comparison of density at two locations, $x = 0.05$ m and $x = 0.1125$ m, along the radial direction. (From Liu et al. (2003), Computers & Fluids, 32(3): 305-322. With permission.)

6.6 Application of SPH to shaped charge simulation

6.6.1 Background

A shaped charge (Walters and Zukas, 1989) is generally a cylinder of high explosive (HE) with a hollow cavity that may assume almost any geometric shape such as a hemi-sphere or a cone in one end with a thin layer of liner, and a detonator at the opposite end (Figure 6.15). When the detonation shock engulfs the lined cavity, the detonation-produced gaseous products at the end with the hollow cavity converge towards the centerline or charge axis of symmetry at an extremely high level of pressure and kinetic energy. The liner material is accelerated under the superimposed extremely high hydrodynamic pressure that leads to the collapse of the cone. The focusing of the detonation products create an intensely localized force, which is capable of creating a deeper cavity on a plate than a plain cylinder of explosive without a hollow cavity, even though more explosive is available in the latter case. If certain criterion is satisfied, the liner can form a jet of very high speed that can be applied to penetrate hard targets, such as tanks.

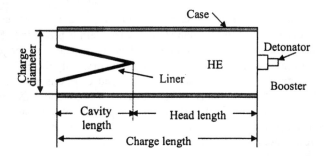

Figure 6.15 Typical configuration of a shaped charge. (From Liu et al., Computers & Fluids, 32(3): 305-322, 2003. With permission.)

The application of the shaped charge involves two stages, the HE detonation process and the later movement of the liner element. The flow of the material element on the liner and the formation of the metal jet can be analyzed by assuming that the strength of the liner is negligible compared with the detonation pressure, and that the liner behaviors as an inviscid and incompressible fluids (Birkhoff et al., 1948; Pugh et al., 1952). The HE detonation process for the shaped charge, which determines the pressure of the focusing detonation products, is however more complicated due to factors such as HE characteristics, HE geometry, interaction of detonation wave and rarefaction wave.

Although the technology of shaped charge was already extensively used in World War II, efficient computer simulation of shaped charges has only been possible recently with the advancement in computer hardware and computational methods. Computer simulation enables parametric studies to be carried out without having to resort to expensive firing trials, and is very useful in designing shaped charges or protection system against shaped charges. Some wave propagation hydro-codes (Cowler et al., 1987; Chou et al., 1983; Hallquist, 1986; Johnson, 1981; Hageman et al., 1975), which were originally developed to solve problems characterized by the presence of shock waves, localized materials response and impulsive loadings, have also been tried to simulate shaped charges. Though some excellent results can and have been obtained for problems with different complex shapes (Liu, Lam and Lu, 1998), some limitations or drawbacks still exist, and prevent further application to shaped charges. One of the most significant aspects is the background mesh inherent in the traditional algorithms and hydro-codes. The grid-based methods are limited in applications in simulating shaped charges due to the difficulties discussed in Chapter 1.

The successful simulations of the one-dimensional TNT slab detonation and the two-dimensional explosive gas expansion provide confidence to further apply the SPH method to simulate shaped charges. Liu et al. (2003c) has applied the SPH method to simulate the shaped charge explosion process. In this section, the application of the SPH method to shaped charge problem is presented. Since the flow of the material element on the liner as well as the formation of the jet can be analyzed using the inviscid, incompressible fluid theory, only the HE detonation as well as the later expansion process of the shaped charge is simulated. The effects of different ignition models, different cavity shapes, and different charge head lengths are investigated in the HE detonation process of the shaped charges. Major shaped charge detonation phenomena are captured numerically.

In the work, interest is concentrated on the HE detonation process of the shaped charge. The movement of the liner element as well as the casing effect is not considered here. For this simplified model, the HE detonation and the focusing of the energy are more complicated than those in the above-presented two examples. The performance of the simplified shaped charge is relevant to such parameters as the charge head length, cavity shape, the detonation ignition models, etc. For a practical shaped charge with liner and case, the material properties and the geometry of the liner and the case also need to be considered, and more involved modeling is required.

Example 6.3 Shaped charge with a conic cavity and a plane ignition

The performance of the shaped charge is related to many parameters. In this example, the detonation is ideally set up as plane ignition. The cavity is of conic shape. As shown from the initial geometry in Figure 6.16, the head length is 0.05 m, the cavity length is 0.15 m, and the conic cavity apex angle is approximately 43.6 degree. This shaped charge geometry is accepted in some engineering and military designs (Zhang, 1976).

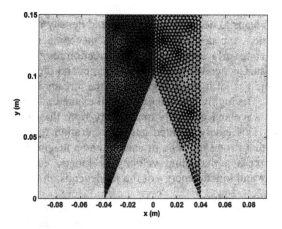

Figure 6.16 Initial geometry and particle distribution for the shaped charge explosion with plane ignition and conic cavity. The left-hand-side (LHS) shows the triangular mesh as well as the corresponding particles generated using the mesh; the right-hand-side (RHS) only shows the generated particles that are actually used in the simulation using the SPH method. (From Liu et al. (2003), Computers & Fluids, 32(3): 305-322. With permission.)

In the simulation, the particles are initially positioned right at the mass center or geometric center of a triangular mesh. As can be seen from Figure 6.16, the left-hand-side (LHS) shows the triangular mesh as well as the corresponding particles generated using the mesh; the right-hand-side (RHS) only shows the generated particles that are actually used in the simulation using the SPH method.

Figure 6.17 shows the velocity vectors at four representative time instants. The length and direction of an arrow represent the magnitude and direction of a velocity vector of a particle; the initial position of the arrow represents the location of that particle. The particle and velocity distribution in the detonation process is very close to what was simulated by Zhang (1976) using the Eulerian-Lagrangian coupling HELP code (Hageman et al. 1975). At 7.2 μs, the detonation wave arrives at the apex of the cone cavity (Figure 6.17*a*). Since the detonation is ideally plane ignited, a uniform planar detonation shock reaches the apex of the conic cavity. The detonation finishes at around 21.6 μs (Figure 6.17*c*) and then expands outwards (Figure 6.17*d*). It can be seen that after the ignition, the detonation produced gas particles expand outside along the direction roughly perpendicular to the surface of the original explosive; the detonation wave moves downward. Since the surrounding outside space of the explosive is assumed to be a vacuum, with the propagation of the detonation wave through the explosive charge, the rarefaction wave propagates also into the gaseous products, and thus leads to decrease of pressure and density in the gas. After 7.2 μs from the ignition, the detonation occurs in the conic region, the gaseous products near the cone surface move along the direction roughly perpendicular to the original cone surface, converge to the centerline, and then form a jet of gas. The pressure of the gas jet is much higher than the already-high pressure of the gaseous products. The highly focused gas jet will expand due to the propagation of the rarefaction wave and gradually move away from the centerline as shown in Figure 6.17*d*. This cavity resulted gas concentration toward the centerline accompanying with a strong focusing of energy, and a magnification of velocity. Note that there exists a maximum velocity or kinetic energy for jet. The maximum velocity and kinetic energy of the jet depends on the configuration of the shaped charge.

Figure 6.18 and Figure 6.19 show the density and pressure evolution in the detonation process at the four corresponding instants. The detonation wave, rarefaction wave and the gas jet can be clearly seen from the figures. During the formation and advancement of the gas jet, the pressure increases to a maximum value located on the centerline. In the expansion process, the pressure gradually decreases due to the inward propagation of the rarefaction wave. At 21.6 μs when the detonation finishes, the maximum pressure is around 10 GPa, while the maximum velocity is approximately 9430 m/s.

Figure 6.20 shows the density, pressure and velocity magnitude contours at 25.2 μs. It is observed that the maximum density, pressure and velocity are all on the centerline, and the particles at the jet tip possess the highest velocity. However, the maximum density, pressure and velocity do not locate at the same position on the centerline. If the shaped charge is lined with a rigid layer of liner, no rarefaction wave will come from outside the liner, and the pressure behind the liner would be built up even higher. Practically it can be expected that, when the detonation wave engulfs the lined cavity, the liner is accelerated under the tremendous pressure built behind it, and under certain condition, forms a powerful metal jet with very large velocity, large density, and hence a high level of energy.

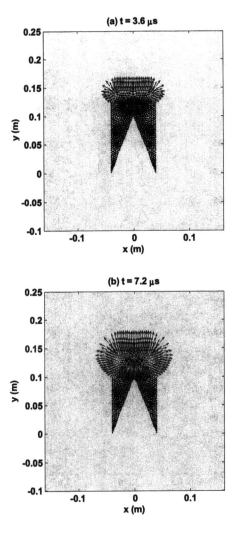

Figure 6.17 Velocity distributions in the detonation process for the shaped charge with a conic cavity and a plane ignition on the upper end. The length and direction of an arrow represent the magnitude and direction of a velocity vector of a particle; the initial position of the arrow represents the location of that particle. (a) The detonation wave advances downwards through the explosive charge before reaching the conic cavity; (b) The detonation wave reaches the apex of the conic cavity; (to be continued.)

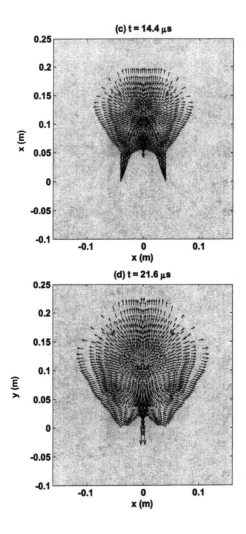

Figure 6.17 Velocity distributions in the detonation process for the shaped charge with a conic cavity and a plane ignition on the upper end. The length and direction of an arrow represent the magnitude and direction of a velocity vector of a particle; the initial position of the arrow represents the location of that particle. (c) The explosive gas converges to the centerline and forms a gas jet traveling downwards at a very high speed; (d) The gas jet advances downwards with a small diverges from the centerline.

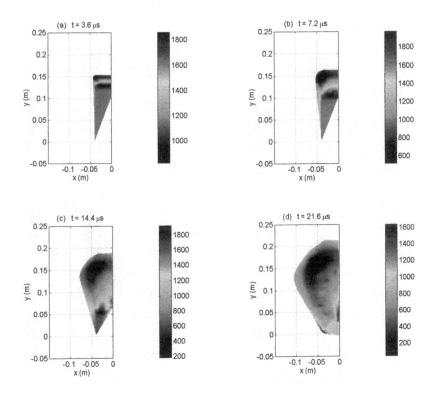

Figure 6.18 Density distributions in the detonation process for the shaped charge with a conic cavity and a plane ignition on the upper end.

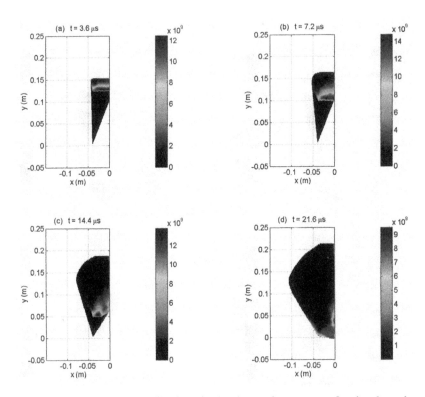

Figure 6.19 Pressure distributions in the detonation process for the shaped charge with a conic cavity and a plane ignition on the upper end.

(a) Density contour (×1.0E+2 Kg/m^3)

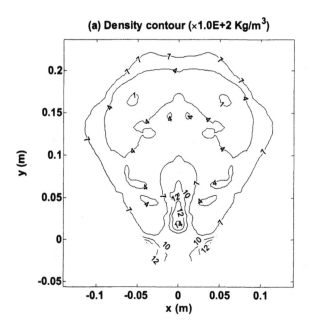

(b) Pressure contour (×1.0E+9 N/m^2)

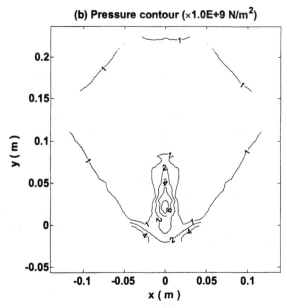

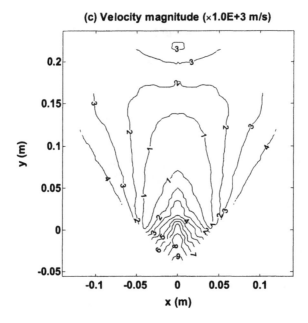

(c) Velocity magnitude (×1.0E+3 m/s)

Figure 6.20 Density, pressure and velocity magnitude contour at 25.2 μs for the shaped charge with a conic cavity and a plane ignition on the upper end.

Example 6.4   Shaped charge with a conic cavity and a point ignition

The plane ignition model in Example 6.3 is not very practical since practical applications of HEs usually involve a detonator, which is approximately a point, and will generate a spherical detonation wave rather than a planar detonation wave through the explosive. The propagation of the spherical detonation wave leads to non-planar profiles when the detonation shock wave reaches the apex of the cavity. It may result in different performance of the gas concentration to the symmetric centerline and the later gas divergence from the centerline. In the simulation, the initial geometry is the same as that in Example 6.3 (Figure 6.16), except that the detonation is ignited from the middle point on the upper end of the charge.

Figure 6.21, Figure 6.22 and Figure 6.23 shows the density, pressure and velocity in the detonation process, respectively, at four time instants of 3.6, 7.2, 14.4 and 21.6 μs. The spherical detonation wave can be clearly seen from these figures. Comparing this point ignition model with the plane ignition model in

Example 6.3, the differences in density, pressure and velocity are substantial at initial stages. At later stages, especially after the detonation wave crosses over the apex of the cone cavity, the differences become smaller. This is because with the increasing distance to the ignition point, and the decreasing arc length in the detonation front, the spherical detonation wave gradually approaches to a planar detonation wave. The differences in density, pressure and velocity for the gaseous products produce differences in the gas focusing performance.

Figure 6.24 shows the density, pressure, and velocity magnitude contours at 25.2 μs. The distributions of density, pressure and velocity are similar to those in Example 6.3. Compared with the counterparts in Example 6.3, the maximum density and pressure obtained using the point ignition model do not differ much. The maximum velocity magnitude is much lower. In the course of gas concentration and divergence, a maximum velocity of around 7770 m/s is observed at a later stage, compared to the maximum velocity of 9430 m/s obtained using the plane ignition model in Example 6.3. This suggests that the means of detonation ignition affect the way, in which the gas focuses.

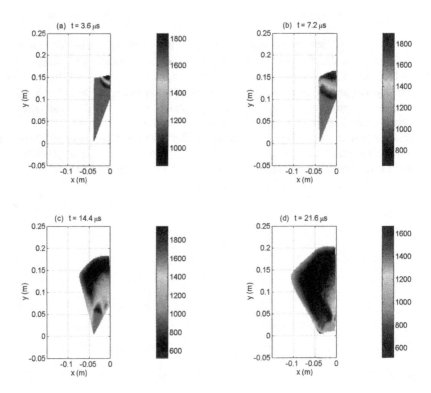

Figure 6.21 Density distributions in the detonation process for the shaped charge with a conic cavity and a point ignition at the middle of the upper end.

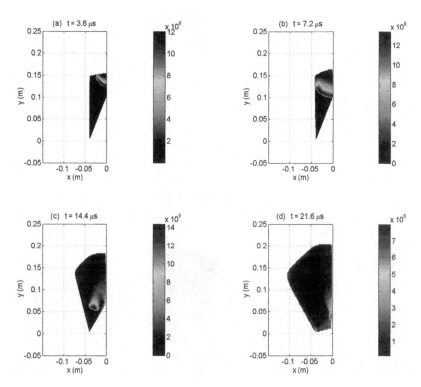

Figure 6.22 Pressure distributions in the detonation process for the shaped charge with a conic cavity and a point ignition at the middle of the upper end.

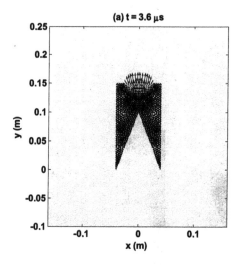

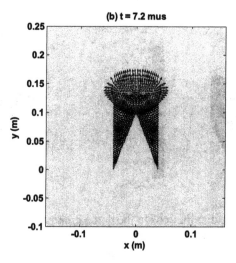

Figure 6.23 Velocity distributions in the detonation process for the shaped charge with a conic cavity and a point ignition at the middle of the upper end. (a) The detonation wave advances downwards through the explosive charge before reaching the conic cavity; (b) The detonation wave reaches the apex of the conic cavity; (to be continued.)

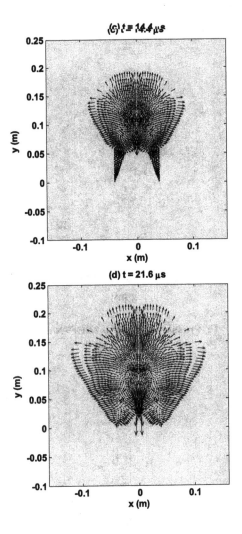

Figure 6.23 Velocity distributions in the detonation process for the shaped charge with a conic cavity and a point ignition at the middle of the upper end. (c) The explosive gas converges to the centerline and forms a gas jet traveling downwards at a very high speed; (d) The gas jet advances downwards with a small diverges from the centerline.

(a) Density contour (×1.0E+2 Kg/m^3)

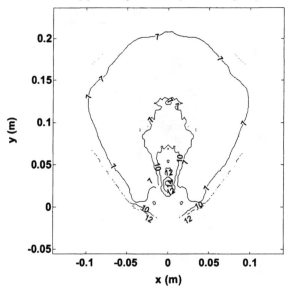

(b) Pressure contour (×1.0E+9 N/m^2)

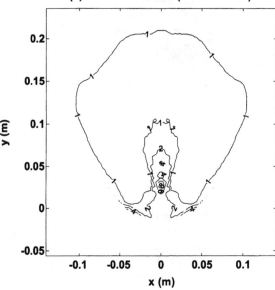

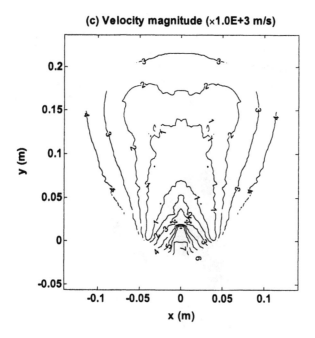

Figure 6.24 Density, pressure and velocity magnitude contour at 25.2 μs for the shaped charge with a conic cavity and a point ignition on the upper end.

Example 6.5   Shaped charge with a hemi-elliptic cavity and a plane ignition

Different cavity shapes for a shaped charge may result in different performances in terms of gas focusing. A shaped charge with hemi-elliptic cavity (Figure 6.25) is investigated here. In this example, the detonation is plane ignited. Figure 6.26 shows the velocity distribution at four instants. It can be seen that before the detonation wave reaches the cavity apex, Figure 6.26 is the same as Figure 6.17. Large differences occur after the detonation wave reaches the cavity apex due to the different cavity contours. Different from the conic cavity, the hemi-elliptic cavity results in inconstant slopes along the cavity contour. Near the cavity apex, the elliptic contour is much smoother than the conic contour. Hence right after the detonation wave reaches the cavity apex, the gas focusing effect is not as strong as the conic cavity case. At later stages, the focusing effect becomes stronger. Density, pressure and velocity magnitude contours at 25.2 μs are plotted in Figure 6.27. These contours are similar in shape to those shown in Figure 6.20. The maximum value of density, pressure and velocity are fairly different from Example 6.3 for the conic cavity. This is because the hemi-elliptic cavity not only changes the cavity contour and the cavity space for focusing gas, but also reduces the gaseous products that take part in the focusing process. The maximum velocity in the whole gas focusing process is around 8183 m/s, which is between the values for the conic cavity cases with a plane-ignition (9430 m/s) and with a point-ignition (7770 m/s).

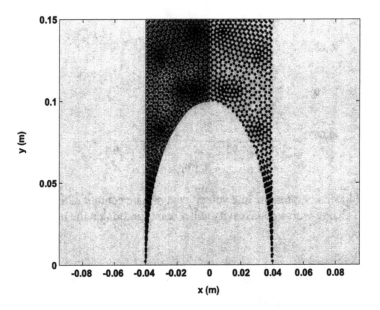

Figure 6.25 Initial particle distribution for the shaped charge explosion with plane ignition and hemi-elliptic cavity. The left-hand-side (LHS) shows the triangular mesh as well as the corresponding particles generated using the mesh; the right-hand-side (RHS) only shows the generated particles that are actually used in the simulation using the SPH method.

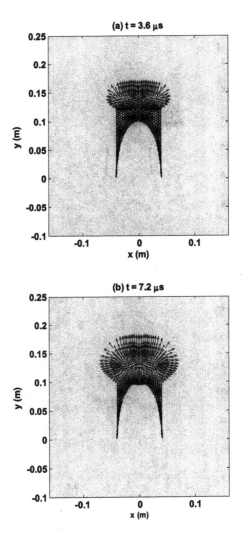

Figure 6.26 Velocity evolution for the shaped charge with a hemi-elliptic cavity and a plane ignition on the upper end. (a) The detonation wave advances downwards through the explosive charge before reaching the conic cavity; (b) The detonation wave reaches the apex of the conic cavity; (to be continued)

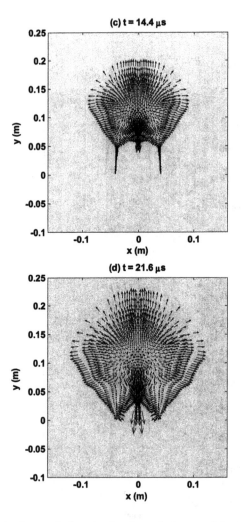

Figure 6.26 Velocity evolution for the shaped charge with a hemi-elliptic cavity and a plane ignition on the upper end. (c) The explosive gas converges to the centerline and forms a gas jet traveling downwards at a very high speed; (d) The gas jet advances downwards with a small diverges from the centerline.

(a) Density contour (×1.0E+2 Kg/m³)

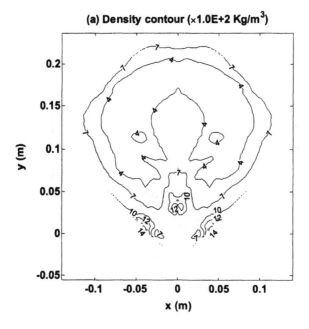

(b) Pressure contour (×1.0E+9 N/m²)

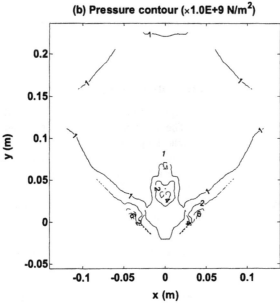

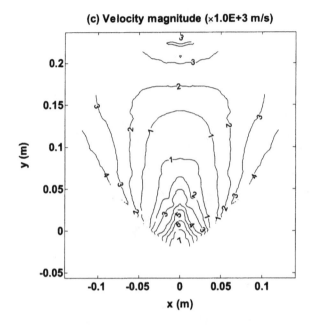

(c) Velocity magnitude (×1.0E+3 m/s)

Figure 6.27 Density, pressure and velocity contour at 25.2 μs for the shaped charge with a hemi-elliptic cavity and a plane ignition on the upper end.

Example 6.6   Effects of charge head length

The charge head length (or the charge length) is another important factor in focusing the gas to form the jet. The influence of the charge head length is investigated by varying the charge head length in Figure 6.16 while keeping the geometry of conic cavity unchanged.   Since the velocity is a typical representation of the gas focusing, the maximum velocity at the point 0.02 m below the lower end of the original explosive charge along the centerline in the gas focusing process is investigated to measure the power of gas focusing.

Figure 6.28 shows the concerned velocity curves with the variation of the charge head length.   The circles represent the results using the plane ignition model; the squares represent the results using the point ignition model.   The charge head length is normalized using the charge diameter, which is in this study equal to the conic cavity diameter. It is seen that, for both the point ignition and plane ignition models, the velocity at the investigated point increases with the increase of the charge head length up to a certain point.   This can be explained as follows.

With the advancement of the detonation wave in the high explosive, the rarefaction wave also propagates to the gaseous product either upwards or sideward. The rarefaction wave reduces the pressure, density and velocity distribution behind the detonation wave. Since the propagation speed of the rarefaction wave is only around one half of the detonation velocity (Zhang, 1976), the influence of the rarefaction wave is restricted to limited areas. Increasing the charge head length means increasing the region not affected by the rarefaction wave, and thus increases the focusing effect. When the charge head length is increased to a certain value so that the gaseous products behind the detonation wave only feel the expansion effect of the rarefaction wave from sideward, but not from upwards, the detonation contour around the conic cavity will be of a conic shape, and the gas focusing process will not be mitigated by the expansion effect of the rarefaction wave. Around a critic value of the ratio of the charge head length to the charge diameter, very little improvement in gas focusing can be achieved.

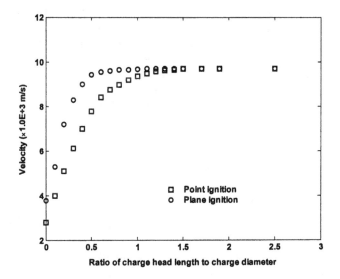

Figure 6.28 Maximum velocities at (0, − 0.02) with different charge head lengths. The charge head length is normalized using the charge diameter. The measure point is 0.02 m below the lower end of the original explosive charge along the centerline.

From Figure 6.28, the critic ratios for the plane ignition model and the point ignition model are estimated around 1 and 1.5, respectively. Since the detonation wave for the point ignition model is of spherical shape rather than the uniform planar shape, it is reasonable that the critic head length is larger than that for the plane ignition model. Increasing the charge head length also means increasing distance from the cavity apex to the ignition point, and the spherical detonation wave gradually approaches to a planar detonation wave, thus the curves become closer at larger charge head length.

The existence of a critical charge head length was also reported by Walters and Zukas (1989) when investigating practical shaped charges with liner. In their discussions, the ratios of the critical charge head length to the charge diameter for planar ignition model and point ignition model were reported to be around 1 and 1.3~1.8, respectively. These values of critical charge head length for practical shaped charge agree well with the simulation results using the SPH method for the simplified shaped charge detonation without liner. This shows that it is the detonation behavior of the explosive in the hollowed shaped charge that determines largely the value of the critical charge head length. This example further reveals that beyond the critical charge head length, the gas focusing effects for planar ignition with planar detonation wave and point ignition with spherical detonation wave should be the same.

In practical applications — though increasing the charge head length up to the critical value — does improve the focusing effect, it is usually desirable to keep the charge head length to the minimum so as to reduce the weight of charge and the length of the entire device. These observations from the simplified shaped charge explosion can be extended to the practical shaped charge with liner and case, and can be used as a general guidance for practical shaped charge design, though perhaps some modifications are needed when considering the liner and case.

6.7   Concluding remarks

The HE explosion process involves a violent chemical reaction which converts the original explosive charge of high energy density into gas at very high temperature and pressure, occurring with extreme rapidity and releasing a great deal of heat. HE explosion simulation is difficult for the conventional grid-based methods due to some special numerical difficulties. Instead, the SPH method is very attractive to simulate HE explosions due to its special features.

In this chapter, the SPH method is applied to simulate the high explosive explosion process. A three-dimensional code is implemented based on the SPH method for the Euler equations with an algorithm of adaptive smoothing length to ensure a proper adaptability of the SPH method to the drastic movement of particles. The method and the code are shown to be robust, computationally efficient, and easy to apply.

Two numerical examples are carried out to examine the method and the code in simulating HE detonation and explosive gas expansion. For the benchmark one-dimension TNT slab detonation, good predictions for the magnitude, location as well as the shape of the detonation wave are obtained. The pressure transients in the expansion process of the explosive gas behind the detonation wave also match the theoretical value fairly well. For the two-dimensional explosive gas expansion problem, the obtained SPH results agree very well with the results from the commercial software package MSC/Dytran.

The SPH method and the code are applied to simulate shaped charge explosions. The effects of different detonation ignition models, different cavity shapes, and different charge length are investigated. The simulation results, though obtained from the simplified shaped charge and need to be adjusted with different specific situations, are revealing for practical applications.

From the numerical simulations and discussions, the following conclusions can be made

1. The SPH method, after suitable modifications together the use of an adaptive smoothing length, can be extended to simulate different HE explosion problems such as arbitrary charge shapes and different detonation ignitions.
2. The SPH method has successfully captured the major features of the shaped charge explosions including the formation of the high velocity gas jet.
3. The SPH results show that the detonation ignition model plays an important role at the earlier stages for the gas jet focusing. The differences resulted from the different ignition models are gradually reduced when increasing the charge head length.
4. Changing the cavity shape leads to different density, pressure and velocity distributions for the gas jet.
5. Increasing the charge head length up to a critic value improves the gas focusing effect. The critic charge head length of the point ignition model is larger than that of the plane ignition model.

Chapter 7

SPH for Underwater Explosion Shock Simulation

In Chapter 6, the SPH method is applied to simulate the complicated explosion process arising from high explosive (HE) detonation. In this chapter, the SPH method is extended to simulate the shock waves generated by underwater explosions.

The formulation of the SPH method used in this chapter is largely the same as those in Chapter 6. The major challenge in this chapter is the presence of multi-materials in the problem domain, which requires special treatments. In order to suit the needs of underwater explosion shock simulation, a *particle-to-particle interface* technique is employed. The interface treatment technique allows the SPH (kernel and particle) approximations among particles form different materials. A special penalty force is applied in the interface treatment technique on the possible penetrating particles. Numerical examples reveal that the SPH method, after properly tuned, can resolve the underwater explosion shocks fairly well.

The SPH method is also used to investigate the real and the artificial detonation models of high explosive as well as their influences to the entire simulation of underwater explosion shock. It is found that beyond the region of 8-10 times the charge radius, both detonation models can get fairly good and close results. In region less than 8-10 times the charge radius, the real detonation model should be employed to obtain more accurate results.

This chapter studies also the water mitigation problems using the SPH method. The water mitigation study using the SPH method is very important in designing high-performance magazines that are robust against accidental explosions. Mitigation simulations with configuration of contact (termed as *contact water mitigation*) and non-contact (termed as *non-contact water mitigation*) water shield have been carried out and are compared with the case without water (explosion in air). It is found that with the presence of water, the peak shock pressure and the equilibrium gas pressure are reduced to different

255

levels depending on the relevant geometrical configuration of the system. An optimal contact water shield thickness is found by the simulation to produce the best mitigation effect for a given high explosive charge. The non-contact water shield, if properly designed, can produce further reduction of peak shock pressure and equilibrium gas pressure.

7.1   Introduction

The underwater explosion (UNDEX) (Cole, 1948; Mader, 1998; Bangash, 1993) produced by the detonation of a submerged high explosive poses a serious threat to the integrity of nearby structures. Issues related to the underwater explosion include

- the detonation process of the high explosive (HE) charge,
- the expansion process of the detonation-produced explosive gas into the surrounding water, and
- the interaction of underwater shocks with the nearby structures.

In the detonation process, the high explosive is converted into gaseous products at very high temperature and pressure through a violent chemical reaction, occurring with extreme rapidity and releasing a great deal of heat. The propagation of the detonation wave through an explosive is so rapid that the gaseous products directly behind the wave front are not in pressure equilibrium with the gas further behind the wave front. As the detonation wave reaches the interface between the explosive and the surrounding water, a high pressure shock wave of step exponential type is transmitted to propagate through the water, followed by a series of bubble pulsation associated with the repeating expansion and contraction of the bubble of the explosive gas. In the entire process of the underwater explosions, some special features such as large deformations, large inhomogeneities, moving material interfaces, deformable boundaries, and free surfaces usually exist.

Underwater explosion has been the subject of experimental and analytical studies since World War II. Analytical or theoretical solutions to underwater explosion problems are only limited to very simple cases. Experimental studies are fairly expensive and very dangerous due to the destructive nature of firing tests. Recently, more and more researches are focused on the numerical simulations using computers. Simulation of underwater explosion problems is a big challenge for the conventional grid-based numerical methods. On one hand, special difficulties such as large distortions, moving material interfaces, deformable boundaries and free surfaces exist; on the other hand, the detonation process of the high explosive in the whole underwater explosion process poses more challenges due to the complexity and larger scale of the problem.

Early theoretical and numerical analyses of underwater explosions were generally based on the following ideal assumptions to simplify the problem (Kirkwood and Bethe, 1942; Penney and Dasgupta, 1942; Brinkley and Kirkwood, 1947).

- The explosive charge is of spherical shape.
- The explosive charge is ignited from the charge center.
- The surrounding water is infinite.

In many numerical implementations, the pressure in the explosive gas was often determined empirically. The interaction between the explosive gas and the surrounding water was not rigorously considered.

In the underwater shock analyses involving fluid structural interaction, the finite element methods were generally used to model the structure behavior and the methods such as doubly asymptotic approximation (DAA) (Geers, 1971; 1978) were used to model the behavior of the shock wave in the surrounding water. The pressure in water was determined empirically. The interaction between the explosive gas and the surrounding water was also not considered.

Recently, some literatures (Brett, 1997; Molyneaux et al., 1994; Liu and Lam et al. 1998; Lam and Liu et al., 1996, Lam et al., 1998; Zong et al., 1998; Gong et al., 1999) tried to use hydro-codes (MSC/Dyna, 1991; MSC/Dytran, 1997) that contain both Lagrangian and Eulerian processors coupled together. In the implementation, the Eulerian processor (mesh) is used to deal with regions where large deformations exist; the Lagrangian processor (mesh) is used to track material interfaces. Computational information is exchanged either by mapping or by special interface treatment between these two types of meshes. This Lagrangian-Eulerian coupling technique is rather complicated and can cause inaccuracy especially in the data exchanging process. Swegle and Attaway (1995) attempted to simulate underwater explosions through coupling the SPH method and the finite element method.

In this chapter, the smoothed particle hydrodynamics is applied to underwater explosion simulations, especially the early time phenomena. Of particular interest is the underwater explosion shock that is the most direct and most important factor that causes the damages to the nearby structures.

The outline of this chapter is as follows.

- In Section 2, the underwater explosion physics and the governing equations are described.
- In Section 3, the basic SPH formulation for underwater explosion simulation is discussed.

- In Section 4, an interface treatment technique suitable for underwater explosion simulations is introduced.
- In Section 5, two numerical examples of different HE charge shapes are presented with comparisons to results from other sources.
- In Section 6, a comparison study is carried out on the real and artificial HE detonation models as well as their influences to the entire underwater explosion process.
- In Section 7, the water mitigation problems, which can be regarded as special cases of underwater explosions, are simulated at different scenarios with some interesting results.
- In Section 8, some remarks and conclusions are given.

7.2 Underwater explosions and governing equations

Underwater explosion event consists of a complicated sequence of events, and can be simplified to the generation of the pressure waves (shock wave and acoustic wave) and the fluid flows produced by the dynamic interaction of the detonation-produced explosive gas and the surrounding water. Immediately after the completion of the HE detonation phase, an underwater explosion can be conceptualized as a shock wave, followed by a series of bubble pulsation associated with the repeating expansion and contraction of the explosive gas bubble. The shock wave, generated when the detonation wave through the high explosive reaches the explosive gas/water interface, travels outwards through the water at a very high speed. The high pressures associated with the shock wave can inflict considerable damage on the nearby targets that it encounters.

7.2.1 Underwater explosion shock physics

The first and most important cause of disturbance to the water in an underwater explosion is the arrival of the HE detonation wave at the interface of water and explosive gas. Immediately upon the arrival of the detonation wave, the extremely high pressure begins to produce an intense pressure wave propagating through the surrounding water, known as *underwater shock wave*, and an outwards motion of the water. The shock wave propagates away into the water very quickly compared to the outwards motion of the water. The dense mass of the explosive gas confined by the surrounding water forms a bubble and begins to expand. The interaction between the inside gas bubble and the surrounding water results in a periodic contraction-expansion of the gas bubble, known as *bubble pulsation*.

The shock wave phenomenon occurs at the earlier stage, and the bubble pulsation occurs at much later stage. This work emphasizes more on the earlier stage related to the generation of shock waves, although the bubble pulsation will also be simulated. The pressure caused by the shock wave is discontinuous and is then followed by a roughly exponential decay. The duration measured for the shock wave is a few milliseconds at most. Once initiated, the disturbance of the shock wave is propagated radially outwards in the water.

As compared to waves of infinitesimal amplitude, the underwater shock wave has some typical characteristics (Cole, 1948).

- The velocity of propagation near the HE charge is several times the sound speed limit (1520 m/s).
- The pressure will fall to acoustic values with the outwards advancement of the shock wave.
- The pressure level in the spherical shock wave falls off more rapidly with distance than the inverse first power law predicted for small amplitudes, but eventually approaches this behavior in the limit of large distances.
- The profiles of the shock wave broaden gradually as the wave spreads out. This spreading effect is most marked in the region of high pressure near the charge.

Underwater explosion phenomena are subject to a number of physical laws and properties, including the physical conditions at the interface of the explosive gas and the surrounding water. The surrounding water has such properties as large density that is approximately 1000 times of the air density, low compressibility, and large sound speed. Owing to the dynamic properties of the water (especially in the regions surrounding the explosive gas), the pressures are generally very high and the wave velocities are dependent on the magnitude of the pressure and the displacement of the water as it progresses. These complications for waves of finite amplitude are expressed in much more difficult mathematical statements than those which suffice to explain the propagation of small amplitude waves whose velocities are practically independent on the magnitude of the pressure.

7.2.2 Governing equations

The physical law of conservation governs the underwater explosion events. Since the HE detonation velocity and the shock propagation velocity in water are rapid, the explosive gas and the water can be assumed to be inviscid, and the entire underwater explosion process is adiabatic. Therefore, the following Euler equation can be used to model the underwater explosion process coupled with a suitable equation of state

$$\begin{cases} \dfrac{D\rho}{Dt} = -\rho \nabla \cdot v \\[2mm] \dfrac{Dv}{Dt} = -\dfrac{1}{\rho} \nabla p \\[2mm] \dfrac{De}{Dt} = -\dfrac{p}{\rho} \nabla \cdot v \\[2mm] p = p(\rho, e) \end{cases} \tag{7.1}$$

where, ρ, e, p, v and t are density, internal energy, pressure, velocity vector and time respectively. The first three equations in equation (7.1) state the conservation of mass, momentum and energy. The fourth equation is the equation of state (EOS). In this work, TNT with a detonation velocity of 6930 m/s is used in the simulation as an example of the HEs.

The derivation of equation (7.1) is briefed in Chapter 4. In equation (7.1), the external forces are not included. For early time phenomena in underwater explosion such as shock generation and propagation, the external forces are generally not important compared with the extremely large internal driven force (Cole, 1948; Zhang, 1976). The always-present gravity is, thus, not considered in the simulation. However, it can be taken into account by simply adding the gravity force into the momentum equation when simulating the later time phenomena, as we have done in Chapter 4 for the water discharge problem.

For the detonation produced explosive gas, the standard Jones-Wilkins-Lee (JWL) (Dobratz, 1981) equation of state can be employed. The JWL EOS for TNT corresponds to a detonation velocity of 6930 m/s and a Chapman-Jouget (CJ) pressure of 21 GPa. The pressure of the explosive gas is:

$$p = A(1 - \frac{\omega \eta}{R_1})e^{-\frac{R_1}{\eta}} + B(1 - \frac{\omega \eta}{R_2})e^{-\frac{R_2}{\eta}} + \omega \eta \rho_0 e \tag{7.2}$$

where η is the ratio of the density of the explosive gas to the initial density of the original explosive charge. e is the internal energy of the high explosive per unit mass. A, B, R_1, R_2 and ω are coefficients obtained by fitting experimental data. The values of the corresponding coefficients are listed in Table 6.1.

The pressure-volume-energy behavior of water is widely studied under different shock loadings of high pressures, densities and temperatures. Water can be modeled as a compressible fluid with the Mie-Gruneisen equation of state, which uses the cubic shock velocity and the fluid particle velocity to determine the pressure of the compressed and expanded water. The Shock Hugoniot experimental data are needed to correlate the equation of the cubic shock velocity and the fluid particle velocity

$$\frac{U_s - C_0}{U_s} = S_1(\frac{U_s}{U_p}) + S_2(\frac{U_s}{U_p})^2 + S_3(\frac{U_s}{U_p})^3 \tag{7.3}$$

where U_s, U_p, and C_0 are the shock wave velocity, the fluid particle velocity and the initial sound speed, respectively. S_1, S_2, and S_3 are the coefficients to determine the slope of the $U_s - U_p$ curve.

Shin et al. (1998) provided a comparison of several approximations from shock Hugoniot experimental data due to the uncertainties in the experimental data for water. In this simulation, the experimental data (listed in Table 7.1) (Steinberg, 1987) is used.

Due to the fact that the behavior of water is very much different when it is under compression and expansion, the EOS for water should depend on the states of the water. The pressure of water in the compression state is given by

$$p = \frac{\rho_0 C^2 \mu[1 + (1 - \frac{\gamma_0}{2})\mu - \frac{a}{2}\mu^2]}{[1 - (S_1 - 1)\mu - S_2 \frac{\mu^2}{\mu+1} - S_3 \frac{\mu^3}{(\mu+1)^2}]^2} + (\gamma_0 + a\mu)e \tag{7.4}$$

in the case of expansion, the pressure of water is

$$p = \rho_0 C_0^2 \mu + (\gamma_0 + a\mu)e \tag{7.5}$$

where ρ_0 is the initial density, η is the ratio of the density after and before disturbance, and $\mu = \eta - 1$. When $\mu > 0$, water is in the compressed state, and when $\mu < 0$, water is in the expanded state. Some material parameters and coefficients in the EOS for water are summarized in Table 7.1.

Chisum and Shin (1997) have provided a procedure to convert the above Gruneisen form of equation of state for water into a polynomial form either in compressed or expanded state. In the compression state

$$p = a_1\mu + a_2\mu^2 + a_3\mu^3 + (b_0 + b_1\mu + b_1\mu^2)\rho_0 e \tag{7.6}$$

in the expansion state

$$p = a_1\mu + (b_0 + b_1\mu)\rho_0 e \tag{7.7}$$

The constants in equation (7.6) and (7.7) were determined, for both compression and expansion states, by matching the terms in equation (7.4) and (7.5) with the Steinberg shock Hugoniot data. Hence, the above equation of state for water can be used over the range where the Mie-Gruneisen EOS is valid. The constants for condensation values in the order of $\mu < 0.8$ are summarized in Table 7.2.

Table 7.1 Material parameters and coefficients in the Mie-Gruneisen EOS for water

Symbol	Meaning	Value
ρ_0	Initial density	1000 Kg/m^3
C	Sound speed	1480 m/s
γ_0	Gruneisen coefficient	0.5
a	Volume correction coefficient	0
S_1	Fitting coefficient	2.56
S_2	Fitting coefficient	1.986
S_3	Fitting coefficient	1.2268

Table 7.2 Material parameters and coefficients in the polynomial EOS for water

Symbol	Value
a_1	2.19×10^9 N/m^2
a_2	9.224×10^9 N/m^2
a_3	8.767×10^9 N/m^2
b_0	0.4934
b_1	1.3937

Air is modeled as an ideal gas, which satisfies the gamma law equation of state, i.e. $p = (\gamma - 1)\rho e$, where γ is the ratio of the specific. In the study, $\gamma = 1.4$. The initial density of air is 1 Kg/m^3, and therefore the initial energy of air is 2.5×10^5 J/Kg, which corresponds to the atmospheric pressure of 1 bar according to the equation of state.

7.3   SPH formulations

Since the explosive gas and the surrounding water are assumed inviscid, and the underwater explosion process is regarded as adiabatic, the Euler equation is the governing equation in the underwater explosion physical process. Using the SPH kernel and particle approximations, a set of possible SPH equations are listed in Table 4.2. In this chapter, the following set of discretized SPH formulae for the Euler equation with artificial viscosity are applied to simulate the underwater explosion process

$$
\begin{cases}
\dfrac{D\rho_i}{Dt} = \displaystyle\sum_{j=1}^{N} m_j (v_i - v_j) \cdot \nabla_i W_{ij} \\[3mm]
\dfrac{Dv_i}{Dt} = -\displaystyle\sum_{j=1}^{N} m_j (\dfrac{p_i}{\rho_i^2} + \dfrac{p_j}{\rho_j^2} + \Pi_{ij}) \nabla_i W_{ij} \\[3mm]
\dfrac{De_i}{Dt} = \dfrac{1}{2} \displaystyle\sum_{j=1}^{N} m_j (\dfrac{p_i}{\rho_i^2} + \dfrac{p_j}{\rho_j^2} + \Pi_{ij}) \left(v_i - v_j \right) \cdot \nabla_i W_{ij} \\[3mm]
\dfrac{Dx_i}{Dt} = v_i
\end{cases}
\tag{7.8}
$$

The Monaghan type of artificial viscosity Π_{ij} (see, Section 4.4.1) is used to resolve the shock and stabilize the calculation. In the simulation of underwater explosion problems, the cubic spline smoothing function (see, Section 3.1) is used. The leapfrog method is used in the time integration for its low memory storage requirement and computational efficiency (see, Section 6.3). Virtual particles (type I) are used to treat the boundaries for the purposes of both improving the accuracy and imposing repulsive force to prevent penetration into the solid wall (see, Section 4.4.8). The smoothing length is evolved in the same procedure as the one in simulating HE explosions, with which only a minimal and necessary number of neighboring particles will contribute to the discrete summations (see, Section 6.4).

7.4   Interface treatment

Underwater explosions involve interfaces of explosive gas with surrounding water and water with outside air. The treatment of interface is important since the computational information is transferred through the interfaces of the media. At the gas/water interface, we have the following continuous conditions

$$p_g = p_w \tag{7.9}$$

$$v_g \cdot n = v_w \cdot n \tag{7.10}$$

where the subscripts g and w represent explosive gas and water, respectively. n is the vector of the outwards normal on the interface. Equation (7.9) states that the pressure on both sides of the interface is the same, and equation (7.10) states that the normal velocities on both sides are equal.

Numerical implementation of equations (7.9) and (7.10) is straightforward in the finite difference method or in the finite element method as long as a predefined mesh is used and the interface is determined using mesh lines or surfaces. However, a special consideration is required for meshfree particle methods such as the SPH method. Due to the free Lagrangian and particle nature of the SPH method, two interface (or contact) particles from different media may move apart, and can even be no longer neighbors (Figure 7.1) in the evolution process.

The material interface treatment is a sore issue in SPH and was ignored in earlier SPH literatures. The interaction or contact from different materials or different bodies is usually handled through the conservation equation with no restriction on the particles from different materials. If the summation allows contributions only from the same material, two particles that are close but of different materials can not be regarded as neighboring particles. Therefore, around the interface region, the simulation suffers from the same particle deficiency problem as for the ordinary boundaries (see Chapter 5). In contrast, if the summation allows contributions of all the particles from different materials, two particles that are close but of different materials can be regarded as neighboring particles. Thus the boundary deficiency problem in the SPH kernel and particle approximations can be reduced. However, this treatment leads to some extent of unphysical penetration or mixture problems. This unphysical penetration is not very serious in most circumstances and typically only one or two layers of particles from different materials penetrate each other. For problems with high intensity loading interactions, the penetration may be fatal to the accuracy and even the execution of the SPH code.

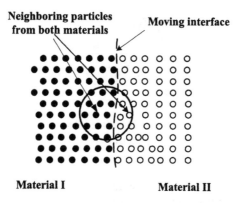

Figure 7.1 Moving material interface between two materials. Some neighboring interface particles at current step may be no longer neighbors in the next time steps.

Monaghan (1992) used the so-called XSPH to treat the interface, in which an average velocity is used to handle the single valued velocity, and no force is generated to resist penetration or sliding. Campbell et al (2000) recently proposed a particle-to-particle contact algorithm that was reported to be effective in simulating high velocity impact problems.

The material interface treatment in the simulation of underwater explosion is more complicated since the particles move more freely due to the violent interaction between the gas and water. Liu et al. (2001a) has developed a particle-to-particle interface algorithm that was shown to be effective both for high velocity impact and underwater explosion problems. In their algorithm, the material interface is handled using kernel summation, which allows for interactions between particles of different materials when solving the governing conservation equation. However, it's not enough since in an underwater explosion, the violent interaction between the high pressure explosive gas and the surrounding water usually leads to unphysical particle penetration or mixing near the interface. Hence, a penalty force is applied to particles from different materials near the interface when they are approaching and tending to penetrate each other. The penetration *pe* is detected (as shown in Figure 7.2) if

$$pe = \frac{h_i + h_j}{2r_{ij}} \geq 1 \qquad (7.11)$$

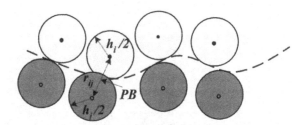

Figure 7.2 Interface particles on the interface of different materials, and particle penetration.

The penalty force employed is similar to the molecular force of Lennard-Jones form and is applied pairwisely on the two approaching particles along the centerline of the two particles.

$$PB_{ij} = \begin{cases} \overline{p}(pe^{n_1} - pe^{n_2})\dfrac{x_{ij}}{r_{ij}^2} & pe \geq 1 \\ 0 & pe < 1 \end{cases} \qquad (7.12)$$

where the parameters $\overline{p}$, n_1, n_2 are taken as 10^5, 6 and 4 respectively. In fact, these parameters can be adjusted to suit the needs of different problems (Liu et al., 2001a), and thus the approaching velocity (in high velocity impact) or driven force (in underwater explosion problems) involved can vary in a very big range. The application of the penalty force combining the summation between the interface particles can well prevent the problems of unphysical penetration in the simulation of underwater explosion, though numerical oscillations near the vicinity of the interface can still exist.

7.5   Numerical examples

Two numerical examples of underwater explosions (UNDEX) are presented here. The first one involves a cylindrical TNT charge surrounded by water with a free circumferential surface, and the second one involves a square TNT charge confined in a frame of four solid boundaries.

Example 7.1   UNDEX of a cylindrical TNT charge

In this example, a cylindrical TNT charge is surrounded by water and then detonated at the center. The radius of the explosive charge is 0.1 m, and that of the problem domain is 0.5 m. The outside water boundary is treated as free surface (Figure 7.3).

Initially, 50 particles are deployed in the radial direction, of which the innermost 10 particles are gas particles and 40 particles are used for water. 60 particles are placed in the tangential direction. The water is initially at atmospheric condition, i.e., a pressure of 101.325 KPa and a temperature of 293 K. Therefore, the initial properties of all particles can be determined using the initial condition. The parameters in the artificial viscosity are the same as those used in the simulation of HE explosions in Chapter 6.

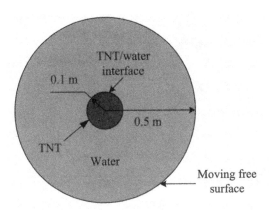

Figure 7.3 Initial geometry of a cylindrical TNT charge in water. 50 particles are deployed in the radial direction, of which the innermost 10 particles are TNT explosive particles and 40 particles are used for water. 60 particles are placed in the tangential direction.

Figure 7.4 shows the pressure distributions in the gas and water as well as the shock waves at the instants of 0.08 ms and 0.12 ms. Corresponding density and velocity distributions are shown, respectively, in Figure 7.5 and Figure 7.6. Good comparisons with MSC/Dytran (1997) are obtained for both the location of the shock wave fronts and the magnitude of the peak of the shock wave. The shock waves resulted from the SPH method are fairly steep, and behave as a step increase followed by an approximately exponential decay. The peak pressure is located immediately behind the shock front and the decay length increases with distance. It is also observed that the pressure behind the shock front obtained by the SPH method is higher that that obtained using the MSC/Dytran (see, Figure 7.4). The same can also be said for the density and velocity, as shown in Figure 7.5 and Figure 7.6.

A numerical blip or oscillation of within several smoothing length can be seen in Figure 7.4. Though this numerical blip is acceptable compared to the magnitude of the shock wave, it may cause errors to the pressure transient especially in the explosive gas. The numerical oscillation in the material interface region is regarded as a characteristic of the SPH method by Campbell (2000). Our investigation found the numerical oscillation appears also in the MSC/Dytran solution, as shown in Figure 7.4, Figure 7.5 and Figure 7.6.

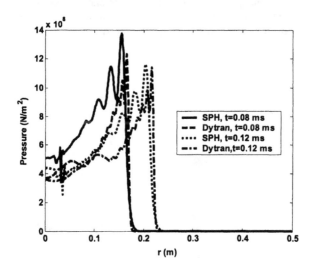

Figure 7.4 Results for the explosion of a cylindrical TNT charge in water. Pressure distributions in the explosive gas and water as well as the shock wave fronts at $t = 0.08$ ms and 0.12 ms. (From Liu et al., Computational Mechanics, 30(2):106-118, 2003. With permission.)

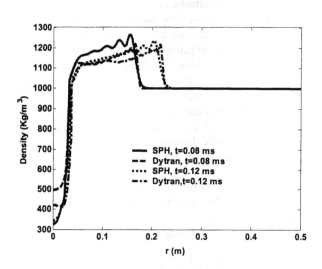

Figure 7.5 Results for the explosion of a cylindrical TNT charge in water. Density distributions in the explosive gas and water at *t* = 0.08 ms and 0.12 ms. (From Liu et al., Computational Mechanics, 30(2):106-118, 2003. With permission.)

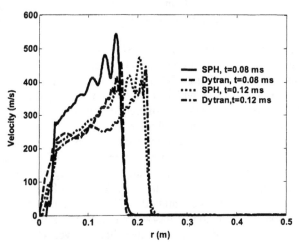

Figure 7.6 Results for the explosion of a cylindrical TNT charge in water. Velocity distributions in the explosive gas and water at *t* = 0.08 ms and 0.12 ms. (From Liu et al., Computational Mechanics, 30(2):106-118, 2003. With permission.)

It has been found that for other numerical methods of discretization type, there will be also numerical oscillations in simulating phenomena of very sharp field variables, such as shock waves. Our numerical experience in carrying out explosion related projects (Liu and Lam et al., 1998; Lam and Liu et al. 1996) indicated that the amplitude and the pitch of oscillation depend on the density of the discretization (mesh). The finer the discretization, the smaller amplitude and the shorter pitch of the oscillation will be. From Figure 7.4, Figure 7.5 and Figure 7.6, we also found that the numerical oscillations on the curves obtained using SPH are larger than those on the curves by MSC/Dytran. This is due to the much finer cell resolution (1500 cells along the radial direction; quasi-one-dimensional model) in the MSC/Dytran simulation, compared to that (50 particles along the radial direction; two-dimensional model) in the SPH simulation.

This numerical oscillation can be reduced by adjusting factors such as the particle resolution, time step, the predefined number of neighboring particles (thus, the smoothing length), the artificial viscosity coefficient, etc. The large inhomogeneities between the high-pressure gas and the water also cause the numerical oscillation. In Campbell's contact algorithm for impact problem (Campbell, 2000), the numerical oscillation was also observed. Compared to the impact problems in which the impacting bodies are of the same material, the numerical oscillation in underwater explosion is more serious due to the large inhomogeneities. Another cause of the numerical oscillation is the penalty forces used in the interface treatment to prevent unphysical penetration. A good model of the penalty force is still under investigation to effectively prevent unphysical particle penetration, and to control the resulted numerical oscillation in a minimal or acceptable level.

Adjustment of the parameters in the SPH simulation may reduce the magnitude of the blip, but cannot completely remove the oscillation. The numerical results presented in Figure 7.4, Figure 7.5 and Figure 7.6 are regarded as our standard SPH solution, in which the total number of particles used is as much as 3000, the predefined number of neighboring particles is the default number of 21 corresponding to the cubic spline smoothing function in two-dimensions, and the time step is decided according to equation (6.11). The effects on the numerical oscillation resulted from the use of doubled particles (6000) in the circumferential direction, doubled neighboring particles (42), and halved time steps are shown in Figure 7.7, Figure 7.8 and Figure 7.9, respectively. Comparing with the standard SPH solution, it is found that, the numerical oscillation in the region of gas/water interface has little effect on the magnitude and the location of the underwater shock wave, which is the most important physics that should be concerned. This may be explained as followings.

The numerical oscillation in the region of gas/water interface occurs only around several smoothing lengths and is a local characteristic of SPH. The shock wave is the physics that represents the global effect of the interaction between gas and water. The local numerical oscillation around the interface is gradually smoothed, and thus will have little effect on the shock wave.

In terms of the computational time, we failed to conduct an accurate and conclusive study, as the computers that run the MSC/Dytran and our SPH code are very much different. In addition the models used in these two codes are also different. Based on our rough estimation, the subjective finding for this example is that the SPH method is around two times faster than the commercial software package MSC/Dytran. As discussed earlier, SPH is a particle method. Once the initial particle positions are determined, the system can evolve in an adaptive manner according to the external force and internal interaction. No numerical integration is required in the entire process of computation. On the other hand, MSC/Dytran is based on the grid-based numerical method (FVM), where numerical integrations over the mesh cells is needed. The numerical integration may be responsible for the poorer computational efficiency.

Note that no exact comparison has been conducted in our study since our SPH code is implemented in PC, while the MSC/Dytran runs on a main frame computer. The parameter settings may also be quite different.

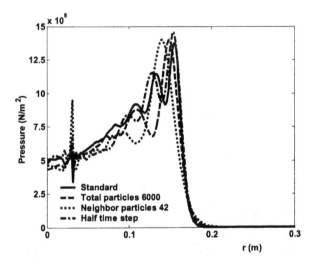

Figure 7.7 Results for the explosion of a cylindrical TNT charge in water. Pressure distributions in gas and water at 0.08 ms for different cases. (From Liu et al., Computational Mechanics, 30(2):106-118, 2003. With permission.)

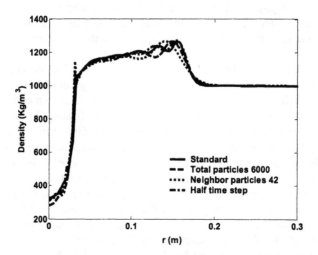

Figure 7.8 Results for the explosion of a cylindrical TNT charge in water. Density distributions in gas and water at 0.08 ms for different cases. (From Liu et al., *Computational Mechanics*, 30(2):106-118, 2003. With permission.)

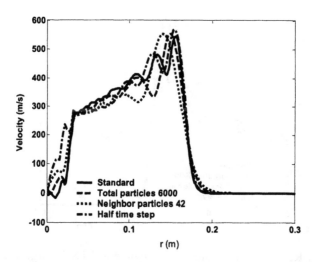

Figure 7.9 Results for the explosion of a cylindrical TNT charge in water. Velocity distributions in gas and water at 0.08 ms for different cases. (From Liu et al., *Computational Mechanics*, 30(2):106-118, 2003. With permission.)

Example 7.2   UNDEX of a square TNT charge

In this example, we consider a standard benchmark problem of UNDEX in two-dimensional space. A square shaped TNT charge (0.1 m × 0.1 m) explodes in water confined in a rigid square frame of 1 m × 1 m. The explosion process and its effects are simulated using the SPH method.

The initial particle distribution used in this simulation is shown in Figure 7.10, where the inner filled circles represent TNT charge of square shape, and others represent water particles that surround the TNT charge. The solid wall boundary conditions are applied in this example and are enforced using virtual particles (type I) as discussed in Chapter 4.

Figure 7.11 shows the pressure distribution at six representative instants in the underwater explosion process. The initial outwards propagating shock wave, reflection wave from the solid wall, explosive gas expansion and later compression can all be seen from these figures. Right after the detonation, a shock wave is generated in the water and propagates outwards. At the same time, the rarefaction wave is also produced within the explosive gas and propagates inwards. With the advancement of the shock wave through the water, the gas bubble expands within the surrounding water. The shock wave reaches the solid wall at around the instant of 200 μs, and then is reflected from the solid wall. This reflection wave propagates inwards and tends to compress the expanding gas. At about 400 μs, the gas bubble reaches to the maximum size and then starts to contract. With the continued process of shock reflection and explosive gas contraction, the gas bubble will reach to a minimal size, and then expands again. This repeating expansion and contraction continue for many circles, and will finally reach the equilibrium at a later time step.

Figure 7.12 shows the velocity distribution in the explosion process. The arrow and length of the arrow represent the direction and the magnitude of the velocity vector, respectively. The starting points of the arrows are the current particle positions. Phenomena such as the shock wave propagation and reflection, water compression and movement, explosive gas bubble expansion and contraction can be observed in the figure. The disturbance in the water by the propagation of the shock wave is obvious. It is noted that with the process of the explosion, the initially square-shaped explosive gas gradually turns to be circular. Since the shock wave reaches and reflects from the square wall at different instants and different positions, the gas bubble size and shape change with respect to the pressure evolutions.

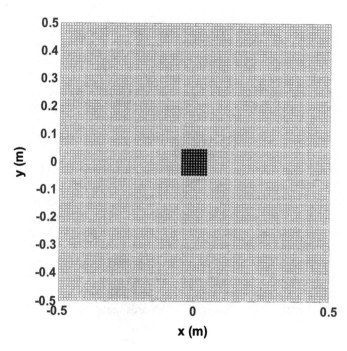

Figure 7.10 Initial particle distributions for the UNDEX of a square TNT charge
(0.1 m × 0.1 m) that explodes in water confined in a rigid square frame of
1 m × 1 m.

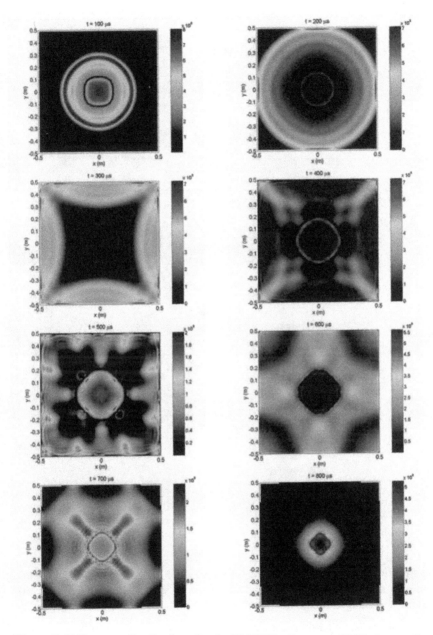

Figure 7.11 Pressure distributions for the UNDEX of a square TNT charge of 0.1 m × 0.1 m confined in a rigid frame of 1 m × 1 m filled with water. (From Liu et al., Computational Mechanics, 30(2):106-118, 2003. With permission.)

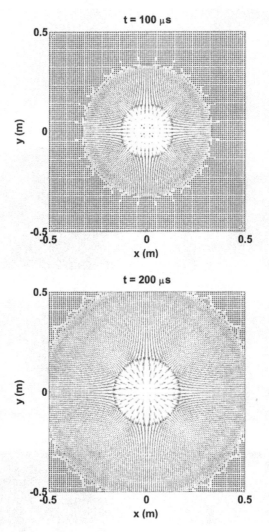

Figure 7.12 Velocity distributions for the UNDEX of a square TNT charge of
0.1 m × 0.1 m confined in a rigid frame of 1 m × 1 m filled with water. (From
Liu et al., Computational Mechanics, 30(2):106-118, 2003. With permission.)

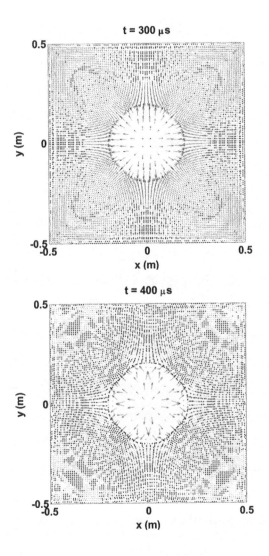

Figure 7.12 Velocity distributions for the UNDEX of a square TNT charge of
0.1 m × 0.1 m confined in a rigid frame of 1 m × 1 m filled with water. (From
Liu et al., Computational Mechanics, 30(2):106-118, 2003. With permission.)

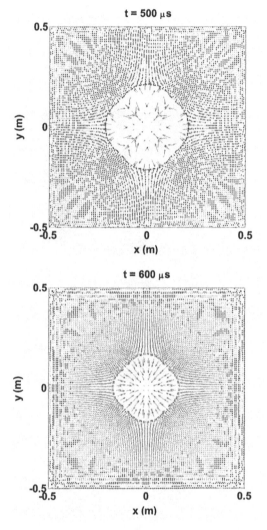

Figure 7.12 Velocity distributions for the UNDEX of a square TNT charge of
0.1 m × 0.1 m confined in a rigid frame of 1 m × 1 m filled with water. (From
Liu et al., Computational Mechanics, 30(2):106-118, 2003. With permission.)

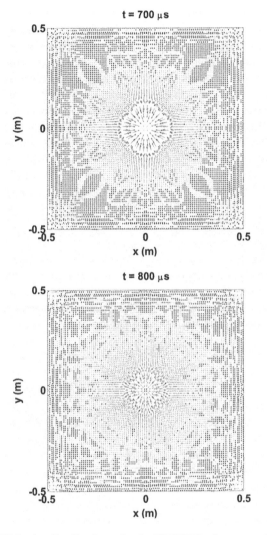

Figure 7.12 Velocity distributions for the UNDEX of a square TNT charge of 0.1 m × 0.1 m confined in a rigid frame of 1 m × 1 m filled with water. (From Liu et al., Computational Mechanics, 30(2):106-118, 2003. With permission.)

Figure 7.13 shows the gas bubble evolution and pulsation in the explosion process. In the figure, the gas bubble size is obtained from the maximum radial coordinate of the explosive gas particle from the gas/water interface. The gas bubble size and shape obtained by SPH agree well with those obtained by MSC/Dytran. The bubble period and size obtained by SPH is a little larger than those by MSC/Dytran. If the gas bubble size is taken as the equivalent radius of the area of the explosive gas, the agreement should be better. Note that in this example, the repeating expansion and contraction of the gas bubble are affected by the repeating shock wave propagation and reflection from the solid wall. While in practical underwater explosions, the movement of the gas bubble is usually fully controlled by the gas pressure inside the bubble and the hydrodynamic pressure from outside (water depth plus atmospheric pressure). If the inner gas pressure is bigger than the outside hydrodynamic pressure, the gas bubble will expand outwards. In contrast, when the outside hydrodynamic pressure is larger than the inner gas pressure, the gas bubble will contract inwards. This outwards expansion and inwards contraction will also last for circles, and finally the gas bubble reaches an equilibrium state. The UNDEX in this confined chamber yields much smaller bubble period, rather than the usual tens of milliseconds' bubble pulsation period observed in practical underwater explosions in infinite media of water at open sea.

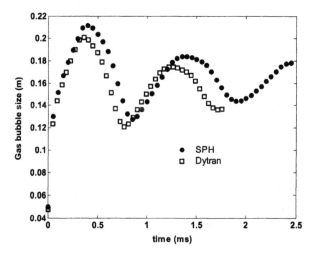

Figure 7.13 Explosive gas bubble evolution and pulsation for the UNDEX of a square TNT charge of 0.1 m × 0.1 m confined in a rigid frame of 1 m × 1 m filled with water. (From Liu et al., Computational Mechanics, 30(2):106-118, 2003. With permission.)

7.6   Comparison study of the real and artificial HE detonation models

Most of the early numerical analyses of underwater explosions employed an *artificial detonation model* of adiabatic explosion. In this model, an ideal gas with the same volume and the same amount of energy as the original explosive charge is considered (Lam and Liu et al. 1996). The detonation process of the explosive is not really simulated. In this artificial detonation model, the detonation process of the HE is neglected with assumptions that the detonation velocity is infinite, and the original HE is replaced by or in a sudden converted to a gas globe of extremely high pressure and temperature. The gas globe possesses the same energy, occupies the same space as the original HE, interacts with the surrounding water, and then produces shock waves as well as bubble pulsation in the water. Due to the assumptions or simplifications on the real detonation process, this artificial detonation model does not give proper pressure profiles in the detonation produced gas products.

A better model (*real detonation model*) may be use to simulate the real detonation process of the HE in the entire underwater explosion simulations. In some numerical simulations (Mader, 1979; Shin and Chisum, 1997; Liu and Lam et al. 1998), the real detonation process of the HE is included into the entire underwater explosion simulations, and thus the pressure profiles in the explosive gas can be numerically determined.

These two models of the detonation process of HEs yield different physics in the explosive gas, and may lead to different results in underwater explosion simulations. However, to the best of the author's knowledge, no other detailed investigation has ever been carried out on the different effects of these two detonation models and their influences on the results of the entire underwater explosion shock simulations.

In this section, the real and artificial detonation models as well as their influences on the entire underwater explosion shock simulations are investigated using the SPH method. The numerical analyses are carried out for the TNT explosion in water with one case of a slab charge and another case of a spherical charge.

Example 7.3   One-dimensional TNT slab

The difference and concern of the two detonation models lies in whether the real detonation process has been included into the entire underwater explosion simulations or the real detonation process is simplified as an artificial detonation condition of adiabatic explosion at a constant volume. In Chapter 6, a one-dimensional TNT slab detonation process was successfully simulated using the SPH method (Example 6.1). In that example, the real detonation model was employed. In this chapter, this TNT slab detonation problem is revisited with the artificial detonation model. Again, the TNT slab is 0.1 m in length.

In the investigation of the real and artificial detonation models, the simulation set up for the TNT slab is the same as that in Example 6.1. Figure 7.14 and Figure 7.15 show the pressure and density along the slab at intervals of 1 μs from 1 to 14 μs. The results are obtained using 4000 particles. The dashed line in Figure 7.14 represents the experimentally determined C-J pressure. For this one-dimensional TNT slab detonation problem, the C-J pressure is 2.1×10^{10} N/m^2. It can be seen from Figure 7.14 that the gas behind the detonation wave front has a pressure profile of a step increase. The pressure of the gas decays exponentially. The peak pressure is located immediately behind the detonation wave front and the decay length increases with the distance traveled by the detonation wave. With the process of the detonation, the detonation pressure converges to the C-J pressure. The detonation shock front is resolved within several smoothing lengths. Further investigation reveals that more particles along the slab result in an even sharper pressure fronts with bigger peak pressures.

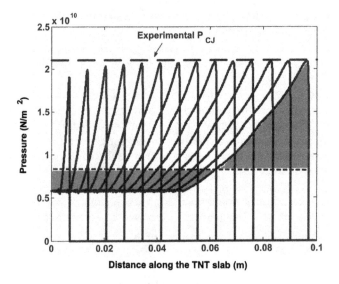

Figure 7.14 Pressure profiles along the one-dimensional TNT slab calculated using the SPH method. The TNT slab is ignited from the left end. The pressure profiles are plotted at 1 μs intervals from 1 μs to 14 μs. The shaded areas are only for the instant of $t = 14$ μs.

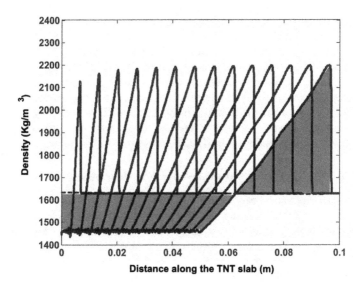

Figure 7.15 Density profiles along the one-dimensional TNT slab calculated using the SPH method. The TNT slab is ignited from the left end. The pressure profiles are plotted at 1 μs intervals from 1 μs to 14 μs. The shaded areas are only for the instant of $t = 14$ μs.

For the simplified artificial detonation, the high explosive is replaced by explosive gas in a sudden, which occupies the same volume with the same energy, and the density also remains the same, as shown by the dotted line on Figure 7.15. It is different from the above calculated density profile that behaves as a step increase followed by an exponential type decay. The gas particles are not in motion at this instant, and hence the particle velocity is zero. The pressure of the gas calculated using the JWL equation of state (equation (7.2)) with $\rho/\rho_0 = 1$ is around 8.3759×10^9 N/m^2 and is shown by the dotted line on Figure 7.14. This value is between the above calculated forward peak pressure and the backward steady pressure. It is much smaller than the peak pressure, but bigger than the steady pressure. Since the detonation process is neglected in this artificial detonation model, no detonation shock will advance along the TNT slab. The pressure suddenly rises to a very high level with its front at the gas/water interface.

Besides the different distribution of the physical variables, the real detonation model also produces the two shaded areas both in Figure 7.14 and Figure 7.15 at different instants. The two shaded areas are the sections intersected by the pressure/density profiles resulted using the two detonation models. The shaded areas in Figure 7.14 and Figure 7.15 are for the results at the instants of 14 μs. It is found that the upper shaded area is approximately equal to the lower shaded area in the density profile shown in Figure 7.15. This is a reflection of the mass conservation in these two models. The upper shaded area in Figure 7.14 is bigger than the lower shaded area. This suggests that when considering the detonation process, the gas will exert a bigger pressure on the surrounding water after the detonation is completed. It is reasonable to presume that the differences of the distribution of the field variables in the high explosive gas will lead to different outcomes when interacting with the surrounding water, and may lead to different behavior of the shock waves in the water.

Example 7.4   UNDEX shock by a TNT slab charge

The TNT slab detonation problem is further extended to underwater explosion problem (with water surrounding the TNT slab charge as shown in Figure 7.16). After the detonation of the high explosive, the explosive gas of high pressure, temperature and velocity tends to move rightwards, and interacts with the surrounding water. For this example of TNT slab detonation, the interacting expansion process of the explosive gas with the outside water is investigated. In the simulation, the problem domain of water is up to 10 m, which is large enough to prevent the effect of the boundary at the right end of the problem domain.

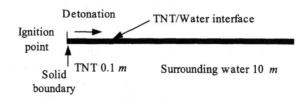

Figure 7.16 Underwater explosion of a TNT slab charge.

Figure 7.17 shows the peak shock pressure in the water at different locations. In the figure, *a* is the length of the original HE charge length (0.1 m). *R* is the distance from the pressure observation location to the detonation end of the charge. Figure 7.18 is a zoom-in figure showing the peak shock pressure for locations from 5 to 15 times the charge length. It can be seen that the peak pressures in both models decrease as the shock waves move away from the explosive gas. The peak pressures decrease in the form of an exponential decay. At locations closer to the charge or explosive gas, the peak pressure from the real detonation model is much bigger than that from the artificial model. As the shock waves advance ahead, the difference becomes smaller. Beyond the range of 8-10 times the charge length, the difference is very small and can be negligible.

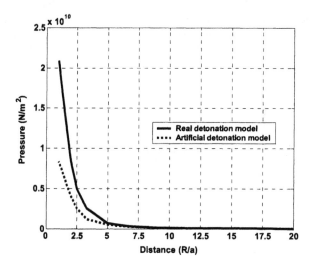

Figure 7.17 Peak pressures at different locations obtained using the SPH method for UNDEX of a TNT slab charge in water.

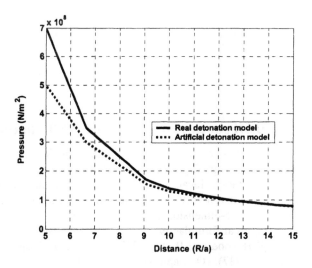

Figure 7.18 Zoom-in figure showing the peak shock pressure at locations from 5 to 15 times the charge length for UNDEX of a TNT slab charge in water.

Figure 7.19 shows the pressure history observed at the location of 0.25 m away from the detonation point, which is 2.5 times the length of the original TNT charge. It is obvious that the peak pressure as well as the pressure curve for the real detonation model is higher than that for the artificial detonation model

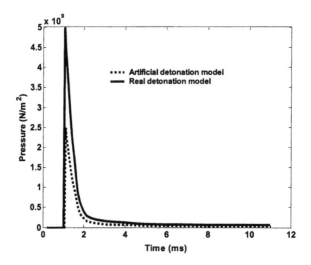

Figure 7.19 Pressure history at the location of R/a=2.5 obtained using the SPH method for UNDEX of a TNT slab charge in water.

Example 7.5   UNDEX shock with a spherical TNT charge

It is observed that the difference of the two detonation models on the entire underwater explosions differs greatly only at earlier stages or nearer locations, and is negligible at very far away locations. To further validate this observation, another numerical example that involves in a spherical TNT charge of 137 Kg is examined. For this quasi one-dimensional spherical charge detonating from the charge center, some empirical formulae exist to predict the peak shock pressure at different locations (Kirkwood and Bethe, 1942; Penney and Dasgupta, 1942; Brinkley and Kirkwood, 1947). One approach is the Penney-Dusgupta theory, which numerically integrate the Riemann equations outwards from the charge. According to the Penney-Dusgupta theory for TNT, the value of the peak pressure is given as a function of shock radius R and the charge weight W in the following form

$$p_m = 2.86 \times 10^7 (W^{\frac{1}{3}}/R) e^{0.108 W^{\frac{1}{3}}/R} \tag{7.13}$$

where p_m is the peak shock pressure.

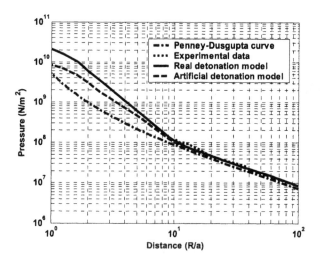

Figure 7.20 Peak pressures at different locations for UNDEX of a spherical TNT charge in water.

Figure 7.20 shows the detailed comparisons of peak pressures obtained from different sources, experimental data from Cole (1948), the Penney-Dusgupta theoretical value, and the SPH simulation results with two detonation models. In the logarithm scaled figure, a is the radius of the original HE charge length; R is the distance from the pressure measure location to the charge center. For the region of $R/a>10$, the Penney-Dusgupta curve can be approximated as a straight line, and agrees well with the experimental data. The peak pressure becomes higher for closer distances. It should be noted that the Penney-Dusgupta curve is only valid for the region of $R/a >10$; for the region of $10> R/a >1$, since the experimental data is not available, Penney-Dusgupta theory needs to be further verified. In the SPH simulation, the peak pressure locations in water can be obtained from the gas/water interface ($R/a=1$) to further distances. The numerical results from SPH simulation also agree well with the experimental data and the Penney-Dusgupta curve for the region $R/a>10$. The peak pressure curves from the two detonation models are both above the Penney-Dusgupta curve, and are closer to the experimental data than the Penney-Dusgupta curve. Similar to the above case of TNT slab explosion in water, the peak pressures in both models decrease as the shock waves propagate ahead. At locations nearer to the charge or explosive gas, the peak pressure for the real detonation model is much bigger than that for the artificial model. Beyond the range of 8-10 times the charge radius, the difference is very small and can be negligible.

7.7   Water mitigation simulation

7.7.1   Background

The detonation of high explosives produces extremely intense shock waves in a very short duration. The blast effects arising from the detonation either in air or underwater are of great significance to nearby personnel and structures. Researchers both from the defense and the academic areas have been looking for ways to effectively mitigate blast effects and thus reduce the consequent damages.

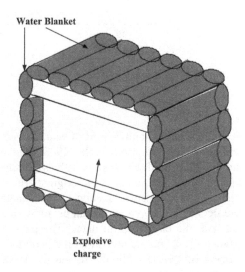

Figure 7.21 An illustration of water mitigation: water held in plastic bags as shelter covering the explosives.

One of the comparatively new concepts is the so-called water mitigation, in which a shield of water is placed in the near proximity of explosives, not necessarily in contact, to change the generation and propagation of the air blast waves. Figure 7.21 illustrates water held in plastic bags as a bomb shelter to mitigate accidental blast effects. A water shield can be used not only as an effective barricade that stops the debris and splinters, but also as an excellent blast mitigator to reduce blast shock, and the internal equilibrium gas pressure if the explosion happens in a confined space. Practically water can be either held in plastic bags for short-term applications such as bomb disposal or held in plastic tanks for long-term applications especially for magazines where accidental events have a big chance to happen.

Water mitigation is a more complicated non-linear wave propagation phenomenon, compared to an ordinary underwater explosion problem or an acoustic water-air interface problem. For ordinary underwater explosion, one computational cell (for grid-based methods) contains at most two materials (explosive gas and water or water and air). In the SPH method a particle may feel the influence from the same kind of particles and at most another kind of particles. On the contrary, for water mitigation problems, one computational cell (for grid-based methods) may contain three materials (explosive gas, water and air), and one particle (for SPH) may interact with up to three kinds of particles. The calculation of the multi-components in the extremely transient water mitigation events is more difficult than the simulation of ordinary underwater explosion. It involves the HE detonation, interaction of the explosive gas, water shield and the outside air. The detonation-produced blast waves propagate outwards in all directions and aerosolize the surrounding water shield. Since the density of the water is much higher than that of the air, when the particles of the explosive gases strike the water particles, the gas particles will be suddenly slowed down due to the momentum exchange between the gas and water particles. The water jacket actually delays the arrival of the shock wave and reduces the magnitude of the peak shock pressure. The aerosolized water expanding with the explosive gas absorbs the detonation energy of the explosive charge, and thus reduces the peak shock pressure. In the case of detonation in a rigidly confined chamber, the equilibrium gas pressure will be also reduced

Many feasibility tests for mitigation problems have been carried out. However, due to the intense interest in this field in the defense related applications, very few documents are publicly available. The NCEL tests (Keenan and Wager, 1992) were conducted at the David Taylor Research Center for the Naval Civil Engineering Laboratory (NCEL). In the tests, 4.67 lb. of TNT explosive were tested in a confined chamber, first without, and then surrounded by water. Both the equilibrium gas pressure and impulse were reported to be reduced by about 90%. In the USACE Huntsville Tests (Marchand et al., 1996), water bags were added around unexploded ordnance in a container. It has reported that the equilibrium gas pressure was successfully reduced by approximately 70%. In the small-scale Alvdalen tests (Forsen et al., 1996), it is also observed that putting water around explosives, either in contact or not, leads to a pressure reduction. While Large-scale Alvdalen tests (Forsen et al., 1997; Hansson and Forsen, 1997) did not show any effect of water mitigation due to other variations in the experimental setup. Therefore, the studies conducted so far have not yet been very conclusive, and the system configuration will have very important influence on the water mitigation effects.

To develop an efficient water mitigation system, more parametric studies are clearly necessary. Numerical simulation provides effective means to reduce greatly the number of expensive and dangerous firing tests. However, the documents related to numerical simulations of water mitigation are also very limited, on one hand, for the sake of confidentiality of actual applications, on the other hand, due to the numerical difficulties inherent in the water mitigation simulations. Challenges such as large deformations, large inhomogeneities, and moving interfaces confront the investigators.

Chong and his colleagues, in a series of papers (Chong et al., 1998a; 1998b; 1999), investigated the water mitigation effects on an explosion inside a vented tunnel system and ascertained the mitigation effects of water in reducing maximum peak shock pressure due to an explosion. Shin et al. (1998) investigated a water mitigation problem in one-dimensional spherical infinite problem domain using the multi-material Eulerian finite volume technique in the software MSC/Dytran. Their numerical results showed that the peak shock pressure generally decreases and the shock arrival time increases. Zhao studied the water mitigation effects on the detonation in a confined chamber (Zhao, 1998) as well as the water effects on shock wave delay in the free field (Zhao, 2001). It is reported that the peak shock pressure is reduced by about 17-46% when explosives are surrounded by water with an amount of 1-5 times the explosive mass. Some other references (Crepeau and Needham, 1998; Malvar and Tancreto, 1998) provided numerical simulations and comparisons with experimental data. The emerged literature usually employs hydrocodes, which generally use domain discretization that is fixed in space, but allow for mass, temperature and pressure transfer across cells (Eulerian mesh).

Liu et al. (2002a) has extended the SPH method to simulate water mitigation problems. In this section, the water mitigation problems are investigated using the SPH method. The objective of the work is, firstly, to probe the feasibility of using SPH to simulate water mitigation problems, and secondly, to investigate water mitigation phenomena as well as the relationship of water mitigation effects and the system geometrical configuration. Water mitigation simulations are carried out for a square shaped TNT charge detonating in a confined square chamber with different setups of water shield and/or air gap.

7.7.2   Simulation setup

In the simulation, a square shaped (0.05 m × 0.05 m) TNT charge detonates and then expands outwards in different cases. Figure 7.22 shows one quarter of the initial geometry of the two-dimensional models. For this planar symmetry problem, $x = 0$ and $y = 0$ represent the symmetric planes. Four observation points are chosen to measure the pressure and shock wave arrival time at the concerned positions. Taking the center point of the geometry as the origin, the locations of the four points as well as their distances to the center points normalized by the explosive thickness (0.025 m) are listed in Table 7.3.

Table 7.3 Locations of the observation points marked in Figure 7.22

Point	Coordinate	Normalized distance*
1	(0, 0.1)	4.0
2	(0, 0.2)	8.0
3	(0, 0.3)	12.0
4	(0.2, 0.2)	11.3

*Normalized by the explosive thickness 0.025 m.

In the simulation, three different cases are modeled for comparison study. The first case (Figure 7.22a) is the explosive detonation in air (without water shield). The second case (Figure 7.22b) is water mitigation with a water shield directly in contact with the explosive charge. The last one (Figure 7.22c) is non-contact water mitigation with an air gap between the explosive and the water shield. In the contact water mitigation, the water shield thickness is varied from 0.01 m to 0.045 m, which corresponds to 0.4 to 1.8 times the explosive thickness. For the case of a non-contact water shield, the mitigation effects for a water shield thickness of 0.025 m and 0.05 m are investigated at the same air gap thickness of 0.025 m. For these three cases, the simulations are carried out in a rigidly confined chamber. Actually, free field water mitigations can also be investigated, and are easier than the confined water mitigation since the SPH method is more natural to treat moving free boundaries. However, in free field water mitigation, only the reducing effect in the peak shock pressure can be investigated. The final equilibrium gas pressure will, with the expansion of the volume, decrease to as low as the atmospheric pressure. In contrast, in confined water mitigation, the two important indices, peak shock pressure and equilibrium gas pressure, can all be investigated. Therefore, this chapter only presents confined water mitigation cases.

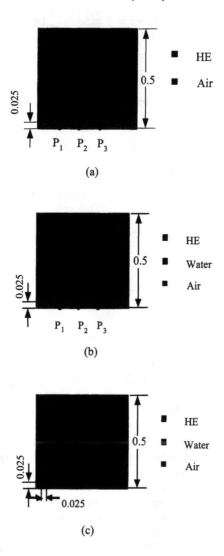

Figure 7.22 One quarter of the geometrical configuration for water mitigation of the blasting effects. Unit: m. (From Liu et al., Shock Waves 12(3):181-195, 2002. With permission.)

7.7.3 Simulation results

Example 7.6 Explosion shock wave in air

An explosion produces blast shock waves in the air that can destroy nearby structures and may lead to losses of lives. Inside a container such as a building it produces a high gas pressure. After the detonation of the explosive charge (TNT), the pressure shock wave advances outwards with rarefaction waves propagating inwards in the inner gas.

Figure 7.23 shows the particle distribution in the explosion process at six representative instants. The configuration of this problem is shown in Figure 7.22a. The small circles and squares represent explosive gas particles and air particles, respectively. The initial particles are exponentially distributed outwards so as to ensure that the gas particles are sufficient though its geometry is much smaller than the entire problem domain. The original square-shaped explosive gas gradually turns to be circular. The air particles near the gas/air interface are highly compressed due to the impulsive driving pressure. At points of the same distance from the origin, the particle velocities around the symmetric planes $x = 0$ and $y = 0$ are higher than those around the symmetric planes $x = y$, while the particle pressure around the symmetric planes $x = 0$ and $y = 0$ are lower than those around the symmetric planes $x = y$.

Figure 7.24 plots the pressure time history for the four observation points. As expected, the calculated pressures at these four points initially stay around 1 bar, later suddenly rise to the peak values, and then quickly decay exponentially. With the propagation of the pressure shock wave, the peak shock pressures decrease and the shock wave becomes obtuse at further locations. At further points, the shock arrives at later stages with a slower decay rate, since the pressures there are lower and the dispersion volume is increased.

Since the chamber is confined by a rigid frame, at very later stages, the shock waves dies off after many cycles of forward and backward reflections from the rigid walls, and interaction of the forward and backward shock waves. The mixture of gas and air gradually stays in a thermodynamic equilibrium of constant gas pressure. In this case, the calculated equilibrium gas pressure is around 48 bar.

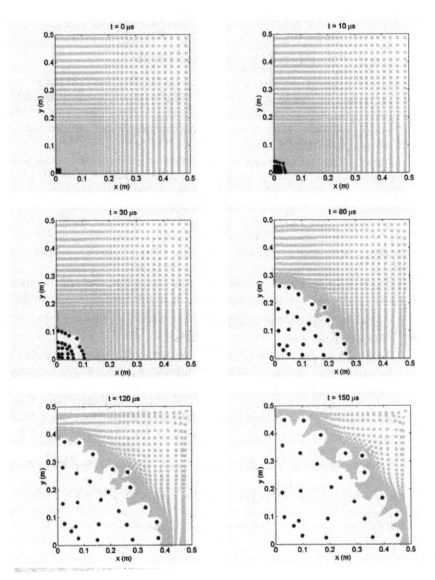

Figure 7.23 Particle distributions in the explosion process in air in a confined space of 1 m × 1 m. One quarter of the model is shown. Circles represent explosive gas particles; squares represent air particles. (From Liu et al., Shock Waves 12(3):181-195, 2002. With permission.)

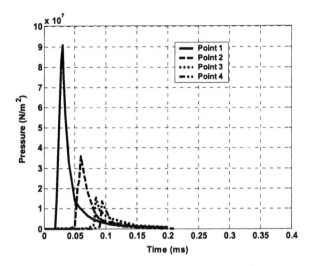

Figure 7.24 Pressure-time histories of the four observation points (see, Figure 7.22) in the explosion process in air. (From Liu et al., Shock Waves 12(3):181-195, 2002. With permission.)

Example 7.7 Contact water mitigation

In this example, the mitigation effect of the water shield directly in contact with the explosive gas is investigated. The configuration of this problem is shown in Figure 7.22b. In the simulation, 100×100 particles are initially deployed exponentially (with the exponential factor 1.01) in the problem domain (one quarter of the geometry). It has also confirmed that more particles produce nearly the same results, the numerical results obtained using 100×100 particles are regarded as the convergent solution.

Different water shield thicknesses are examined. Figure 7.25 and Figure 7.26 show the peak shock pressure and shock arrival time at the four observation points. The peak shock pressure and shock arrival time are normalized by the corresponding peak shock pressure and shock arrival time without water at the same points. The abscissa is the normalized water shield thickness by the initial explosive TNT thickness 0.025 m. As anticipated, the shock arrival time increases with the increase of the water shield thickness. It can be seen from Figure 7.25 that putting water around explosive does lead to mitigations of the peak shock pressures at all the sampled locations. The mitigation effect is stronger at nearer points and weaker at points further from the explosive.

Varying the water shield thickness gives rise to different mitigation behavior. For a certain observation point, increasing water shield thickness up to a certain value results in a gradually decreasing peak shock pressure. The mitigation effect of peak shock pressure reaches the maximum when the normalized water shield thickness approaches one (or 0.025 m in absolute value). Beyond this value, increasing water shield thickness, on the contrary, leads to a worse mitigation effect.

In fact, since the density of the water is much higher than that of air (approximately 1000 times), when the explosive gas particles strike the water particles, the gas particle velocities will be suddenly slowed down with a rapid deceleration due to the momentum exchange. In the interaction process of the explosive gas and water particles, the water particles are compressed and heated. The momentum and energy exchange between the different particles mitigate the blast, and reduce the peak shock pressure. On the other hand, the water shield, as a barricade and a mitigator to the shock, confines the outwards burst of the gas particles. This confining effect actually raises the peak shock pressure and acts as an offset to the mitigation effect. When increasing the water shield thickness, the momentum and energy exchange gradually slow down while the confining effect becomes stronger. Therefore, an optimal value of water shield thickness exists, on either side of which the mitigation effect is weaker. In other words, increasing water does not necessarily always produce better mitigation effects. For the contact water mitigation case considered here, the normalized optimal water shield thickness is 1 (or 0.025 m in absolute value), where the surrounding mitigation water weighs 1.84 times the TNT charge. This finding is interesting for practical applications, and may provide a reference criterion for the design of water mitigation systems after taking into account the specific situations.

Figure 7.27 and Figure 7.28 show the density and pressure evolution at four time instants for the calculated optimal water shield thickness. The shock wave and rarefaction wave can be clearly seen from the figures. Similar to Example 7.6 (explosion in air without water), the initially square-shaped explosive gas and water particles gradually turn to be circular. With the propagation of the shock, the water particles are compressed into a thinner layer. The pressure decay rates around the symmetric planes $x = 0$ and $y = 0$ are greater than those around the symmetric planes of $x = y$.

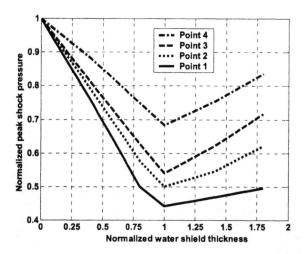

Figure 7.25 Normalized peak shock pressure using different water shield thicknesses. (From Liu et al., Shock Waves 12(3):181-195, 2002. With permission.)

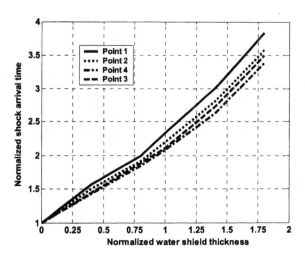

Figure 7.26 Normalized shock arrival time using different water shield thicknesses. (From Liu et al., Shock Waves 12(3):181-195, 2002. With permission.)

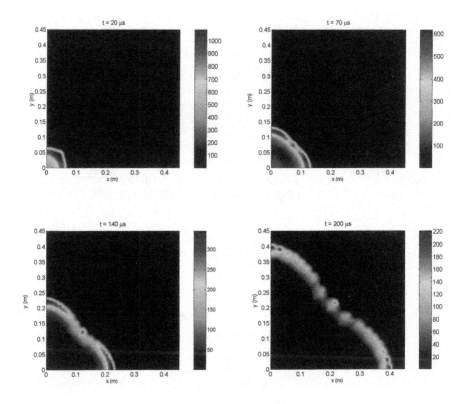

Figure 7.27 Density evolution at $t = 20$, 70, 140, and 200 µs for the contact water mitigation with water shield thickness of 0.025 m. (From Liu et al., Shock Waves 12(3):181-195, 2002. With permission.)

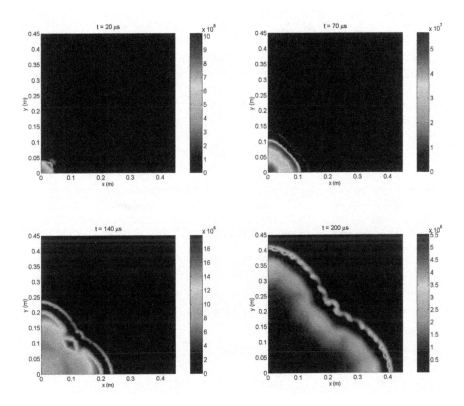

Figure 7.28 Pressure evolution at t = 20, 70, 140, and 200 µs for the contact water mitigation with water shield thickness of 0.025 m. (From Liu et al., Shock Waves 12(3):181-195, 2002. With permission.)

The final equilibrium gas pressure is determined by the thermodynamic equilibrium of the constituents of the explosive gas, water and air in the confined chamber. The equilibrium gas pressure at different water shield thicknesses normalized by 0.025 m is listed in Table 7.4. The equilibrium gas pressure reduces for the given water shield thickness. The maximum reduction occurs when the normalized water shield thickness is 1, and the equilibrium gas pressure is around 17.8% of the equilibrium gas pressure of the explosion in air. This can also be explained from two aspects. Water absorbs and dissipates the released energy from the explosive gas when being compressed and heated. More water means better energy absorption effect and thus lower equilibrium gas pressure. On the other hand, more water occupies more space, which in reverse leads to higher gas pressure. Making good use of these influences from the two aspects can produce the optimal effect by selecting an optimal water shield thickness.

Table 7.4 Equilibrium gas pressure for the contact water mitigation

Normalized Water shield thickness	Equilibrium gas pressure (E+5 N/m^2)	Normalized equilibrium gas pressure*
0	48.3	100%
0.4	35.6	73.7%
0.8	28.4	58.8%
1.0	8.6	17.8%
1.4	15.8	32.7%
1.8	26.8	55.5%

* Normalized by the equilibrium pressure without water.

Example 7.8   Non-contact water mitigation

When an air gap is placed in between the explosive and the water, the mitigation effect may be different from the above example of contact water mitigation. The relevant geometrical configuration of the air gap to those of the water shield and the explosive charge greatly changes the peak shock pressure at a certain location as well as the final equilibrium gas pressure. In this example, an air gap with the normalized thickness of 1 (0.025 m) is employed with two different normalized water shield thicknesses of 1 (corresponding to 0.025 m) and 2 (corresponding to 0.05 m).

Figure 7.29 and Figure 7.30 show the density and pressure evolutions at the instants of 20, 70, 150 and 250 µs with the normalized water shield thickness of 1. Comparing Figure 7.29 with Figure 7.27 and Figure 7.30 with Figure 7.28, it is found that the shock arrival time is obviously delayed. The peak shock pressures are further reduced especially for positions nearer to the explosive. The final equilibrium gas pressure is reduced to around 8 bar, that is around 16.6% of that in Example 7.6 (explosion in air without water).

Figure 7.31 and Figure 7.32 show the density and pressure evolution at the time instants of 20, 100, 220 and 350 µs with the normalized water shield thickness of 2. Though the shock arrival times are further delayed, the peak shock pressures, in reverse, are larger than those in the case of contact water mitigation. The calculated final equilibrium gas pressure is 11.4 bar, that is around 23.6% of the equilibrium gas pressure without water mitigation.

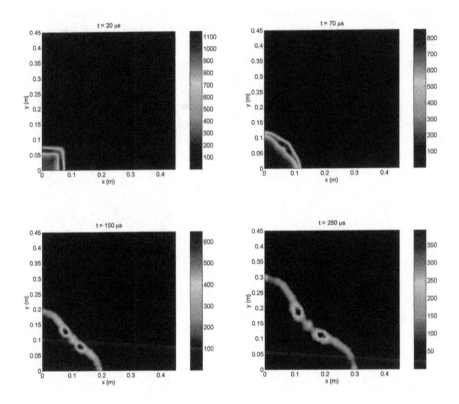

Figure 7.29 Density evolution at t = 20, 70, 150, 250 µs for the non-contact water mitigation with the normalized air gap thickness of 1 and the normalized water shield thickness of 1. (From Liu et al., Shock Waves 12(3):181-195, 2002. With permission.)

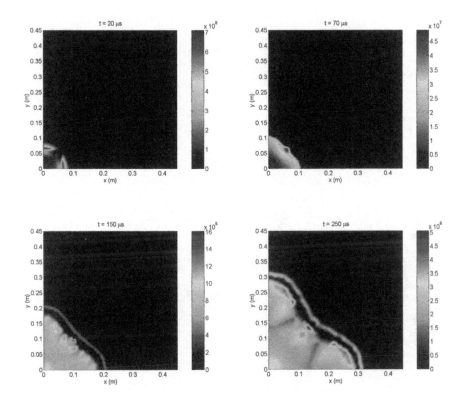

Figure 7.30 Pressure evolution at *t* = 20, 70, 150, 250 μs for the non-contact water mitigation with the normalized air gap thickness of 1 and the normalized water shield thickness of 1. (From Liu et al., Shock Waves 12(3):181-195, 2002. With permission.)

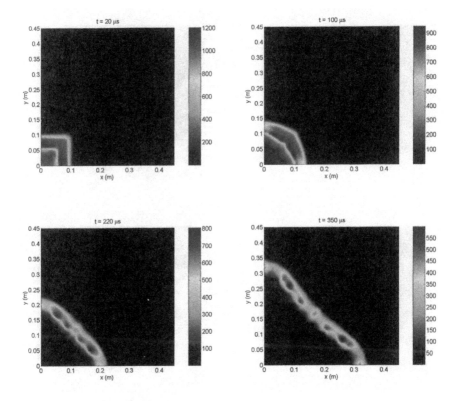

Figure 7.31 Density evolution at t = 20, 100, 220, 350 μs for the non-contact water mitigation with the normalized air gap thickness of 1 and the normalized water shield thickness of 2. (From Liu et al., Shock Waves 12(3):181-195, 2002. With permission.)

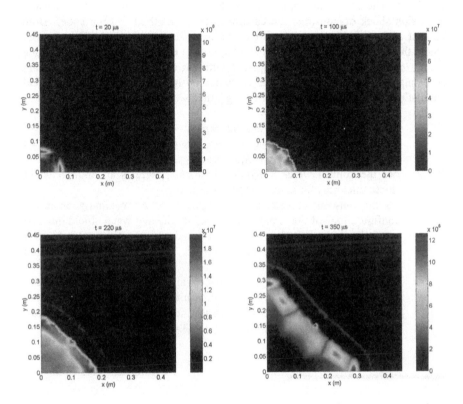

Figure 7.32 Pressure evolution at t = 20, 100, 220, 350 µs for the non-contact water mitigation with the normalized air gap thickness of 1 and the normalized water shield thickness of 2. (From Liu et al., Shock Waves 12(3):181-195, 2002. With permission.)

7.7.4 Summary

As a relatively new concept in mitigating the blast effects arising from HE detonations, water has been increasingly used so as to reduce the resultant damages. In this chapter, the SPH method is applied to simulate water mitigation problems. The successful applications of SPH to HE explosions and underwater explosion shock demonstrate the capability of the method and the code. Both contact and non-contact water mitigations are investigated with different water shield thicknesses. The simulation results, though obtained from the small-scale simple cases, and need to be adjusted with different specific situations, are revealing for practical applications of water mitigation. From the numerical simulations and discussions, the following conclusions can be drawn:

1. Putting a layer of water around explosive, either in direct contact or separated by another layer of air gap, generally mitigates the peak shock pressure as well as the final equilibrium gas pressure.
2. For the contact water mitigation problems, there is an optimal water shield thickness for a given explosive charge.
3. For the non-contact water mitigation problems, the relevant geometrical configurations of the explosive charge, air gap and water shield must be carefully investigated. A properly designed non-contact water mitigation system can produce further mitigation effects than the contact water mitigation system.
4. Either for the contact or non-contact water mitigation, the mitigation effect in the final equilibrium gas pressure is much more obvious than that of the peak shock pressure. In this study, the maximum reduction of the equilibrium gas pressure is up to 83.4%.

7.8   Concluding remarks

Underwater explosions consist of complicated sequence of physical processes, and usually involve large deformation, large inhomogeneities, moving interfaces and free surfaces. Simulation of underwater explosions is usually a big challenge for the conventional numerical methods due to the special features inherent in the underwater explosion process.

This chapter presents the application of the SPH method to simulate underwater explosions. Of particular interest is the underwater explosion shock that is the first and the most important cause of the damages to nearby targets.

In order to suit the needs of underwater explosio₁ shock simulation, an algorithm is proposed to treat the material interface. The particle-to-particle interface technique allows the kernel and particle approximations among particles form different materials/media. It applies a special penalty force to particles that tend to pass over each other. This interface treatment technique, not only implements the interface condition through kernel and particle approximations, but also can effectively prevent unphysical particle penetration between different media.

Numerical examples have demonstrated that the SPH method, after appropriately tuned, can resolve the underwater explosion shocks fairly well. Though there are some oscillations around the gas water interface, the local oscillations do not influence much on the global shock physics. The major early time phenomena in underwater explosion can be well observed. Since the water depth does not have much influence on the early time phenomena in underwater explosions, the gravity is not taken into account in the simulation. It is obvious that it may not be reasonable to expect the simulations to be able to capture both the strong shock wave present at early times in the calculation, but also the late time effects due to acceleration of gravity and bubble buoyancy. The ability to accurately model the late time phenomena in the underwater explosion events may require extensive method development and fine-tuning of numerical artifacts. Moreover, the amount of computer time required to reach such late times with explicit dynamics calculation is considerable. Modeling water as incompressible fluids rather than compressible fluids may be more suitable for late-time phenomena. The current investigations show that although the SPH method needs to be further tuned for late-time calculations, it is effective in simulating early time underwater explosion physics such as shocks.

In this chapter, the real and artificial detonation models of HE as well as their influences to the entire underwater explosion shock simulations are also investigated using the SPH method. The study is carried out with numerical examples of slab and spherical TNT charges. It is found that the two different detonation models leads to different profiles of physical variables along the high explosive gas, and yields different impulses to the surrounding water. The real detonation model produces stronger impulse than the artificial detonation model. Since the real detonation model gives the proper profiles of physical variables in the explosive gas before interacting with water, it leads to more reasonable results. For the artificial detonation model, neglecting the detonation process and simply replacing the original HE charge with explosive gas yield lower peak pressure. The difference in peak pressure is rather large at closer distances, and become smaller with the advance of the shock wave, and finally turns to be negligible after the region of 8-10 times the charge radius (or equivalently charge length). It can be concluded that for numerical simulations of underwater explosions, beyond the region of 8-10 times the charge radius, both two detonation models can get fairly good and close results; in region nearer than 8-

10 times the charge radius, the real detonation model rather than the artificial detonation model should be employed to obtain more reliable results. This finding is roughly the same as those from our past experience in carrying out explosion related projects using grid-based methods (Liu and Lam et al. 1998; Lam and Liu et al. 1996). The use of the real detonation model will of course require more computational efforts.

This chapter also presents an investigation of water mitigations that are employed to reduce the blast effects. Both contact and non-contact water mitigations are investigated with different water shield thicknesses. Some interesting observations have been obtained from the investigation. These observations, though obtained from the small-scale simple cases, and need to be adjusted with different specific situations, are revealing for practical applications of water mitigation. The SPH method is very a useful alternative tool for investigating the water mitigation effects for explosives.

Chapter 8

SPH for Hydrodynamics with Material Strength

In the previous chapters, the SPH method has been applied to different types of computational fluid dynamics (CFD) problems without considering the physical strength of the media or material. This chapter presents the application of the SPH method to hydrodynamics with material strength. In addition, the adaptive smoothed particle hydrodynamics (ASPH) is introduced. The constitutive model and equation of state for materials with strength are incorporated into both the SPH and ASPH equations. Two examples of applications, a cylinder impacting on the rigid surface and a high velocity impact of a cylinder on a plate, are presented using both the SPH and ASPH.

8.1 Introduction

Solids under extreme situations behave like fluids. Two typical situations are high velocity impact (HVI) and penetration. In HVI, the kinetic energy of the system dominates and forces the solid material to deform extremely and the material actually "flows". In the events of penetration, the materials can even be broken into pieces that "fly", in addition to the extremely large deformation.

One significant and challenging task has been focused on the simulation of hydrodynamics with material strength in high velocity impacts and penetration to study the effects of projectiles impacting upon space assets (satellites, space stations, shuttles). Since HVI and penetration processes usually involve large deformations, the simulation is generally difficult for traditional grid-based numerical methods such as the FEM and FDM, as discussed in detail in Chapter 1.

The SPH method is, on the other hand, a particularly suitable candidate for these types of problems. The difference is the presence of the material strength that requires proper treatments.

Libersky and his co-workers carried out the pioneering work of applying the SPH method to problems including HVI, fracture and fragmentation (Libersky et al., 1991; 1993; 1995; Randles and Libersky, 1996; Randles et al., 1995a; b). The group from the Applied Physics Division of Los Alamos National Laboratory (LANL) has modeled high velocity impacts ranging from the very small size (femtogram-scale projectiles that have been accelerated by a Van de Graaf machine to create craters in various types of targets) to the very large size (Shoemaker-Levy comet impact with the planet Jupiter). Attaway, Pampliton and Swegle et al. in the Sandia National Laboratories (SNL) have worked in coupling the SPH processor with a transient-dynamics FEM code, PRONTO, in which high-strain areas that typically tangle or break conventional finite element meshes are resolved using the SPH method (Plimpton et al., 1998; Brown et al., 2000). Benz and Asphaug (1993; 1994; 1995) extended the SPH method to the simulation of fracture of brittle solids. Johnson et al. (1993; 1996a; 1996b), who proposed a normalized smoothing function (NSF) for axisymmetric problems based on the condition of uniform strain rate, have made outstanding contributions in the application of SPH to impact problems.

There are advantages to use SPH together with finite element, finite difference or finite volume grids. A number of workers have coupled SPH with grid-based methods though at present there is no standard method for the coupling and contact between particles and grids (Attaway et al., 1994; Johnson, 1994; Johnson et al., 1996a). Due to its success in HVI simulations, the SPH method has been incorporated into some commercial software packages as one of the key processors (Century Dynamics, 1997; Lacome and Gallet, 2002; Johnson et al., 1993).

This chapter presents some work in hydrodynamics with material strength using SPH and its variation, the adaptive smoothed particle hydrodynamics (ASPH). The chapter is outlined as follows.

- In Section 2, the physics of hydrodynamics with material strength is addressed.
- In Section 3, the conventional SPH formulation for hydrodynamics with material strength is introduced.
- In Section 4, the tensile instability problem in the SPH method is discussed.
- In Section 5, The ASPH method and its formulation for hydrodynamics with material strength as well as its similarities and differences from the SPH method are briefed.

- In Section 6, two numerical examples, the impact of a plate against a rigid surface and the impact of a cylinder on a plate, are investigated using both the SPH and ASPH code.
- In Section 7, some remarks and conclusions are given.

8.2   Hydrodynamics with material strength

8.2.1   Governing equations

In high velocity impact (HVI) problems with material strength, shock waves propagate through the colliding bodies, which often behave like fluids. Analytically, the equations of motion and a high-pressure equation of state are the key descriptors of the dominated material behavior. Material strength is significant only at the late stages of this energy driven problem and may often be treated with a simple incremental elastic, perfectly plastic model with an appropriate value of flow stress obtained from dynamic experiments (Zukas, 1982, 1990). The governing equations for hydrodynamics with material strength are the conservation equations of continuum mechanics.

$$
\begin{cases}
\dfrac{D\rho}{Dt} = -\rho\dfrac{\partial v^{\beta}}{\partial x^{\beta}} \\[2mm]
\dfrac{Dv^{\alpha}}{Dt} = \dfrac{1}{\rho}\dfrac{\partial \sigma^{\alpha\beta}}{\partial x^{\beta}} \\[2mm]
\dfrac{De}{Dt} = \dfrac{\sigma^{\alpha\beta}}{\rho}\dfrac{\partial v^{\alpha}}{\partial x^{\beta}} \\[2mm]
\dfrac{Dx^{\alpha}}{Dt} = v^{\alpha}
\end{cases}
\tag{8.1}
$$

where the scalar density ρ, and internal energy e, the velocity component v^{α}, and the total stress tensor $\sigma^{\alpha\beta}$ are the dependent variables. The spatial coordinates x^{α} and time t are the independent variables. The summation in equation (8.1) is taken over repeated indices, while the total time derivatives are taken in the moving Lagrangian frame. The detailed process that leads to equation (8.1) can be found in Chapter 4.

8.2.2   Constitutive modeling

Material models such as strength, fracture, and fragmentation are important in modeling impacts. The total stress tensor $\sigma^{\alpha\beta}$ in equation (8.1) is made up of two parts, one part of isotropic pressure p and the other part of viscous shear stress τ (see, Section 4.2)

$$\sigma^{\alpha\beta} = -p\delta^{\alpha\beta} + \tau^{\alpha\beta} \tag{8.2}$$

Note that in fluid mechanics, for Newtonian fluids, the stress is proportional to the *strain rate* through the *dynamic viscosity*. In solid mechanics, the constitutive model, in general, permits the stress to be a function of *strain* and *strain rate*. For the anisotropic shear stress, if the displacements are assumed to be small, then the stress rate is proportional to the strain rate through the *shear modulus*.

$$\dot{\tau}^{\alpha\beta} = G\bar{\varepsilon}^{\alpha\beta} = G(\varepsilon^{\alpha\beta} - \frac{1}{3}\delta^{\alpha\beta}\varepsilon^{\gamma\gamma}) \tag{8.3}$$

where G is the shear modulus, $\dot{\tau}$ is stress rate, and $\varepsilon^{\alpha\beta}$ is the strain rate tensor defined as

$$\varepsilon^{\alpha\beta} = \frac{1}{2}\left(\frac{\partial v^{\alpha}}{\partial x^{\beta}} + \frac{\partial v^{\beta}}{\partial x^{\alpha}}\right) \tag{8.4}$$

$\bar{\varepsilon}^{\alpha\beta}$ is the traceless part of $\varepsilon^{\alpha\beta}$.

Comparing equation (8.3) with equation (4.16), it is seen that the unit of the shear modulus in solid mechanics is Pa. It is different from that of the dynamic viscosity in fluid mechanics, $Pa \cdot s$.

In order to get the material frame indifferent strain rate, the Jaumann rate is adopted here with the following constitutive equation as

$$\dot{\tau}^{\alpha\beta} - \tau^{\alpha\gamma}R^{\beta\gamma} - \tau^{\gamma\beta}R^{\alpha\gamma} = G\bar{\varepsilon}^{\alpha\beta} \tag{8.5}$$

where $R^{\alpha\beta}$ is the rotation rate tensor defined as

$$R^{\alpha\beta} = \frac{1}{2}\left(\frac{\partial v^{\alpha}}{\partial x^{\beta}} - \frac{\partial v^{\beta}}{\partial x^{\alpha}}\right) \tag{8.6}$$

The provisional von Mieses flow stress J is computed using the shear stress

$$J = \sqrt{\frac{3}{2}\tau^{\alpha\beta}\tau^{\alpha\beta}} \tag{8.7}$$

In the perfectly plastic yield model, the yield strength is constant if the second stress invariant $J_2 = \sqrt{\frac{1}{2}\tau^{\alpha\beta}\tau^{\alpha\beta}}$ exceeds the known stress J_0. The shear stresses have to be scaled back to the yield surface

$$\tau^{\alpha\beta} = \tau^{\alpha\beta}\sqrt{J_0/3J_2} \tag{8.8}$$

8.2.3 Equation of state

The pressure-volume-energy behavior for the material is computed from the equations of state (EOS). The Mie-Gruneisen equation for solids in (Libersky et al., 1993; Zukas, 1990) can be used

$$p(\rho,e) = (1-\frac{1}{2}\Gamma\eta)p_H(\rho)+\Gamma\rho e \tag{8.9}$$

where the subscript "H" refers to the Hugoniot curve, Γ is the Gruneisen parameter, and

$$\eta = \frac{\rho}{\rho_0}-1 \tag{8.10}$$

where ρ_0 is the initial density.
In equation (8.9),

$$p_H = \begin{cases} a_0\eta + b_0\eta^2 + c_0\eta^3, & \eta > 0 \\ a_0\eta & \eta < 0 \end{cases} \tag{8.11}$$

The constants a_0, b_0 and c_0 can be computed form the linear shock velocity relation

$$U_S = C_S + S_S U_p \tag{8.12}$$

where U_S is the shock velocity, U_p is the material particle velocity, C_S is the constant in the linear relation between the shock-velocity and the particle-velocity, and S_S is the slope.

$$a_0 = \rho_0 C_S^2 \tag{8.13}$$

$$b_0 = a_0 \left[1 + 2(S_S - 1)\right] \tag{8.14}$$

$$c_0 = a_0 \left[2(S_S - 1) + 3(S_S - 1)^2\right] \tag{8.15}$$

8.2.4 Temperature

In this work, a simplified method for computing the temperature T as suggested by Zukas (1990) is implemented

$$T = T_0 + \frac{e - e_0}{mC_v} \tag{8.16}$$

where T_0 is the initial temperature, e_0 is the initial energy, which is usually set to zero at the start of the computation, C_v is the constant specific heat, and m is the mass.

8.2.5 Sound speed

The sound speed of solids under three-dimensional stresses and strain is much more complicated than that of liquids, and is not commonly defined. One possible choice is (Hallquist, 1998)

$$c = \sqrt{\frac{4G}{3\rho_0} + \left.\frac{\partial p}{\partial \rho}\right|_{entropy}} \tag{8.17}$$

Equation (8.17) is a combination of the uniaxial strain elastic sound speed and the isentropic fluid sound speed. When the hydrostatic pressure is zero, we have

$$c = \sqrt{\frac{4G}{3\rho_0}} \tag{8.18}$$

When the isentropic fluid sound speed is neglected, we have

$$c = \sqrt{\left.\frac{\partial p}{\partial \rho}\right|_{entropy}} \tag{8.19}$$

8.3   SPH formulation for hydrodynamics with material strength

Deriving the conventional SPH formulation for hydrodynamics with material strength is very much similar to deriving the SPH formulation for the NS equations governing fluid mechanics. The difference is on the specification of the material, which states the relationship of the stress, strain and strain rate. For convenience, some key SPH equations are rewritten here (see, Section 4.3 for more details).

For the density evolution, either the summation density approach

$$\rho_i = \sum_{j=1}^{N} m_j W_{ij} \tag{8.20}$$

or the continuity density approach

$$\frac{D\rho_i}{Dt} = \sum_{j=1}^{N} m_j (v_i^\beta - v_j^\beta) \frac{\partial W_{ij}}{\partial x_i^\beta} \tag{8.21}$$

can be used.

The advantages and disadvantages of these two approaches for evolving density are also discussed in Chapter 4. It is noted that using the summation density approach, only particle mass and position are required to in the computation of the current step. For the continuity density approach, besides particle mass and position, particle velocity is also necessary in the computation.

The momentum can either be evaluated using equation

$$\frac{Dv_i^\alpha}{Dt} = \sum_{j=1}^{N} m_j \frac{\sigma_i^{\alpha\beta} + \sigma_j^{\alpha\beta}}{\rho_i \rho_j} \frac{\partial W_{ij}}{\partial x_i^\beta} \tag{8.22}$$

or according to

$$\frac{Dv_i^\alpha}{Dt} = \sum_{j=1}^{N} m_j (\frac{\sigma_i^{\alpha\beta}}{\rho_i^2} + \frac{\sigma_j^{\alpha\beta}}{\rho_j^2}) \frac{\partial W_{ij}}{\partial x_i^\beta} \tag{8.23}$$

Taking into account equation (8.2), equations (8.22) and (8.23) can be further expanded by using the shear strain rate tensor and rotation rate tensor. On the other hand, the strain rate tensor and the rotation rate tensor are given by

$$\varepsilon_i^{\alpha\beta} = \frac{1}{2} \sum_{j=1}^{N} \left(\frac{m_j}{\rho_j} v_{ji}^{\alpha} \frac{\partial W_{ij}}{\partial x_i^{\beta}} + \frac{m_j}{\rho_j} v_{ji}^{\beta} \frac{\partial W_{ij}}{\partial x_i^{\alpha}} \right)$$

(8.24)

$$R_i^{\alpha\beta} = \frac{1}{2} \sum_{j=1}^{N} \left(\frac{m_j}{\rho_j} v_{ji}^{\alpha} \frac{\partial W_{ij}}{\partial x_i^{\beta}} - \frac{m_j}{\rho_j} v_{ji}^{\beta} \frac{\partial W_{ij}}{\partial x_i^{\alpha}} \right)$$

(8.25)

where $v_{ji}^{\alpha} = v_j^{\alpha} - v_i^{\beta}$. Therefore, after $\varepsilon^{\alpha\beta}$ and $R^{\alpha\beta}$ for particle i and j have been calculated, the shear stress tensor $\tau^{\alpha\beta}$ and total stress tensor $\sigma^{\alpha\beta}$ can be calculated. The acceleration can subsequently be calculated in a nested summation on equations (8.22) and (8.23).

Similarly, the energy evolution can be expressed as

$$\frac{De_i}{Dt} = \frac{1}{2} \sum_{j=1}^{N} m_j \frac{p_i + p_j}{\rho_i \rho_j} (v_i^{\beta} - v_j^{\beta}) \frac{\partial W_{ij}}{\partial x_i^{\beta}} + \frac{1}{\rho_i} \tau_i^{\alpha\beta} \varepsilon_i^{\alpha\beta}$$

(8.26)

or

$$\frac{De_i}{Dt} = \frac{1}{2} \sum_{j=1}^{N} m_j \left(\frac{p_i}{\rho_i^2} + \frac{p_j}{\rho_j^2} \right)(v_i^{\beta} - v_j^{\beta}) \frac{\partial W_{ij}}{\partial x_i^{\beta}} + \frac{1}{\rho_i} \tau_i^{\alpha\beta} \varepsilon_i^{\alpha\beta}$$

(8.27)

Finally, considering the artificial viscosity effect Π_{ij} and artificial heating H_i (see, Section 4.4), a set of SPH formulation for hydrodynamics with material strength can be summarized as follows.

$$\begin{cases} \dfrac{D\rho_i}{Dt} = \sum_{j=1}^{N} m_j (v_i^\beta - v_j^\beta) \dfrac{\partial W_{ij}}{\partial x_i^\beta} \\[3mm] \dfrac{Dv_i^\alpha}{Dt} = -\sum_{j=1}^{N} m_j \left(\dfrac{\sigma_i^{\alpha\beta}}{\rho_i^2} + \dfrac{\sigma_j^{\alpha\beta}}{\rho_j^2} + \Pi_{ij} \right) \dfrac{\partial W_{ij}}{\partial x_i^\beta} \\[3mm] \dfrac{De_i}{Dt} = \dfrac{1}{2} \sum_{j=1}^{N} m_j \left(\dfrac{p_i}{\rho_i^2} + \dfrac{p_j}{\rho_j^2} + \Pi_{ij} \right)(v_i^\beta - v_j^\beta) \dfrac{\partial W_{ij}}{\partial x_i^\beta} + \dfrac{1}{\rho_i} \tau_i^{\alpha\beta} \varepsilon_i^{\alpha\beta} + H_i \\[3mm] \dfrac{Dx_i^\alpha}{Dt} = v_i^\alpha \end{cases} \tag{8.28}$$

Equation (8.28) can be integrated using some standard methods such as Leapfrog method as described in Chapter 6.

It is worth noticing that the calculation of work done by the traceless deviatoric stress in the above energy resolution equation is only valid in the elastic range. In the situation where the plastic yielding is dominant, one must calculate incremental plastic work during every time step and incorporate its contribution into energy equation.

8.4   Tensile instability

When using the SPH method for hydrodynamics with material strength, one numerical problem called tensile instability (Swegle et al., 1995; Balsara; 1995; Dyka and Ingel, 1995; Dyka et al., 1997) arises. The tensile instability is the situation that when particles are under tensile stress state, the motion of the particle becomes unstable. It could result in particle clumping or complete blowup in the computation. As pointed out by Swegle (1995), the tensile instability neither depends on the artificial viscosity, no on the time integration scheme. The tensile instability problem is typically not a question for problems with equations of state where no tensile stress generates.

In a one-dimensional von Neumann stability analysis, Swegle et al. (1995) identified a criterion for being stable or instable in terms of the stress state and the second derivative of the smoothing function, i.e., a sufficient condition for the unstable growth is

$$W_{\alpha\alpha} \sigma^{\alpha\alpha} > 0 \tag{8.29}$$

where $W_{\alpha\alpha}$ is the second derivative of the smoothing function.

In the SPH method, the cubic spline smoothing function (illustrated in Figure 8.1) is often employed. The initial smoothing length is usually set to be equal to the particle spacing. Under such a circumstance, the first nearest neighbor particles are at $r/h = 1$; and the next nearest neighbor particles are at

$r/h = 2$. As can be seen from Figure 8.1 that the second derivatives of the cubic spline function from $r/h = 1$ to $r/h = 2$ are always positive. It therefore is expected that, according to equation (8.29), the SPH method with the cubic spline function would be stable in a compressed state but could be unstable in a tensile state.

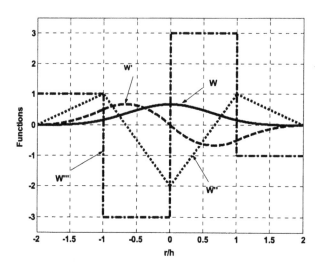

Figure 8.1 The cubic spline function and its first three derivatives.

Several remedies have been proposed to improve or avoid such tensile instability. Morris (1996) suggested using special smoothing functions since the tensile instability is closely related to the second order derivative of the smoothing function. Though successful in some cases, they do not always yield satisfactory results generally. Chen and his co-workers proposed the corrective smoothed particle method (CSPM) (1999c), which improves the tensile instability. Recently Monaghan and his colleagues (Monaghan, 2000; Gray et al. 2001) proposed an artificial force to stabilize the computation.

Note that the tensile instability problem is rooted mainly at the particle approximation that converts the integration over the support domain into a summation only over a finite number of particles. The situation is very much similar to the so-called "node integration" in the implementation of the element free Galerkin method (EFG) (Beissel and Belytschko, 1996). This kind of insufficient sampling point for integration can lead to numerical instability problem (see, Section 2.2.3 in this book and Example 6.2 in the monograph by Liu (2002)).

In the EFG method, the instability is restored by adding stabilization terms in the Galerkin weak form. Another method to solve this problem is to somehow make use of the information at additional points in the support domain, rather than use only these particles. Dyka et al. (1995; 1997) first introduced additional stress points other than the normal particles in a one-dimensional algorithm aimed at removing the tensile instability in SPH. The stress points were also shown to be stable in tension and contributed considerably to the accuracy in wave propagation problems. Later, Randles and Libersky (2000) extended Dyka's approach to multi-dimensional space by staggering the SPH particles and stress points so that there are essentially an equal number in each set of points.

Randles and Libersky (2000) pointed out that, the tensile instability for problems involving material strength generally is latent. The growth rate for damages in solid continuum models is often much faster than that for the development of the tensile instability.

8.5   Adaptive smoothed particle hydrodynamics (ASPH)

8.5.1   Why ASPH

Previous applications of SPH to impacts with material strength are based on isotropic smoothing function with a scalar smoothing length, which may evolve with space and time. For HVI problems with strong shock waves and extremely large deformations, this isotropic kernel may seriously mismatch the generic anisotropic volume changes. In this section, we present our attempt to implement the adaptive smoothed particle hydrodynamics (ASPH) and its applications to HVI simulations (Chin et al., 2002a, b).

Different from the conventional SPH method that employs isotropic kernel with scalar smoothing length, the adaptive smoothed particle hydrodynamics (ASPH) uses anisotropic kernel with a smoothing length tensor. Bicknell and Gingold (1983) first utilized ellipsoidal SPH kernels to study the tidal flattening of stars involved in close encounters with black holes. Later in the two major papers regarding ASPH (Shapiro et al., 1996; Owen et al., 1998), Shapiro, Owen, Martel and Villumsen systematically presented the developments, tests, mathematical prescriptions and the implementation of the ASPH method in both two and three dimensions. The investigation led to an ASPH formulation (Shapiro et al., 1996) in which the scalar smoothing length h of standard SPH is replaced by a generalized smoothing tensor $\mathbf{H}$. The nine components of $\mathbf{H}$

define three vectors that correspond to the three axes of the smoothing ellipsoid for each particle. A later alternative mathematical formalism was derived for evolving the ASPH smoothing scale with a local, linear coordinates transformation (Owen et al., 1998).

In the two ASPH papers, Shapiro et al. have identified the following generic problem with standard SPH that led to the development of ASPH. In the SPH method with spatially and temporally variable smoothing length, the smoothing length h is usually adjusted in proportion to $\rho^{-1/D}$, where ρ is the mass density and D is the number of dimensions. This is adequate only for isotropic volume changes, but is seriously mismatched to the anisotropic volume changes that occur in problems in astrophysics. In general, the local mean inter-particle spacing varies in time, space as well as *direction*. A series of numerical examples including two-dimensional Zeldovich pancake collapse, two-dimensional Sedov blast wave, and Pseudo-Keplerian disks were given in (Owen et al., 1998) for illustration. Along the direction of compression, the smoothing length should vary in proportion to ρ^{-1}, rather than $\rho^{-1/D}$ of D-dimensional standard SPH. The smoothing length in the transverse direction should not vary at all with density. Otherwise, the standard SPH method will lose neighbor information in the transverse direction. The ASPH method replaces the isotropic smoothing algorithm of conventional SPH, which uses spherical interpolation kernels characterized by a scalar smoothing length that varies spatially and temporally according to the local variations of the density. This is implemented by an anisotropic smoothing algorithm that uses ellipsoidal kernels characterized by a different smoothing length along each axis of the ellipsoid and varies these three axes so as to follow the value of the local mean spacing of nodes surrounding each node, as it changes in time, space and direction. By deforming and rotating the ellipsoidal kernels so as to follow the anisotropy of volume changes local to each node, ASPH adapts its spatial resolution scale in time, space and direction. This significantly improves the spatial resolving power of the method over that of the standard SPH for the same number of nodes.

8.5.2   Main idea of ASPH

For SPH applications, the corresponding kernel and particle approximations are carried out within the influencing domain of the smoothing kernel function, which is specified by the smoothing length. For the conventional SPH method, since the smoothing length is a scalar quantity associated with each discrete SPH particle, each particle samples a spherical volume in isotropic nature. This assumption of isotropic smoothing may not be optimal especially for cases with direction dependent features such as shock waves. As discussed in Chapter 4 when generalizing the constructing conditions for smoothing kernel function, the smoothing function in conventional SPH can be expressed as a function of

$\kappa \equiv \dfrac{r}{h}$, such that $W(r,h) = W(\dfrac{r}{h}) = W(\kappa)$, where r is the position vector, $r = |r|$, and κ is known as the normalized position vector. Extending this idea into ASPH, a linear transformation G is used to map from real position space to normalized position space ($r \rightarrow G$). In comparison with SPH, this relation is given as follows

$$\text{SPH: } \kappa = \frac{r}{h} \tag{8.30}$$

$$\text{ASPH: } \kappa = G \cdot r \tag{8.31}$$

$$\kappa = |\kappa| \tag{8.32}$$

The G tensor has the units of the inverse of length. The spherical smoothing of SPH is generalized to a smoothing in elliptical volumes in two dimensions and ellipsoidal volumes in three dimensions. Placing such a restriction upon G implies that it must be a real, symmetric matrix. An ellipsoidal G tensor, taking the place of the conventional SPH scalar smoothing length h , is associated with each particle. The conventional SPH can be regarded as a special simplified case, where the G tensor is diagonal and each diagonal element is equal to $1/h$.

The spatial gradients of the smoothing kernel in both SPH and ASPH are

$$\text{SPH: } \nabla W(\kappa) = \frac{\partial W(r/h)}{\partial r} = \frac{1}{h} \frac{r}{r} \frac{\partial W}{\partial r} \tag{8.33}$$

$$\text{ASPH: } \nabla W(\kappa) = \frac{\partial W(Gr)}{\partial r} = G \frac{\kappa}{\kappa} \frac{\partial W}{\partial \kappa} \tag{8.34}$$

The interpolation symmetry problem arising from variable smoothing scales (see Chapter 4 for details) in ASPH is similar to that in SPH.

$$\text{SPH: } W(\kappa_{ij}) = \frac{1}{2} \Big[W(\kappa_i) + W(\kappa_j) \Big] \tag{8.35}$$

$$\nabla W(\kappa_{ij}) = \frac{1}{2} \Big[\nabla W(\kappa_i) + \nabla W(\kappa_j) \Big] \tag{8.36}$$

$$\kappa_i = \frac{r_{ij}}{h_i} \ , \ \kappa_j = \frac{r_{ij}}{h_j} \tag{8.37}$$

$$\text{ASPH: } W(\kappa_{ij}) = \frac{1}{2}\Big[W(\kappa_i) + W(\kappa_j)\Big] \tag{8.38}$$

$$\nabla W(\kappa_{ij}) = \frac{1}{2}\Big[\nabla W(\kappa_i) + \nabla W(\kappa_j)\Big] \tag{8.39}$$

$$\kappa_i = G_i r_{ij} \ , \ \kappa_j = G_j r_{ij} \tag{8.40}$$

For the smoothing function, what constructed in Chapter 4 in analytical form can all be employed. For example, the cubic spline function in ASPH for D dimensions is expressed as

$$W(\kappa) = \alpha_D \times \begin{cases} 1 - \dfrac{3}{2}\kappa^2 + \dfrac{3}{4}\kappa^3 & 0 \le \kappa < 1 \\ \dfrac{1}{4}(2-\kappa)^3 & 1 \le \kappa < 2 \\ 0 & \kappa \ge 2 \end{cases} \tag{8.41}$$

$$\nabla W(\kappa) = \alpha_D G \frac{\kappa}{\kappa} \begin{cases} -3\kappa + \dfrac{9}{4}\kappa^2 & 0 \le \kappa < 1 \\ -\dfrac{3}{4}(2-\kappa)^2 & 1 \le \kappa < 2 \\ 0 & \kappa \ge 2 \end{cases} \tag{8.42}$$

$$\alpha_1 = \frac{2}{3}|G| \ , \ \alpha_2 = \frac{10}{7\pi}|G| \ , \ \alpha_3 = \frac{1}{\pi}|G| \tag{8.43}$$

Similar to the smoothing length evolution in SPH, the G tensor in ASPH also varies spatially and temporally. Owen et al. (1998) gave the full details of evolving the G tensor in both two and three-dimensions.

In order to stabilize the ASPH scheme, the G tensor needs to be replaced periodically by an averaged G', which for D dimensions is

$$\left\langle G_i^{-1} \right\rangle = \frac{\sum_j G_j^{-1} W_{ij}}{\sum_j W_{ij}} \tag{8.44}$$

$$G' = |G| \left| \left\langle G_i^{-1} \right\rangle \right| \left\langle G_i^{-1} \right\rangle \tag{8.45}$$

For the kernel and particle approximation, as long as the quantities are expressed in terms of the normalized position vector κ rather than the explicitly using h, the SPH and ASPH dynamic equations are identical. The standard time integrators like Leapfrog can also be used in the time integration. Note that in ASPH, a multiplicative correction factor by Balsara (1995) is employed in the Monaghan type artificial viscosity Π_{ij} (see, Section 4.4.1) to restrict the excessive heating resulted from the spurious shear viscosity,

$$\tilde{\Pi}_{ij} = \frac{1}{2}(f_1 + f_2)\Pi_{ij} \tag{8.46}$$

$$f_i = \frac{|\nabla \cdot v_i|}{|\nabla \cdot v_i| + |\nabla \times v_i| + 0.0001 \dfrac{c_i}{\bar{h}_{12i}}} \tag{8.47}$$

$$\bar{h}_{12i} = \frac{1}{2}(h_{1i} + h_{2i}) \tag{8.48}$$

where the third term in the denominator is used to prevent numerical singularity. For shear free flows, where $\nabla \cdot v \neq 0$ and $\nabla \times v = 0$, $\tilde{\Pi}_{ij} = \Pi_{ij}$. For pure shear flows, where $\nabla \cdot v = 0$ and $\nabla \times v \neq 0$, $f = 0$, therefore the artificial shear viscosity is removed.

8.6　Applications to hydrodynamics with material strength

A series of numerical tests for simulating hydrodynamics with material strength have been carried out using SPH and ASPH (Liu et al. 2001a; Chin et al., 2002a;

b). Presented herein are two typical cases. One is a cylinder impacting on rigid surface; another is a cylinder impacting on a plate.

Example 8.1   A cylinder impacting on a rigid surface

In this case, an Armco iron cylinder (actually a plate in two dimensions) traveling at 221 m/s impacts on a rigid surface. The plate is 2.546 cm long, and 0.760 cm wide. The motion is normal to the rigid surface. Johnson and Hollquist (1998) ever examined the usefulness of cylinder impact test data to determine constants for various constitutive models. Libersky et al. (1991; 1993) also numerically simulated this case using the SPH method. In our work, The Armco iron plate is modeled as an elastic-perfectly plastic material in Section 8.2. The Mie-Gruneisen equation of state is employed. The material properties of the Armco iron and the parameters for the Mie-Gruneisen equation of state are listed in Table 8.1.

For the simulation with both the SPH and ASPH, there are 67 particles along the length and 20 particles along the diameter. The rigid surface is simulated using 19 layers of ghost particles mirrored on the other side of the rigid surface. There are a total of 1340 real particles and 380 ghost particles. The cylinder is initially in contact with the rigid surface. In the ASPH simulation, the G tensor field is smoothed every 2 time steps, while the initial h_1 and h_2 are equal to 1.0 times the initial particle spacing in each dimension.

Table 8.1 Material properties for the Armco iron

$\rho_0(Kg/m^3)$	$C_S(m/s)$	S_S	μ (GPa)	Γ	J_0 (MPa)
7850	3630	1.8	80	1.81	500

Figure 8.3 comparatively shows two sets of the particle distribution obtained using the SPH and ASPH for easy comparison. It is seen that the SPH results reach the steady state when the material stops deforming after around 90 µs. In the steady state, the length and diameter of the cylinder are 1.925 cm and 2.404 cm, respectively. It can be seen from the figure that there are voids formed around the centre of the plate near the impact end. Some particles from each side have lost contact with the other particles. This is the well-known phenomenon called numerical fracture caused by the tensile instability.

The ASPH results also reach the steady state after around 90 µs. The length and diameter of the cylinder in the steady state are 1.945 cm and 2.245 cm respectively. The particle distribution clear shows that the particles are flattened along the direction of the impact. It is necessary to smooth the G tensor field because without smoothing the **G** tensor filed could be strongly disordered. The voids become bigger if the smoothing frequency becomes be smaller. Numerical fracture due to tensile instability is not solved by the ASPH method, even though the **G** tensor adapts according to the anisotropy of the volume changes. Only certain smoothing frequency gives satisfactory results. There is a trend of turning points with regard to the length, diameter, number of time steps, the fraction of artificial viscosity energy, and energy conservation as the smoothing frequency changes. Note that the energy conservations of the ASPH cases are better than the SPH case. One possible reason is the smaller time steps due to the anisotropy of the kernel.

It is found that the ASPH case generally predicts less residual kinetic energy than corresponding SPH case. The ASPH results also clearly show the anisotropy of the deformations. However, it is found that the parameters chosen have a great effect on the results. Much work is still needed to find out the optimal set of parameters.

Figure 8.2 shows the CPU time elapsed for both the SPH and ASPH simulation. It is seen that the elapsed CPU times for the ASPH simulation is much longer than those for the SPH simulation. There are some possible reasons. Firstly, in the ASPH simulation, smoothing the **G** tensor every several steps is quite time-consuming. Secondly, the anisotropy of the volume change leads to a compression of the smoothing length in some direction, which yields a smaller component of the smoothing length in that direction. The time step used in the integration is closely related to the smallest smoothing length component, and therefore becomes very small to be efficient.

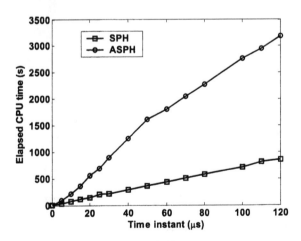

Figure 8.2 Elapsed CPU time vs. the event time instant for both the SPH and ASPH simulation of a cylinder impacting on a plate.

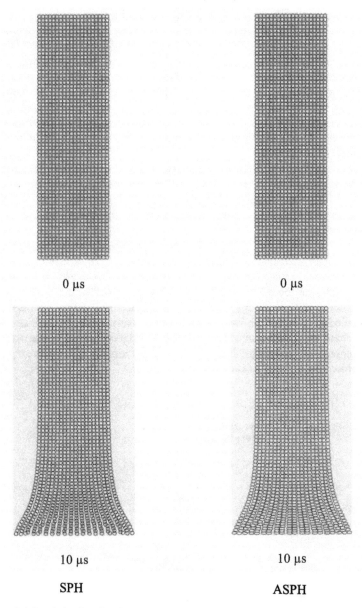

0 µs   0 µs

10 µs   10 µs

SPH   ASPH

Figure 8.3 Particle distributions obtained using SPH and ASPH for the case of a cylinder impacting on a rigid plate. (to be continued)

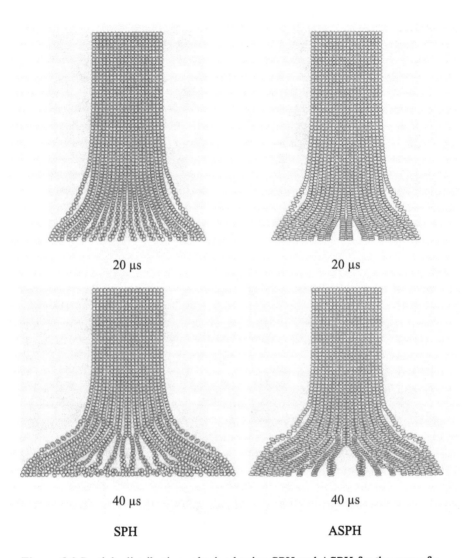

20 μs   20 μs

40 μs   40 μs

SPH   ASPH

Figure 8.3 Particle distributions obtained using SPH and ASPH for the case of a cylinder impacting on a rigid plate. (to be continued)

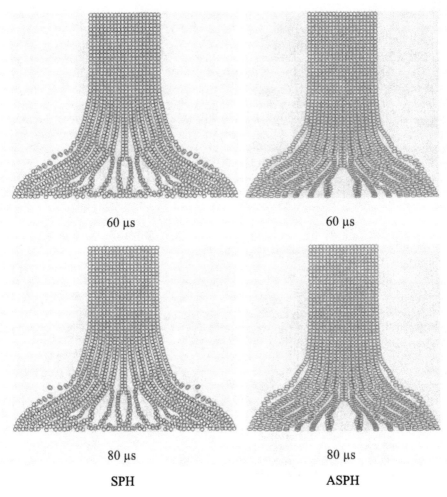

60 µs   60 µs

80 µs   80 µs

SPH   ASPH

Figure 8.3 Particle distributions obtained using SPH and ASPH for the case of a cylinder impacting on a rigid plate. (to be continued)

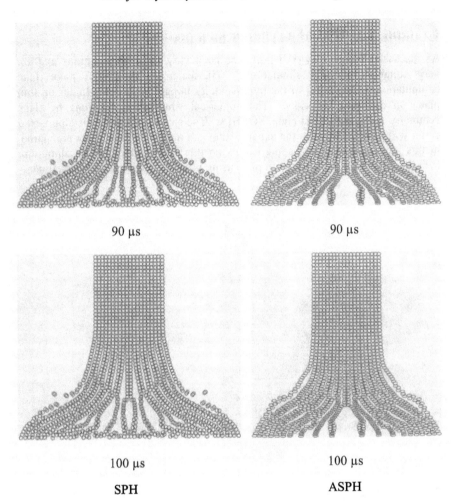

90 μs   90 μs

100 μs   100 μs

SPH   ASPH

Figure 8.3 Particle distributions obtained using SPH and ASPH for the case of a cylinder impacting on a rigid plate.

Example 8.2 HVI of a cylinder on a plate

As discussed before, the HVI problems have very large deformations and are very suitable for SPH simulations. Hiermaier et al. (1997) have done computational simulation of the hypervelocity impact of an Al-cylinder on thin plates of different materials. The numerical simulations are done in plane symmetry using their SPH code, SOPHIA. The shapes of the debris clouds and crater widths at 20 μs from the experimental and numerical results are compared. In this section, we present some results on this HVI problem of an aluminum cylinder penetrating an aluminum plate using both the SPH and ASPH methods.

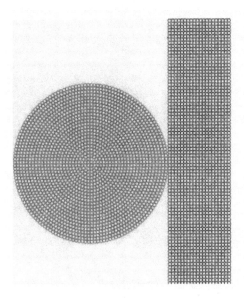

Figure 8.4 Initial particle distribution in the vicinity of the contact area for both the SPH and ASPH simulation of a cylinder impacting on a plate.

In the simulation, the cylinder is of 1.0 cm diameter. The plate is 0.4 cm thick. The plate length is of 10 cm. The particles are all initialized as squares of 0.02 cm in side dimensions. The particles in the cylinder, which is an infinite cylinder in plane symmetry, are arranged in circumferential rings as this gives a most realistic representation of the geometry. The particles in the plate are arranged in a rectangular Cartesian array (Figure 8.4). There 500 particles along the length and 20 particles along the thickness of the plate. There 1956 particles in the cylinder and 10000 particles in the plate, for a total of 11956 particles. The cylinder is initially in contact with the center of the plate. The problems are run with the plate free of constraints. The impact speed of the cylinder is 6180 m/s. The problems are run to 20 μs after the impact. The environmental and initial temperature of the cylinder and plate are set to 0 °C. Again the elastic-perfectly plastic yield model and the Mie-Gruneisen equation of state are employed. The material properties for aluminum and the parameters for the Mie-Gruneisen equation of state are listed in Table 8.2.

Table 8.2 Material properties for the aluminum used in the penetration simulation

ρ_0 (Kg/m^3)	G (GPa)	e_0 (MJ/Kg)	e_s (MJ/Kg)	e_{sd} (MJ/Kg)
2710	27.1	5	3	15

For the simulation with SPH, the smoothing length varies both spatially and temporally. The initial smoothing length is set to be equal to the particle spacing. The results of some key overall dimensions of the objects are shown in Table 8.3 together with the experimental data. It is noted that the experimental data is obtained from Hiermaier et al. (1997), in which the crater diameter was given to be 2.75 cm when including crater lip, while 3.45 cm excluding crater lip. Therefore, the mean value of 3.1 cm is taken as the experimental crater diameter, as shown in Table 8.3. Another point is that, in Hiermaier et al. (1997), instead of giving the experimental results of debris cloud length and width, only the ratio of debris cloud length to width was given. The SPH simulation results and the experimental data are in an acceptably agreement.

It must be pointed out that different parameter setting in the SPH simulation may influence the numerical results. For instance, if the initial smoothing length is set to be 1.5 or 2.0 times the particle spacing, the resultant crater diameter, debris cloud length and width will slightly change.

Figure 8.5 shows the particle distributions obtained by the SPH simulation. The symmetry of the problem is well preserved. Figure 8.6 and Figure 8.7 show the close-up view of the particles near the penetrated edge of the plate and the frontal region of the debris cloud, respectively. It is seen that the outer layer of the debris cloud is made of the material from the plate, and the inner layer of the debris cloud is made up of the material from the cylinder.

Table 8.3 SPH and ASPH results of key overall dimensions of an aluminum cylinder penetrating an aluminum plate using both the SPH and ASPH methods. The results are compared with experimental data

	Crater diameter (cm)	Debris cloud length (cm)	Debris cloud width (cm)	Ratio of debris cloud length to width
Experimental data	3.1	NA	NA	1.39
SPH Results	3.25	10.74	7.99	1.34
ASPH Results	3.55	10.44	7.35	1.42

For the simulation with ASPH, the initial h_1 and h_2 are equal to 1.0 times the initial particle spacing in each dimension. Again, the G tensor field is smoothed every 2 time steps. The ASPH results are also shown in Table 8.3. It is found that the ASPH seems deviate much from the experimental data compared with the SPH results when the same parameters are used. Further investigations reveal that the energy deviation in ASPH simulation is bigger than that in SPH.

Figure 8.8 shows the particle distribution obtained by the ASPH simulation. Figure 8.9 and Figure 8.10 show the close-up view of the particles near the penetrated edge of the plate and the frontal region of the debris cloud, respectively. In comparison with the SPH simulation, the orientation and anisotropy of the deformation of the particles can be clearly seen. This is an obvious advantage of using ASPH for HVI problems, which are with inherent anisotropic deformations. With orientation and anisotropy of volume changes, the ASPH method gives better predictions than SPH method (see the ratio of debris length to width).

An obvious disadvantage of the ASPH is that the computational expense of ASPH is much higher than that of SPH. This is because in ASPH simulation, the anisotropic volume change leads to greatly reduced smoothing length in the compressed direction, which determines the time step used in the simulation. In order for the ASPH computation to reach the same interested time instant, a larger number of smaller time steps is necessary.

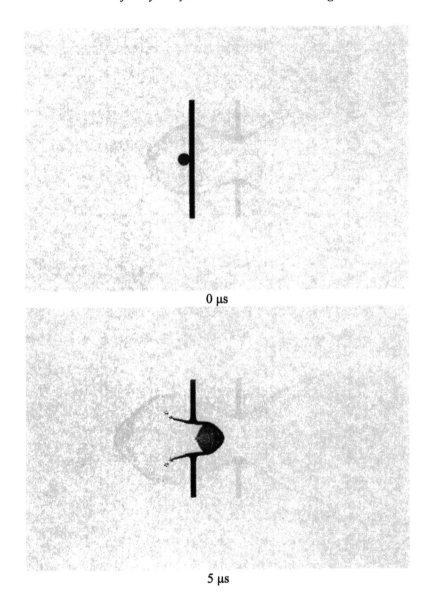

0 μs

5 μs

Figure 8.5 Particle distributions obtained using the SPH method for the case of an aluminum cylinder penetrating an aluminum plate. (To be continued.)

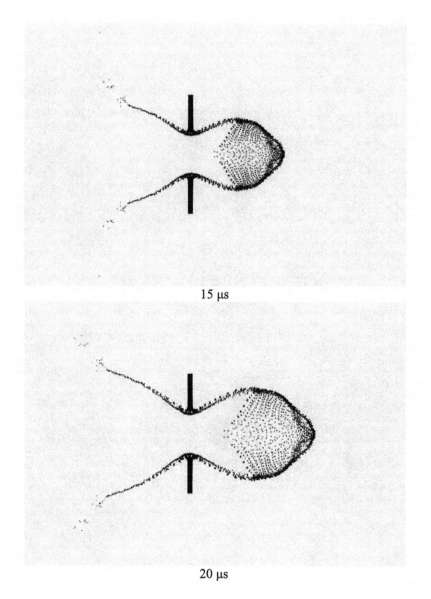

15 µs

20 µs

Figure 8.5 Particle distributions obtained using the SPH method for the case of an aluminum cylinder penetrating an aluminum plate.

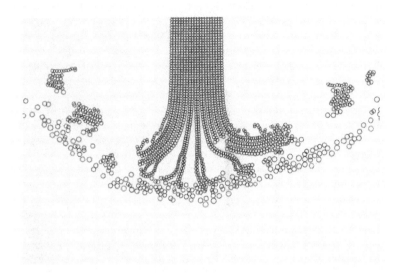

Figure 8.6 Close-up view of the particle distributions near the penetrated edge of the plate for the case of an aluminum cylinder penetrating an aluminum plate at 20 μs.

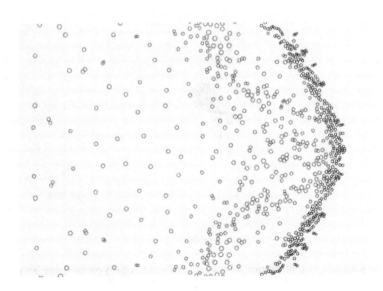

Figure 8.7 Close-up view of the frontal region of the debris cloud at 20 μs.

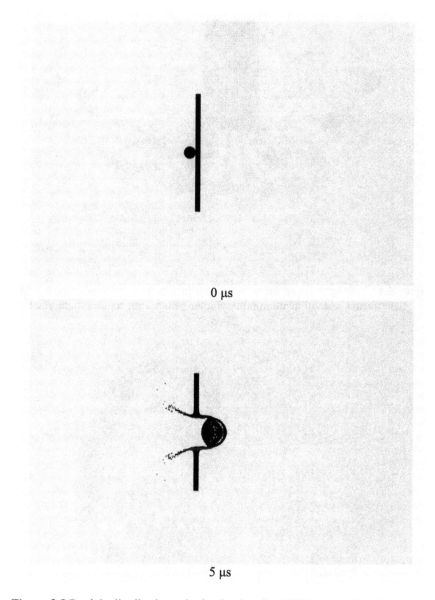

0 μs

5 μs

Figure 8.8 Particle distributions obtained using the ASPH method for the case of an aluminum cylinder penetrating an aluminum plate. (To be continued.)

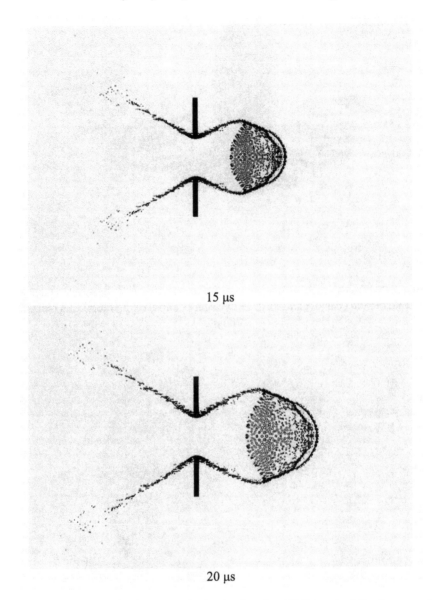

15 µs

20 µs

Figure 8.8 Particle distributions obtained using the ASPH method for the case of an aluminum cylinder penetrating an aluminum plate.

Figure 8.9 Close-up view of the particle distributions near the penetrated edge of the plate for the case of an aluminum cylinder penetrating an aluminum plate at 20 μs.

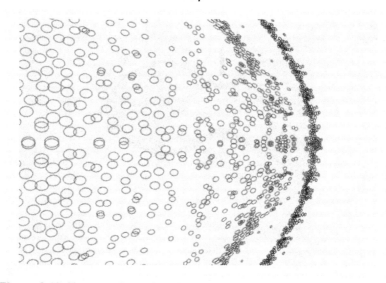

Figure 8.10 Close-up view of the frontal region of the debris cloud at 20 μs.

8.7   Concluding remarks

Simulating hydrodynamics with material strength is one of the attractive applications of the SPH method and its variations. Large deformations inherent in the hydrodynamic process can be handle relatively well in the simulation. Existing constitutive model and equation of state can be readily incorporated with the corresponding SPH formulations. In the work presented in this chapter, the tensile instability problem is not treated since tensile instability in solid dynamics generally remains latent, and the growth rate for damage in solid continuum models is often much faster than that for the development of the numerical tensile instability. However, this tensile instability problem needs to be improved for either SPH or ASPH to realize their full potential.

Adaptive smoothed particle hydrodynamics (ASPH), as a general approach with ellipsoidal smoothing tensor, can capture features of anisotropic deformations. Its spatial and temporal evolution is similar to the smoothing length in SPH. The resultant ASPH dynamic equations are also very much similar to SPH equations. Theoretically, ASPH should have better accuracy than the original SPH method especially for hydrodynamic problems with inherent anisotropic deformations.

It is also found that practically the parameters chosen in ASPH have a great effect on the numerical results. Much work still needs to be done in order to find out an optimal set of parameters. It should be noted that ASPH improves the approximation accuracy, but it does not improve the tensile instability problem. The biggest challenge of the SPH method for simulating hydrodynamics with material strength may be the tensile instability. One disadvantage of ASPH is that the time step may be much smaller than that in SPH due to the reduced smoothing length in the compressed direction. This leads to a larger number of smaller time steps, and hence more computational effort in ASPH than that in SPH.

Chapter 9

Coupling SPH with Molecular Dynamics for Multiple Scale Simulations

9.1   Introduction

Recent development of micro and nano systems has been attracting the attention of researchers in many different areas. Flows in micro and nano mechanics systems are often very complex in nature, and usually involve multi-scales and multi-phases. One of the major outstanding challenges in the simulation of complex fluid flows is the necessity of a systematic frame, which bridges the gap between nano, micro, meso and macro scales for physics on multiple scales. Coupling length scale (CLS) is very important for such simulations and is usually implemented by combining different approaches to account for the different phenomena that dominate. One example is the flows in nano and microfluidic devices. Modeling the flows in micro devices with molecular dynamics (MD) is impractical since the usual atomistic MD simulations are limited to very small length scale over very short times. Application of the macro continuum numerical methods such as the finite element method (FEM), finite difference method (FDM) and finite volume method (FVM) is invalid for the atomistic regions due to the continuum assumptions. Coupling the atomistic molecular dynamics with the continuum methods tends to be a good approach for multiple scale computations. In the coupling practice, MD is employed for atomistic regions with inhomogeneities and complex features, and a continuum approach is used for other regions.

Coupling atomistic and continuum simulation for solids is well investigated and widely practiced (Broughton, 1999; Rudd and Broughton, 1999; Smirnova, 1999). It is especially popular in fracture mechanics for simulating micro crack initiation and propagation. In the simulations, MD is applied for the nano sized region, while some well developed numerical methods (e.g. FEM) in other regions with larger scales. The coupling is implemented by some kind of handshaking algorithm to treat the interface region.

For fluid flows, there are only limited cases in the atomistic and continuum coupling simulation. O'Connell and Thompson (1995) proposed a MD-continuum hybrid technique by constraining the dynamics of the fluid atoms in the vicinity of the MD-continuum interface. The validity of the hybrid technique was demonstrated by the Couette flow problem using an overlap region mediating between a particle ensemble and an explicit FDM approximation for the incompressible Navier-Stokes equation. Hadjiconstantinou and Patera (Hadjiconstantinou and Patera, 1997; Hadjiconstantinou, 1999a; b) coupled the microscopic and macroscopic length scales by extracting the molecular solution in the vicinity of the contact line as the input of the boundary conditions (BCs) for separate FEM computations. This approach is generally used for simulating steady-state flows.

Garcia and his co-workers (1999) have developed a sophisticated adaptive mesh refinement (AMR) algorithm, which embeds the Direct Simulation Monte Carlo (DSMC) (Bird, 1994; Pan et al., 1999; 2000; 2002) within a continuum method at the finest level of an AMR hierarchy. Due to the nature of DSMC, this approach is more suitable for rarefied systems. This paper also provides a comprehensive review on literatures related to hybrid approaches. Flekkoy et al. (2000) have constructed another hybrid model for combined particle and continuum dynamics, which is symmetric in the sense that the fluxes of the conserved quantities are continuous across the particle-field interface. Aktas and Aluru (2002) proposed a coupling approach, in which DSMC was coupled with a scattered point based finite cloud method for solving the Stokes equations for continuum fluids. The combination is implemented by an overlapped Schwarz alternating method with the Dirichlet-Dirichlet type BCs. Qian et al. (2001) presented a new combined molecular dynamics/continuum approach with account of both the non-bonded and bonded interactions to model C_{60} in nanotubes.

In order to make the thermodynamics and transport properties across the interface between the two descriptions continuous, the hybrid models are either implemented by providing BCs from one description to another, or through some kind of average to map the field variables from one description to another.

Different from the previous CLS approaches, a novel combination of smoothed particle hydrodynamics (SPH) method with MD is presented in this chapter for multi-scale fluid dynamic problems. Some of the results are from the work by Liu et al. (2002b). SPH is applied to the Navier-Stokes equation in the continuum region. MD is employed in atomistic regions where complex flow features prevent a continuum description of the fluid. Since both SPH and MD are all practically particle methods, this hybrid particle-atom (SPH fluid particles and MD atoms) coupling approach is very flexible. The momentum exchange between SPH particles and MD atoms is realized through particle-atom interactions.

9.2 Molecular dynamics

9.2.1 Fundamentals of molecular dynamics

The molecular dynamics (MD) was first introduced in 1957 by Alder and Wainright (1957) to study the solid-fluid transition in a system composed of hard spheres interacting by instantaneous collisions. Gibson et al. (1960) first employed the continuous repulsive interaction potential in the MD simulation of radiation damage in a Cu target. In 1964, Rahman (1964) carried out the first MD simulation using the Lennard-Jones potential to describe both attractive and repulsive interaction in a system of 864 Argon atoms. The first MD simulation for a realistic system of liquid water was done by Stillinger and Rahman (1974).

As one of the most important and most widely used particle methods, molecular dynamics has rapidly found more and more applications in both engineering and science. In the study of liquids, molecular dynamics is used to investigate the transport properties such as viscosity and heat conductivity either using equilibrium or non-equilibrium techniques (Allen and Tildesley, 1987; Rapaport, 1995). It is also being employed to investigate fluid dynamics with the purpose of studying complex fluid behavior. In the study of solids, molecular dynamics are frequently used to investigate surface effects (Wu et al., 2003) and material defects (Ciccootti et al., 1987; Meyer and Pontikis, 1991, Wu et al., 2002), from point defects (vacancies, interstitials etc.), linear defects (dislocations etc.) to planar defects (grain boundaries, stacking faults etc.). Investigations on the fracture/crack initiation and propagation by molecular dynamics are also providing some profound insights (see, e.g., Ruth and Lynden, 1994; Xu and Liu, 2002). Molecular dynamics is currently finding a rapidly increasing number of applications in biology, chemistry and medicine since it facilitates the study of dynamics of large macromolecules including biological systems such as proteins, nucleic acids (DNA, RNA), and membranes (Fraga et al., 1995; Allen and Tildesley, 1993). Drug design in pharmaceutical industry involves repeatedly testing the properties of a molecule on the computer without expensive real synthesis.

In principle, for a molecular system with the presence of interacting nuclei and electrons, it is necessary to solve the SchrÖdinger equation and find a total wave function which tells the complete behavior of the system. The complexity in solving the SchrÖdinger equation is greatly simplified using the Born-Oppenheimer approximation, which treats the dynamics of the nuclei and electrons separately due to the very large difference of the masses of the nuclei and electrons. The Born-Oppenheimer thus leads to the following two steps. The first step is to calculate the motion of electrons for fixed nuclei by solving the electronic Schrödinger equation for a specific set of nuclear variables. The second step is to calculate the motion of the nuclei by solving the nuclear Schrödinger equation with the energy obtained from the solution of the electronic Schrödinger equation as the interactomic potential for the interaction of the nuclei. There are various strategies to obtain the interatomic potential of a molecular system as a function of nuclear coordinates. The most fundamental approach is to calculate the potential energy from first principles by directly solving the electronic Schrödinger equation. Since this first principle or *ab initio* molecular dynamics is quite expensive, simplifications are usually necessary to approximate the most time consuming parts in the *ab initio* approach by choosing various empirical parameters. This is a semi-empirical approach, which uses analytical, semi-empirical potential. To further simplify the problem, one can assume an analytical potential function with properly tuned parameters, which can reproduce a set of experimental data or the exact solutions using more accurate approaches such as *ab initio* approach. This gives the so-called empirical potential functions in the classic molecular dynamics. Using inverse techniques (Liu and Han, 2003), Xu and Liu (2003) have constructed an inter-atomic potential using a genetic algorithm.

Molecular dynamics method is a deterministic technique. Since after the initial positions and velocities of the atoms are given and the potential is known, the evolution of the system in time is in principle completely determined. On the other hand, molecular dynamics simulation can also be used as a statistical mechanics method since it generates a set of configurations at the microscopic level that are distributed according to the statistical distribution functions. The connection of the microscopic information to macroscopic observables such as pressure, energy, heat capacities, etc., requires the use of the treatments in the statistical mechanics (Hoover, 1991). Molecular dynamics simulations provide the means to solve the equation of motion of the atoms and evaluate the corresponding mathematical formulas. Statistical mechanics provides the rigorous mathematical expressions that relate the macroscopic properties to the distribution and motion of the atoms and molecules of the system. According to statistical mechanics, a macroscopic property observable in experiments is the average of the corresponding ensemble. The ensemble is a collection of all possible phase states of the system with different microscopic states but an identical macroscopic state. The fundamental axiom of statistical mechanics, the *ergodic hypothesis*, which is based on the idea that if the system is allowed to evolve in time infinitely, the system will eventually pass through all possible states, describes that the ensemble average equals to the time average obtained by molecular dynamics simulation. Experimentally observable structural, dynamic and thermodynamic properties can therefore be calculated using molecular dynamics simulation if the run time is sufficiently long enough to generate enough representative configurations of the system.

9.2.2 Classic Molecular Dynamics

Equations of motion

In classic molecular dynamics, the equation of motion is based on Newton's second law. After giving the initial conditions (initial atomic positions and velocities), the interaction potential is used for deriving the forces among all the atoms. The derived forces will be used to integrate the equation of motion so as to yield new atomic positions and velocities. For a particular particle i, the Newton's equation of motion is given by

$$F_i = m_i a_i \tag{9.1}$$

or

$$F_i = m_i \frac{dv_i}{dt} \tag{9.2}$$

or

$$F_i = m_i \frac{d^2 x_i}{dt^2} \tag{9.3}$$

where F_i is the force exerted on atom i, m_i is the mass of atom i, a_i, v_i and x_i are the acceleration, velocity and position of atom i respectively. For a system with a total number of N atoms, the force on atom i at a given instant can be obtained from the inter atomic potential $u(x_1, x_2, ... x_N)$, which is in general a function of the position vector x of all the atoms

$$F_i = -\nabla_i u(x_1, x_2, ... x_N) \tag{9.4}$$

Interaction potential function

The interaction potential function is therefore important in a MD simulation since it determines the forces exerting on the atoms and hence the way in which the atomic system evolves in time. The choice of interaction potential is usually problem dependent, and should consider both the accuracy and computational speed. Excluding the external force field effects, the interaction potential can be categorized into two classes: pair potential and multi-body potential (Allen and Tildesley, 1987). The multi-body potential is very computationally expensive, since it needs to consider the interaction between the multi atoms, which results in multiple nested summations. The pair potential, as the most important contribution, depends only on the magnitude of the separation of the pair of atoms i and j, $r_{ij} = |x_i - x_j|$. The pair potential gives a remarkably good description of most of the problems simulated, and sometimes if necessary, average multi-body effects can be partially included by defining an effective pair potential. Therefore, if only taking account of the pair potential, the total potential energy of the system of N atoms is then given by

$$u(x_1, x_2, \ldots x_N) = \sum_{i}^{N-1} \sum_{j>i}^{N} u(r_{ij}) \tag{9.5}$$

There are many kinds of pair potential, each with its advantages and disadvantages. Some simple examples are listed as follows (see, e.g., Allen and Tildesley, 1987).

The hard sphere potential (Figure 9.1) is given by

$$u^{HS}(r_{ij}) = \begin{cases} \infty, & r_{ij} < r_c \\ 0, & r_{ij} \geq r_c \end{cases} \tag{9.6}$$

where r_c is a cutoff distance.

The square well potential (Figure 9.2) has the form of

$$u^{SW}(r_{ij}) = \begin{cases} \infty, & r_{ij} < r_{c1} \\ -\varepsilon, & r_{c1} \leq r_{ij} < r_{c2} \\ 0, & r_{ij} \geq r_{c2} \end{cases} \tag{9.7}$$

where r_{c1} and r_{c2} are two cutoff distance.

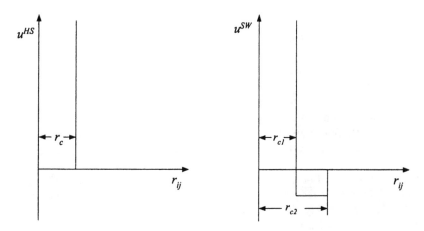

Figure 9.1 The hard sphere potential. **Figure 9.2** The square well potential.

The soft-sphere potential (Figure 9.3) with a repulsion parameter γ can be written as

$$u^{SS}(r_{ij}) = \varepsilon(\frac{\sigma}{r_{ij}})^{\gamma} \tag{9.8}$$

where ε governs the strength of the interaction; σ defines a length scale.

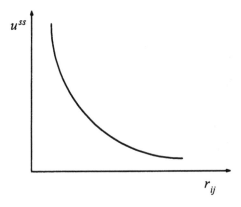

Figure 9.3 The soft sphere potential.

Some other examples include

- the ionic potential, which takes account of the Coulomb interaction of charges;
- the Morse potential, which is more suitable for cases when attractive interaction comes from the formation of some kind of chemical bond;
- the Bunkingham potential, which provides a better description of strong repulsion due to the overlap of the closed shell electron clouds.

Many other forms of potentials can be found in (Meyer and Pontikis, 1991). Xu and Liu (2003) have recently developed an approach to establish the inter-atomic potentials by inversely fitting of experimental data using molecular dynamics simulations and the inter-generation projection genetic algorithm.

Among all the pair potentials, the most widely used is the Lennard-Jones (LJ) potential (Figure 9.4) that can be written as

$$u^{LJ}(r_{ij}) = 4\varepsilon[(\frac{\sigma}{r_{ij}})^{12} - (\frac{\sigma}{r_{ij}})^{6}], \quad r_{ij} \leq r_c \tag{9.9}$$

where r_c is a cutoff distance, ε governs the strength of the interaction, and σ defines a length scale.

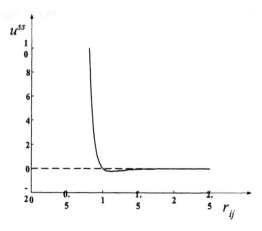

Figure 9.4 The LJ potential.

The LJ potential has a long distance attractive tail (term $-1/r_{ij}^6$), which represents the van der Walls interaction due to electronic correlations, a negative well of depth ε, and a strongly repulsive core (term $1/r_{ij}^{12}$) arising from the non-bonded overlap between the electron clouds. MD simulations using the LJ potential can give a reasonably good agreement with the experimental properties of liquid argon if proper parameters are chosen. The LJ potential is popular in MD simulations when the objective is to model a general class of effects, which only require a reasonably accurate potential.

Time integration

After the potential function is determined, the Newton's equation of motion expressed in equations (9.1) and (9.3) can be numerically integrated in time to obtain the time-dependent atomic velocities and positions. There are numerical integration algorithms available for partial differential equations, e.g. Euler's scheme, Runge-Kutta scheme, leapfrog etc. A good time integration algorithm should conserve energy and momentum, be computationally efficient and has a time step Δt as large as possible. A most popular way of time integration is to march position and velocity with Taylor series expansion based on another time step and often with some transformations.

$$x(t+\Delta t) = x(t) + v(t)\Delta t + \frac{1}{2}a(t)\Delta t^2 + \dots \qquad (9.10)$$

$$v(t+\Delta t) = v(t) + a(t)\Delta t + \dots \qquad (9.11)$$

Applying forward and backward Taylor series expansion to position and then combining the resultant expressions together result in the following famous Verlet algorithm.

$$x(t+\Delta t) = -x(t-\Delta t) + 2x(t) + a(t)\Delta t^2 \qquad (9.12)$$

The velocities in the Verlet algorithm, which are necessary in computing kinetic energy so as to check the conservation of the total energy, do not appear in the algorithm explicitly. One variant of the Verlet algorithm, the velocity Verlet algorithm is to introduce velocities into the algorithm

$$x(t+\Delta t) = x(t) + v(t)\Delta t + \frac{1}{2}a(t)\Delta t^2 \qquad (9.13)$$

$$v(t + \Delta t) = v(t) + \frac{1}{2}\Delta t[a(t) + a(t + \Delta t)] \qquad (9.14)$$

Another popular integration scheme is the leapfrog algorithm, which staggers velocity and position at different instants (see, Chapter 6).

Boundary conditions

A typical molecular dynamics simulation is very limited in the number of atoms. In order to simulate the bulk material properties with small number of atoms, it is necessary to use the periodic boundary condition (Figure 9.5). With the periodic boundary condition, an atom that leaves the specified simulation region through a particular bounding face immediately reenters the region through the opposite face. An atom lying within the cutoff distance from a boundary interacts with atoms in an adjacent copy of the system, or equivalently with atoms near the opposite boundary. This wraparound effect of the periodic boundary condition should be taken into consideration in both the integration of the equations of motion when moving the atoms and the interaction computations between interacting atoms. As shown in Figure 9.5, in the integration process, it is necessary to check if an atom has moved out outside the region. If so, its coordinates must be adjusted to bring it back inside from the other side of the boundary. The checking and adjustment are: if $x_i \geq LX/2$, replace x_i with $x_i - LX$, and if $x_i \leq -LX/2$, replace x_i with $x_i + LX$. Similar checking and adjustment also apply to the interaction calculation process. The only difference is to replace the coordinate of a single atom with the position vector difference between two atoms.

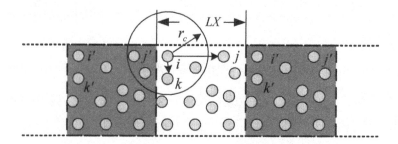

Figure 9.5 Periodic boundary condition in one direction for a molecular dynamics simulation.

It is an important aspect in molecular dynamics simulation to describe the interaction of atoms in the MD computational cell with surrounding environment. This interaction of MD atoms with surrounding environment involves more complex boundary condition (BC), e.g. free surface BC, rigid solid BC, non slip BC, etc. Implementing these boundary conditions usually requires special methods for temperature and pressure control, which exchange heat and work between the MD computational cell and the environment. To sufficiently consider environmental effects on the motion of the MD atoms, the surrounding environment has been gradually incorporated into the MD simulation. The reality is, however, any MD simulation is always limited by the number of atoms that the computer can accommodate.

To save the computational expense, continuum approaches such as the finite element and finite difference methods are used to simulate the surrounding environment by solving the governing equations of continuum theorem. The molecular dynamics is applied to the atomic-sized regions for accuracy, whereas the continuum approaches are applied to other peripheral regions for efficiency. Coupling length scale (CLS) has been one of the hottest research topics for multi scale simulations. In solid mechanics, MD is usually combined with FEM for simulating micro crack initiation and propagation (Broughton et al., 1999). In fluid mechanics, it is attempted to couple MD with FDM for simulating complex fluid dynamics (O'Connell and Thompson, 1995). Coupling length scale bridges the gap between nano, micro, meso and macro scales for physics on multiple scales, and has great potential in modern engineering and science. In this chapter, some discussions and preliminary results on the simulation of multi-scale fluid flows using MD (for atomic region) and meshfree particle method of SPH (for larger scale region) will be presented.

9.2.3 Classic MD simulation implementation

A typical molecular dynamics simulation consists of four sequential stages: initialization, equilibration, production and analysis state.

1. **Initialization:** For the first run of an MD simulation, it is necessary to initialize the coordinates of the atoms, their velocities and the target temperature for the simulation. Typically the atoms are initially placed in a regular lattice spaced to give the desired density. The initial velocities are assigned with random directions and a fixed magnitude. It is preferred to initialize the velocity with the appropriate Maxwell-Boltzman distribution for the specified temperature. However, the usual rapid equilibration renders the careful fabrication of a Maxwell-Boltzman distribution unnecessary. Initialization of atom velocities is subject to a number of conditions:
    a) There is no overall momentum in any Cartesian direction;
    b) For a non-periodic calculation, there is no angular momentum with respect to any of the Cartesian axes;

       c) The total kinetic energy is appropriate to the temperature specified.

2. **Equilibration**: After the initialization of the MD simulation, it should take some period of time for the system to achieve equilibrium before collecting data. This equilibration involves the achievement of the correct partitioning of energy between kinetic and potential energy, as well as attaining a Maxwell-Boltzmann velocity distribution corresponding to the concerned temperature. The time needed for equilibration to occur is variable as it depends on the nature and size of the system being run. During this stage, the velocities are normally scaled to maintain a proper temperature. Equilibration of the system can be accelerated by first starting the simulation at a higher temperature and later cooling by rescaling the velocity.

3. **Production**: After the MD system reaches to the equilibrium state, it usually takes another period of time to collect data. In this production stage after equilibration, no velocity scaling for temperature control is involved, while the trajectories are written out in some interval to the external file for later analysis. The number of time steps in this production stage usually depends on the nature of the problem and the purpose of the simulation.

4. **Analysis**: This stage involves analyzing the information stored in the trajectory file in the production stage. The trajectory file usually contains the absolute Cartesian coordinates, the velocities and various observables of the system, such as the energy, temperature, pressure, etc. Depending on the different purpose of the simulation, the trajectory information can be extracted and employed to analyze either the material properties or other physical characteristics. The analysis of the trajectory information can be related to experimental observables.

9.2.4   MD simulation of the Poiseuille flow

As mentioned above, molecular dynamics has been widely used to different areas with various applications. In fluid mechanics, molecular dynamics techniques combined with proper inter atomic potential function can be used either to predict the properties of fluids, or to model complex flow phenomena. Presented here is a molecular simulation of the Poiseuille flow, which is a classic benchmarking problem and important in engineering and science. The Poiseuille involves flow between two stationary plates, which is driven by some kind of force, and finally reaches to the equilibrium state after some time (Figure 9.6).

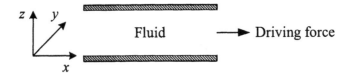

Figure 9.6 The Poiseuille flow in two parallel infinite plates driven by an external force.

The classic molecular dynamics with LJ potential is used to simulate this Poiseuille flow of liquid argon. In the simulation, periodic boundary condition applies to the x and y direction, while reflection solid boundary condition (Rapaport, 1995) applies to the z direction. The MD simulation is carried out using a leapfrog integrator with a time step of $0.005\,\tau_0$, where $\tau_0 = \sqrt{m\sigma^2/\varepsilon}$ is the characteristic time scale for molecular motion. The parameters used in the MD simulation are as follows. The characteristic length, energy and time scales of liquid argon are $\sigma = 3.4$ A, $\varepsilon/k_B = 120$ K, and $\tau_0 = 2.161 \times 10^{-12}$ s. The fluid temperature is constant at $T = 1.2\,\varepsilon/k_B$, where k_B is the Boltzmann constant. The fluid density is initially given as $\rho\sigma^3 = 0.80$. The flow is driven by a uniformly distributed external force of $g = 0.1\sigma/\tau_0^2$ in the x direction to maintain a low shear rate.

Figure 9.7 shows the velocity profile along the z coordinate. The results are interpolated from 50 layers in the z direction. The velocity magnitude and the z coordinate are nondimensionalized by the maximal velocity and the thickness of the entire problem domain in the z direction respectively. For the Poiseuille flow, there is analytical solution when using constant viscosity. The solid line in Figure 9.7 corresponds to the analytical solution of the Poiseuille flow with a dynamic viscosity of $\mu = 2.2(m\varepsilon)^{1/2}/\sigma^2$. This analytical solution is very much close to the MD solution, in which no viscosity is directly used in the simulation. Therefore, matching the MD result with the analytical solution can give rise to the dynamic viscosity of the fluids in the MD simulation. This approach of matching the analytical solutions for the classic Poiseuille flow with the MD results (velocity and temperature) is usually used to predict the fluid transport properties such as viscosity and heat conductivity (Rapaport, 1995).

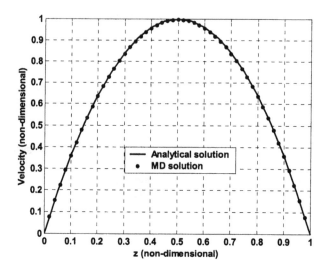

Figure 9.7 Velocity profile of the Poiseuille flow.

9.3   Coupling MD with FEM and FDM

The handshaking algorithm is the most important part in the hybrid atomistic continuum combination. It determines the consistency of the transport properties and field variables around the handshaking interface area.

In the early approaches of coupling MD with FEM, the computational domain was divided into two parts without any overlap region (Figure 9.8). The atoms are usually treated as a node of the boundary elements. This treatment directly provides the boundary condition for both sides, but usually leads to property inconsistency in the boundary. Later, alternative MD-FEM coupling approach has been developed that employs a special region, in which the FEM elements and MD atoms are overlapped (Smirnova et al., 1999) (Figure 9.9). The implementation of boundary conditions for the FEM is to assign values on the overlapped nodes from the averaged properties over the overlapped atoms. The boundary conditions for the MD are to distribute the FEM force according to the Boltzman distribution to each atom.

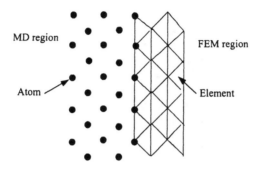

Figure 9.8 MD-FEM CLS simulation without overlap region.

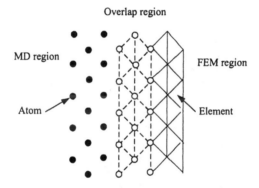

Figure 9.9 MD-FEM CLS simulation with overlap region.

For the approaches of coupling MD with FDM, since MD is a moving particle method and FDM is based on the Eulerian grid, which is fixed in space. In a certain instants, there are atoms moving in or out of the FDM region. How to insert or remove atoms from the FDM domain is quite difficult. Similarly, two different handshaking treatment techniques have been used, one without overlap region (Figure 9.10), another with an overlap region (O'Connell and Thompson, 1995; Flekkoy et al., 2000) (Figure 9.11). If there is no overlap region, the fluid properties around the interface area are usually inconsistent. If with overlap region, since the atoms move in the overlap grid, it is not easy to calculate the mass, momentum flux in a grid cell.

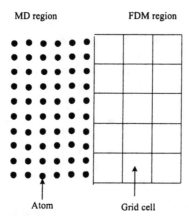

Figure 9.10 MD-FDM CLS simulation without overlap region.

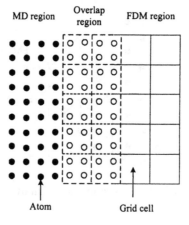

Figure 9.11 MD-FDM CLS with overlap region.

9.4   Coupling SPH with MD

This section discusses algorithms that couples SPH with MD. These methods were studied by Liu et al. (2002b)[1].

---

[1] Comments and suggestions from Professor W. K. Liu on this work are sincerely appreciated.

As a meshfree Lagrangian particle method, SPH has much in common with the MD in the particle sense and therefore seems well suited for coupling with MD to simulate nano systems with multi-scale physics. The SPH method is used to give the continuum solution. For handshaking the interface, two different possible models can be employed, model I with overlap region, and model II without overlap region.

9.4.1 Model I: Dual functioning (with overlapping)

In this model, the domain is divided into three regions (Figure 9.12) according to different characteristics, one region for MD simulation with a potential cutoff distance r_c for every atoms, another region for ordinary SPH particle simulation with a smoothing dimension of κh , and a layer of transitional SPH particles placed between the MD region and the ordinary SPH region. Each SPH particle has its corresponding smoothing length representing the influencing area and length scale of the particle. For the ordinary SPH particle region, the length scale of the particles is graded down to the order of the atomic lattice size when approaching the interface area. The length scale of the transitional SPH particles is the same as the atomic lattice size. The transitional SPH particles interact with neighboring transitional and ordinary SPH particles that are within the influencing area of κh . They act also as virtual atoms which interact with the real and virtual atoms that are within the potential cutoff distance r_c. This dual role of the transitional SPH particle or virtual atoms acts as some kind of overlapping, or the layer of transitional SPH particles is overlapped with MD atoms. The influencing area of κh of the transitional SPH particles is not necessarily equal to the potential cutoff distance r_c. The width of the transitional SPH region is around 4σ .

Therefore for model I, if the LJ potential force corresponding to equation (9.9) is used, the equations of motion for the real atoms are

$$\frac{Dv_i^\alpha}{Dt} = -\frac{1}{m}\sum_{j\neq i}^{N_{md}^r}\frac{\partial u(r_{ij})}{\partial x_i^\alpha} - \frac{1}{m}\sum_{j=1}^{N_{md}^v}\frac{\partial u(r_{ij})}{\partial x_i^\alpha} \tag{9.15}$$

where m is the mass of the MD atoms; v is the velocity vector. The first term in the RHS comes from the contribution of real atoms. The second term comes from the molecular interaction with virtual atoms (or transitional SPH particles). N_{md}^r and N_{md}^v are the number of the interacting real and virtual atoms within the cutoff distance r_c .

The equations of motion for the virtual atoms (or transitional SPH particles) are

$$\frac{Dv_i^\alpha}{Dt} = -\frac{1}{m}\sum_{j=1}^{N_{md}^r}\frac{\partial u(r_{ij})}{\partial x_i^\alpha} - \frac{1}{m}\sum_{j\neq i}^{N_{md}^v}\frac{\partial u(r_{ij})}{\partial x_i^\alpha}$$

$$+ \sum_{j=1}^{M_{sph}^o}m_j\frac{s_i^{\alpha\beta}+s_j^{\alpha\beta}}{\rho_i\rho_j}\frac{\partial W_{ij}}{\partial x_i^\beta} + \sum_{j=1}^{M_{sph}^t}m_j\frac{s_i^{\alpha\beta}+s_j^{\alpha\beta}}{\rho_i\rho_j}\frac{\partial W_{ij}}{\partial x_i^\beta}$$

(9.16)

where $s^{\alpha\beta}$ is the total stress. The Greek superscripts α and β are used to denote the coordinate directions. m_i and ρ_i are mass and density associated with particle i. W is the smoothing function (see, Chapter 4). It is seen that the force exerting on the virtual atoms consists of four parts, the molecular interaction with the real atoms (N_{md}^r), and the virtual atoms (N_{md}^v); SPH interaction with the ordinary SPH particles (M_{sph}^o), and the transitional SPH particles (M_{sph}^t).

The equations of motion for the ordinary SPH particles are

$$\frac{Dv_i^\alpha}{Dt} = \sum_{j=1}^{M_{sph}^o}m_j\frac{s_i^{\alpha\beta}+s_j^{\alpha\beta}}{\rho_i\rho_j}\frac{\partial W_{ij}}{\partial x_i^\beta} + \sum_{j=1}^{M_{sph}^t}m_j\frac{s_i^{\alpha\beta}+s_j^{\alpha\beta}}{\rho_i\rho_j}\frac{\partial W_{ij}}{\partial x_i^\beta}$$

(9.17)

Therefore, the force exerting on the ordinary SPH particles consists of two parts, the interaction with the ordinary SPH particles (M_{sph}^o), and the transitional SPH particles (M_{sph}^t).

Virtual atoms or transitional SPH particles $r_c = kh$

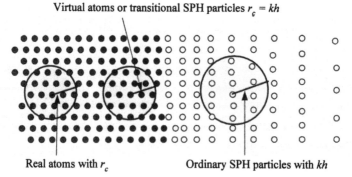

Real atoms with r_c Ordinary SPH particles with kh

Figure 9.12 MD-SPH coupling: handshaking with overlap.

9.4.2 Model II: Force bridging (without overlapping)

In model II, the computational domain consists of a MD region and a SPH region (Figure 9.13). Similar to model I, there is also a layer of transitional SPH particles. The SPH particles in this transitional layer lie around the interface as the neighbors of the MD atoms, and are in the same length scale as the atoms. The influencing area of these neighboring SPH particles is the interaction potential cutoff distance of MD atoms. The thickness of this layer of SPH particles is around 4 σ. With the increasing distance from the MD region, the SPH particles are gradually coarse-grained from finer distribution so as to improve the computational efficiency.

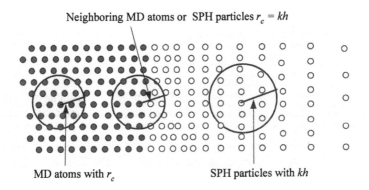

Neighboring MD atoms or SPH particles $r_c = kh$

MD atoms with r_c SPH particles with kh

Figure 9.13 MD-SPH coupling: handshaking without overlap.

The handshaking is implemented by allowing interaction between neighboring MD atoms and SPH particles. For atoms near the interface, they not only feel the influence from other atoms, but also experience interaction with neighboring SPH particles. For SPH particles near the interface, they may also experience forces from the other SPH particles and neighboring atoms. The interactions between MD atoms and the interactions between SPH particles are treated traditionally. The interaction between atoms and neighboring SPH particles can be implemented using pairwise forces, which are exerting on the centerline of the neighboring pair of the MD atom and the SPH particle. The pairwise forces are equal in magnitude, but opposite in direction. It is convenient to take the pairwise force as some kind of potential force (e.g., LJ potential) within with a cutoff distance.

If the LJ potential force corresponding to equation (9.9) is applied, for the MD atoms in neighboring with the SPH particles, the equations of motion are

$$\frac{Dv_i^\alpha}{Dt} = -\frac{1}{m}\sum_{j\neq i}^{N_{md}}\frac{\partial u(r_{ij})}{\partial x_i^\alpha} - \frac{1}{m}\sum_{j=1}^{N_{sph}}\frac{\partial u(r_{ij})}{\partial x_i^\alpha} \tag{9.18}$$

The first term in the RHS comes from the contribution of the molecular interaction. The second term comes from the interaction with the neighboring SPH particles. N_{md} is the number of interaction atoms within the cutoff distance r_c; N_{sph} is the number of neighboring SPH particles within the cutoff distance r_c.

For the SPH particles in neighboring with atoms, the equations of motion are

$$\frac{Dv_i^\alpha}{Dt} = \sum_{j=1}^{M_{sph}} m_j \frac{s_i^{\alpha\beta} + s_j^{\alpha\beta}}{\rho_i \rho_j}\frac{\partial W_{ij}}{\partial x_i^\beta} - \frac{1}{m_i}\sum_{j=1}^{M_{md}}\frac{\partial u(r_{ij})}{\partial x_i^\alpha} \tag{9.19}$$

Similarly, the first term in the RHS comes from the contribution of the SPH particle interactions. The second term comes from the interaction with the neighboring MD atoms. M_{sph} is the number of the neighboring SPH particles within the smoothing length, M_{md} is the number of the interaction atoms within the influencing area κh, which is equal to the cutoff distance r_c for these transition SPH particles.

Since the smoothing area of the particles in the transitional region is equal to the cutoff distance of the molecular interaction, the force exerting on an atom by an SPH particle and the force on the SPH particle by the atom are equal in magnitude but opposite in direction. This ensures the momentum conservation during the interaction of atoms and SPH particles. Through the interaction of the atoms and the neighboring SPH particles, the momentum and energy are exchanged. Due to the average nature of the SPH method, the thermodynamic and transport properties around the interface should be continuous. It is clear that this handshaking technique is carried out on two types of moving particles, i.e. MD atoms and SPH particles. It can be extended to unsteady flows with momentum and energy exchange, and is very suitable to solve the complex flows with moving interfaces.

9.4.3   Numerical tests

To demonstrate the validity of the MD/SPH CLS algorithm, the unidirectional Poiseuille flow and Couette flow of liquid argon are tested. The Poiseuille involves flow between two stationary plates, which is driven by some kind of force, and finally reaches to the equilibrium state after some time. The geometry of flow system is shown in Figure 9.14. The central part around the symmetric line is modeled with MD, while the upper and lower parallel parts close to the wall are modeled with SPH. The system measures $13.6\sigma \times 8.5\sigma \times 85\sigma$. Periodic boundaries are imposed in the x and y direction. Non-slip boundary condition is imposed in the z direction. The thickness of the central MD region is 17.0σ. A total of 1600 atoms are distributed within the FCC lattice. This thickness of the upper and lower SPH region is 34σ each. Excluding the boundary virtual SPH particles, a total of 1664 SPH particles are employed in each of the upper and lower SPH region. After the layer of transitional SPH particles, other SPH particles are distributed exponentially with the increasing distance from the MD region.

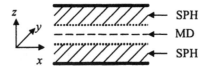

Figure 9.14 Geometry for simulating the Poiseuille flow.

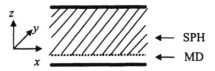

Figure 9.15 Geometry for simulating the Couette flow geometry.

The Couette flow involves flow between two initially stationary parallel infinite plates after one of the plates suddenly moves at a certain constant velocity. The geometry of flow system is shown in Figure 9.15. In this work, the upper plate is moving. The region close to the upper plate is modeled with SPH, and the region close to the lower plate is modeled with MD. The system measures $13.6\sigma \times 8.5\sigma \times 85\sigma$. Periodic boundaries are imposed in the x and y directions, and the non-slip boundary condition is imposed in the z direction. The thickness of the MD region is 8.5σ. A total of 800 atoms are distributed within the FCC lattice. The thickness of the SPH region is 76.5σ. A total of 2184 SPH particles are employed in each of the upper and lower SPH region.

After the layer of transitional SPH particles, other SPH particles are distributed exponentially with the increasing distance from the MD region.

The MD simulation is carried out using a leapfrog integrator with a time step of $0.005\,\tau_0$, where $\tau_0 = \sqrt{m\sigma^2/\varepsilon}$ is the characteristic time scale for molecular motion. The parameters used in the MD simulation are as follows. The characteristic length, energy and time scales of liquid argon are $\sigma = 3.4$ A, $\varepsilon/k_B = 120$ K, and $\tau_0 = 2.161\times10^{-12}$ s. The fluid temperature is constant at $T = 1.2\,\varepsilon/k_B$, where k_B is the Boltzmann constant. The fluid density is initially given as $\rho\sigma^3 = 0.80$. The constant temperature T is to be maintained by weakly coupling the y component of the molecular velocity to a thermal reservoir (Grest and Kremer, 1986). For the SPH region, the dynamics of the fluid is modeled using the Navier-Stokes equation with a constant viscosity of $\mu = 2.2(m\varepsilon)^{1/2}/\sigma^2$ and density of $\rho = 0.80/\sigma^3$. The equation is integrated using the leapfrog method with a time step of $0.005\,\tau_0$. For the Poiseuille flow, the flow is driven by the external force of $\rho = 0.1\sigma/\tau_0^2$ in the x direction to maintain a low shear rate. For each SPH particle i, the external force $g = 0.1(\sigma/\tau_0^2)(m_i/m)$ applies to move the flow. For the Couette flow, the upper plate is moving at a constant velocity of $v = 2\sigma/\tau_0$.

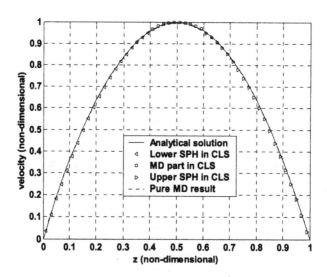

Figure 9.16 Velocity profiles for the Poiseuille flow obtained using different methods.

The two models to treat handshaking interface are tested. In our experience, model II gives better results. Figure 9.16 and Figure 9.17 show the results from the analytical solution, pure MD solution and the MD/SPH coupling with model II handshaking. The results are interpolated from 50 layers in the z direction. The velocity magnitude and the z coordinate are normalized by the maximal velocity and the thickness of the entire computational domain in the z direction, respectively. The close agreement between the analytical solution and the pure MD simulation results verifies the validity of the MD/SPH code. It is seen that, for either the Poiseuille flow or the Couette flow, the results obtained by combining MD and SPH are approaching the analytical solution fairly well. The results around the interfacing region are quite continuous.

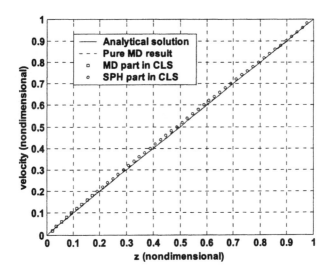

Figure 9.17 Velocity profiles for the Couette flow obtained using different methods.

9.5   Concluding remarks

This chapter presents a novel approach to couple the meshfree particle method of SPH with molecular dynamics. The handshaking interface is treated in such manner that allows interaction between the SPH particle and the MD atoms. Since MD is applied to atomistic region, and SPH is a continuum approach with meshfree, Lagrangian particle nature, this coupling length scale is very attractive for solving multiple scale physics. Due to the common particle nature of MD and SPH, this novel CLS approach will be useful for studying complex flows including studies of convection, coalescence, spreading and wetting, instability in boundary lubrication, and moving contact flows. The numerical test on the Poiseuille flow and Couette show the preliminary success of this new atomistic continuum CLS approach.

Two handshaking models of MD-SPH CLS have been presented in detail in this chapter. The dual function model uses an overlap region where the atoms and SPH particles possess dual functions, so as to achieve the purpose of exchanging information. The force bridging model achieves the information exchange by allowing both the atoms and SPH particles in the common domain to exert forces to their counterparts. Our example tests suggest that the force bridging model works better inters of accuracy and stability.

Chapter 10

Computer Implementation of SPH and a 3D SPH Code

Two main purposes of developing meshfree methods are
1. to relieve the users or the analysts from the trivial and time-consuming task of meshing, and
2. to solve problems with extremely large deformation and even breakage.

The computer implementations of meshfree methods are, in general, more difficult than that in a grid/mesh-based method, simply because there is no predefined grid/mesh to use in establishing the discrete system equations. Among the meshfree methods, SPH method is relatively simple to code. A detailed discussion on issues and techniques related to computer implementation of a meshfree method has been given by Liu (2002) in his recent monograph.

In this chapter, issues related to the computer implementation of meshfree particle methods are discussed. Computer implementations of meshfree particle method over serial and parallel computers are also briefly addressed. The source code of a standard serial 3D SPH code is provided. The main features of the SPH code, detailed descriptions and the source code of the related subroutines are also provided.

The programs demonstrate most of the concepts and techniques related to the SPH method. The code can be easily extended to other corrective or modified versions of SPH with proper treatment either on the kernel approximation or the particle approximation. The 3D SPH code can solve compressible flows with real viscosity, and can be readily modified for applications to hydrodynamics with material strength with a proper constitutive model and an equation of state. As a structured code, it can be readily modified or imported to other user-application subroutines according to users' specific requirements. Two benchmark problems, a 1D shock tube and a 2D shear driven cavity are provided to test the accuracy and efficiency of the attached SPH source code.

The purpose of releasing the source code is to save readers' time in developing their own SPH code. Readers are free to use in part or the entire code at their own risks, as long as a proper reference and acknowledgement are given. The authors apologize for not being able to provide technical support, but welcome comments and suggestions from the users. We will try to improve the code in the future, whenever there is a chance.

10.1   General procedure for Lagrangian particle simulation

From the earliest computer, MANIAC in the Los Alamos Laboratory, computer hardware has undergone a rapid and revolutionary advancement.   Modern computers range from microcomputers, to workstations, and to the extremely high performance mainframes. The rapid development of computer hardware is still on the way, with the introduction of some specialized features such as pipeline and array parallel processors. With the aid of the rapidly developing computer hardware, the meshfree particle methods have increasingly been a class of very important and powerful computational techniques valuable for studying complex interacting systems from atomistic scale to astronomic scale. Numerical simulations using particle methods have been carried out for physical and engineering problems involving electronics, gas dynamics, structural systems, and astrophysics.

Meshfree particle simulation in particular represents an important branch of computer simulations.   From the atomistic molecular simulation to the continuum meshfree particle methods, and to the particle methods applied in astrophysics, the computer simulations and implementation share some common features. Meshfree particle methods generally solve the dynamic equations of motion, in forms of Newton's equation, Euler equation, N-S equation or equations in other possible forms of physical laws. The particle interactions are simulated using either some kind of potential or weight function. Note that a particle usually does not interact globally with all other particles. Even if the global interaction between all particles is physical, this global interaction is usually treated in most particle methods as local interaction with some kind of truncation to reduce computational efforts. In the local interaction scenario, each particle has its corresponding support and influence domains.   For some meshfree particle method like SPH, the support and influence domains may vary spatially or/and temporally (see, Chapter 2 and Chapter 4). Due to the local interaction feature, an efficient neighboring particle-searching algorithm is necessary to improve the computational efficiency, such as the bucket searching algorithm (Liu and Tu, 2001; 2002), link-list search and tree search algorithms (see, Chapter 4). Meshfree particle simulations are usually carried out in the

Lagrangian frame as an initial value problem. Standard time integration schemes can be employed to obtain field variables in the next time step. The time step in the time integration is closely related to the particle resolution (see, Chapter 4).

The code implementation of a technique is a mixture of engineering, physics, computational science, and mathematics to convert the numerical procedures into a usable tool for performing complex tasks. In serial computing with a scalar computer, there is a logical flow of control through the program. Though different methods for different physics may need some specific considerations, programming for the Lagrangian particle methods of various kinds usually share many common features.

10.2   SPH code for scalar machines

As a Lagrangian particle method, SPH possesses the typical features in carrying out simulations. The SPH method approximates a function and its spatial derivatives through averaging or summation over neighboring particles. The basic SPH methodology and the accompanied algorithms for various numerical aspects of SPH result in some special features in the SPH coding. These special features are generally involved under the main loop of time integration process, including the smoothing function and derivative calculation, particle interaction calculation, smoothing length evolution, spatial derivative estimation, artificial viscosity, artificial heat, boundary treatment, etc. A typical procedure for SPH simulation includes

1. Initialization module, which includes the input of the initial configuration of the problem geometry (dimensions and boundaries), discretization information of the initial geometry of particles, material properties, time step and other simulation control parameters. This initial setup process can either be loaded from external files or generated by the code itself, depending on the complexity of the problem. CAD/CAE tools may be used in helping to create the geometry and/or the initial particle distribution.

2. Main SPH processor, which contains the major modules in the SPH simulation, and is implemented in the time integration module. Standard methods such as Leapfrog, predictor-corrector and Runge-Kutta can be employed, which may have different advantages and disadvantages. Following modules need to be included into the time integration process:

    2.1 Generation of boundary (virtual or ghost) particles.

    2.2 Nearest neighboring particle searching (NNPS) module. Different approaches such as the direct all-pair search, linked-list search algorithm, tree search algorithm, and bucket search algorithm can be employed (see, Section 4.5).

2.3 Calculating the smoothing function (for the summation density approach) and its derivatives from the generated information of interaction particle pairs.

2.4 Updating density if the summation density approach is used.

2.5 Calculating the artificial viscous force.

2.6 Calculating the internal forces arising from the particle interactions. Note that the particle pressure is obtained from the density and energy through an equation of state.

2.7 Calculating the external forces if necessary.

2.8 Calculating the change of momentum, energy and density (if using the continuity density approach).

2.9 Updating smoothing length for the next time step.

2.10 Updating particle momentum, energy and density; Updating particle position and velocity; checking the conservation of the energy and momentum.

2.11 Applying boundary conditions.

3. Output module. When the time step reaches to a prescribed one or at some interval, the resultant information for a computational state is saved to external files for later analyses or post-processing.

The structure of a typical SPH code is schematically shown in Figure 10.1.

10.3 SPH code for parallel machines

10.3.1 Parallel architectures and parallel computing

The success of computational science to accurately describe and model the real world has helped to fuel the ever-increasing demand for cheap computing power. Parallel computing offers a way to tackle the problem with a cost-effective manner to achieve high performance. Parallel computing is concerned with producing the same results using multiple processors as many as possible in parallel to minimize the *execution time* of the program. The problem to be solved is divided among the available processors. Dividing the problem up in a sensible and efficient manner is critical to achieve a good performance on parallel machines. The aim of an efficient code is to keep all the processors busy (load-balancing) while minimizing the amount of communication (usually the bottleneck in parallel calculation). There is of cause a tradeoff between the communication load and the processing load for solving the problem.

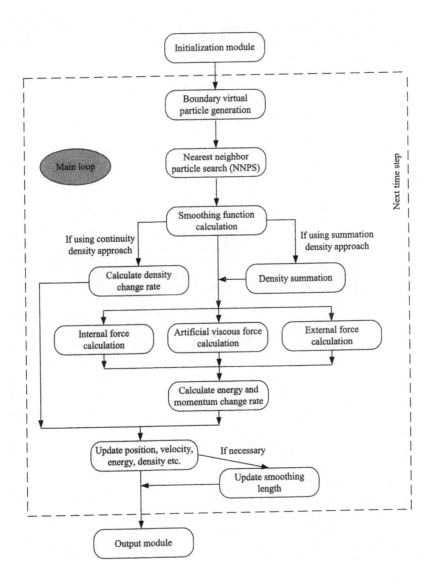

Figure 10.1 Structure of a typical serial SPH code.

Parallel system architecture is far from simple. Among the factors to be taken into account are:

1) whether the individual processors all carry out the same operation during each operation, or whether they are able to act independently;
2) whether each process has its own private memory, or all share a common memory, or both;
3) whether the processors communicate with one another by passing messages across a communication network, or through common memory;
4) the nature and topology of the communication network, if any.

Table 10.1 provides some types of the parallel architectures of computers (Yukiya and Jun, 1999).

Table 10.1 Categorization of parallel architectures

	Shared Address Space	Individual Address Space
Centralized memory	SMP	N/A
Distributed memory	NUMA	MPP

SMP (symmetric multi processor, e.g. SGI Origin) architecture uses shared system resources such as memory and I/O subsystem that can be accessed equally from all the processors. SMP machines have a mechanism to maintain the coherency of data held in its local caches. A single operating system controls the SMP machine and schedules process and threads on processors so that the load is balanced. MPP (Massively Parallel Processors) architecture (e.g. IBM SP, clustered PC) consists of nodes connected by a network, with each node in the network having its own processors, memory, I/O, and operating system. NUMA (Non-uniform Memory Access) architecture behaves between SPM and MPP, since each node in the network not only has its own local memory, but also shares a global memory. A single operating system controls the whole system.

Parallel computing is quite different on different parallel architectures. Writing a parallel program for SMP machines is much easier than that for MPP machines, which needs message-passing interface (MPI) between processors. SMP machines are best fit for multiple threaded programming using Fork-Join model through sharing the available resources. For SMP machines, there are even compilers that can parallelize programs automatically or with the aid of compiler directives given to the user (Yukiya and Jun, 1999). The executables generated by such compilers run in parallel using multiple threads, and those threads can communicate with each other by using shared address space without explicit message passing statements.   It is clear that the control over parallelization is limited. Instead, parallelizing code by using MPI and running it on MPP machines are more flexible since it is the programmer rather than the compiler who determines when, where and how to parallelize, pass message and divide tasks.

10.3.2 Parallel SPH code

Though the SPH method is quite attractive in many applications, high speed is required. Current simulations on workstations (scalar machines) are usually limited to around 100,000 particles and may require months of machine time to model a problem for a meaningful time duration, especially when a supplicated version of SPH (such as the ASPH detailed in Chapter 8) is used. Parallelization is therefore very important to reduce the execution time of a SPH code.

After the invention of the SPH method, different parallel computing techniques have been employed to enhance the computational performance of the SPH method. Asphaug and Olson (1998), during their research on impacts into asteroids and planets, developed a parallel SPH tree code on CRAY T3E. Their work on SPH parallelization has been carried out for a couple of years and is still on the way. Dave et al. (1997), based on the TreeSPH code (Hernquist and Katz, 1989), developed a parallel version of TreeSPH code. Plimpton and his co-workers (1998) proposed some algorithms for contact detection and smoothed particle hydrodynamics for parallel transient dynamics simulations. Their work was further carried out with more parallel strategies (Brown et al. 2000). The implementation of parallelization is based on the transient dynamics analysis code PRONTO, which incorporates FEM with SPH. Many complex crash and impact problems were simulated by using the presented parallel strategies. Lia and Carraro (2000) developed a parallel TreeSPH code for simulating the formation of an X-ray galaxy cluster. Speith et al. (2000) also presented a distributed implementation of SPH for simulations of accretion disks. Chin and his co-workers (Chin, 2001; Chin et al., 2002a; 2002b) have developed shared memory parallel programs for the SPH and ASPH codes using the industry standard OpenMP C/C++ Application Program Interface (API). In this section, only the very basic parallel techniques are briefly discussed.

An important step for parallel programming is to decompose the potential parallel or partition a serial code through different ways among the multi processors. For the SPH code, the act of partitioning has been tested on the particles involved, on the domain simulated and on computation of disparate functions.

Particle-based decomposition

If the partition is based on the particles involved, a group of particles or even each particle can be assigned to a particular processor for the duration of the simulation, irrespective of its spatial location. This partition is conceptually simple, but large amounts of communication are required to handle interactions between particles assigned to each individual processor. Since the interaction process in the SPH method is frequently calculated, the amount of communication can be quite considerable, and may become the bottleneck limiting the efficiency of the entire computation.

Domain-based decomposition

The partition can also be based on the simulation domain, which subdivides the problem domain and assigns each processor a particular sub-region. All the particles that are in a given sub-domain at some moment in time reside in the processor responsible, and when a particle moves between sub-domains all the associated variables are explicitly transferred from one processor to another. This partition is frequently used in molecular dynamics simulations in which the simulation domain is some kind of lattice, and the number of atoms moving from one sub-domain to another is very few.

In this parallel implementation, the particles distributed in the entire domain can be divided into small particle blocks, with each particle block having its own processor as shown in Figure 10.2. Not all the required neighboring particles will necessarily reside in the local data block of the same processor. In order to access data from neighboring particle blocks, each particle blocks overlap particles with interaction. As the particles near the edge of each particle block, communication is required to transfer data from one processor to another. Therefore each processor is prepared to send copies of particle information from the edges of its own particle block, out to neighboring particle block, and thereafter to update the information in the overlap region.

Operation-based decomposition

The partition can be also readily implemented based on the different operations in the SPH computations. As can be seen from the structure of a typical SPH code for serial computers (Figure 10.3), the code can be divided into two types of subroutines/operations. The first one is naturally serial in time, and the second type consists of many program units that are naturally in parallel level, such as the parallel internal, artificial and external force computation.

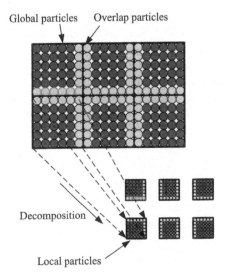

Figure 10.2 Illustration of SPH particle decomposition for parallel computing.

The SPH operation based partition is easy to be implemented by assigning one processor to process the serial subroutines, while assigning one processor to execute each parallel SPH operations. Since there are no dependencies between different runs, the number of processors that can be used is limited only by the number of runs to be performed. The time for computing the parallel functions will be the execution time of the most time consuming run in the parallel sets. The total execution time for a typical SPH run is the sum of the time for serial computation, and the time for parallel computations. Clearly, this trivial parallelism can be exploited to provide almost linear speed-up if runs take a similar length of time. As each run is completely independent, no communications is required between processes and this technique is most effective on cluster systems, which could be of poor communication performance. One problem inherent in this approach is that its efficiency is significantly limited by the speed of the sequential part in the SPH code, since according to the Amdahl's Law, the performance improvement to be gained using parallelism is limited by the fraction of the serial code.

One special point in SPH code parallelization is the treatment of the loops used for calculating the interaction between nearest particles. The interaction loops are frequently encountered and are one of the most time consuming parts in a typical SPH simulation. The nested interaction loop, when in the parallelization, should miss minimal cache and minimal communication overhead. If the total number of interaction is *NIAC*, the total interaction loop is divided into parts using one processor to carry out summation on each part. As shown in Figure 10.3, if *NP* processors are employed in the interaction calculation with $NIAC = q \times (NP-1) + r$, where q and r are the quotient and remainder, respectively, the interaction loop can be divided into the *NP* processors. Processors from 1 to *NP*-1 process q times interaction, while the last processor processing only r times of interaction and gathering all the interaction information from other processors. Therefore, the CPU time for each processor can be well balanced, and the total elapsed time for the whole program is minimized.

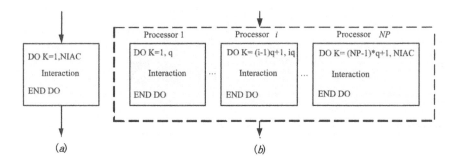

(a)   *(b)*

Figure 10.3 Parallelization of interaction loop for the SPH approximation.

We have implemented a parallelized version of SPH code. The decomposition is basically based on the operations and then each interaction loop in each of the basic functions is parallelized to different portions, except for some serial work carried out in the pre- and post-analyses. The outline of the parallel SPH implementation is

I. Serial pre-analysis
    Input initial particle information and other necessary data for the computation;
II. Parallel computing on each processor
    A. Perform the nearest neighboring particle search, get the total number of interaction pairs, *NIAC*;
    B. Calculate the smoothing function, with a parallelized interaction loop;
    C. Density summation with a parallelized interaction loop;
    D. Parallel calculating the internal, artificial, external force with a parallelized internal interaction loop;

E. Parallel updating the position, velocity, energy and the smoothing
length of each particle;
III Serial post-analysis
Output the particle information for post-processing.

Parallel performance of the SPH code

The performance of the parallel SPH code is tested on a benchmark problem of
the one-dimensional TNT slab detonation, which was described in Chapter 6.
The simulation was carried out both on an IBM SP2 and on a PC (Pentium 755).
Table 10.2 shows the comparison of the elapsed time at different scenarios for
performing one typical SPH time step. The elapsed time in one step is obtained
from the elapsed time in 1000 steps divided by the total number of steps (1000).

Table 10.2 Comparison of the elapsed time

Number of particles	IBM (8 processor)	IBM (1 processor)	Pentium 755 (1 CPU)
4000	0.110 s	0.315 s	0.285 s
8000	0.224 s	0.595 s	0.550 s

It is seen that the elapsed time for 8 processors is around one third of that for
only one processor. It is noted that since our SPH parallel decomposition is
based on the different function units with internal parallel interaction loop, the
CPU time spent at each processor may not be balanced very well. However, the
function decomposition greatly reduces the communication overhead.
Considering the fact that this simple parallel SPH code still has some serial parts,
the achieved parallel efficiency is quite satisfactory for such a simple parallel
code.

10.4   A 3D SPH code for solving the N-S equations

In this section, a sample serial SPH code is described. The main features of this
sample code, conventions for naming variables in FORTRAN, and some more
detailed descriptions are given below. The programs herein demonstrate most of

the concepts and techniques related to the conventional SPH methodology. It is hoped that these programs can serve to help to understand the SPH algorithms, and can be used to solve practical CFD problems.

Readers are free to use in part or the entire source code, as long as a proper reference and acknowledgement are given. However, the authors make no warranty whatsoever, expressed or implied, that the programs contained in this book are free of errors, or that these programs will meet the specific requirements for any particular application. In no event shall the authors become liable to any party including the users of the materials contained in the book (the programs, codes, methods, algorithms, etc.) for any loss, including but not limited to the loss of time, money, or goodwill, which may arise form the use of these materials. The authors apologize for not being able to provide technical support, but welcome comments and suggestions from the users.

10.4.1 Main features of the 3D SPH code

The sample SPH code has following main features:

1. The code is based on the SPH techniques presented in this book. For the sake of generality and simplicity to avoid possible confusion, some treatments for the simulation of special phenomena (explosion, impact, penetration, etc.) are not listed in the code.
2. The code can be easily extended to other corrective or modified versions of SPH with proper treatment either on the kernel approximation or the particle approximation.
3. The code is implemented in three-dimensional space. Therefore, 1-, 2- and 3-dimensional simulations are possible.
4. The code can solve compressible flows with real viscosity. The shear stress and strain rate can all be calculated using the corresponding SPH approximations (see, Chapter 4).
5. The code can be employed to simulate incompressible flows using the concept of artificial compressibility through a properly selected equation of state (see, Chapter 4).
6. The code can be readily modified for application to hydrodynamics with material strength with a proper modification on the constitutive model and the equations of state (see, Chapter 8).
7. The code is implemented and tested in the environment of Visual Fortran Standard 5.0 under Microsoft Windows in standard F77 plus F90 for time subroutines. Since standard F77 and F90 are used, the programs can be implanted to other platforms with these Fortran compilers.

8. The SPH code only includes the key components for implementing the SPH algorithms. Simple operations such as the input consistency check and conservation (mass, momentum, energy) check are not included.

9. The code is structured. Users can readily modify or import their own application subroutines according to their specific requirements.

10.4.2 Conventions for naming variables in FORTRAN

For easier understanding of the programs given later, we have adopted the following conventions for naming FORTRAN variables and subroutines:

1. In all the programs, all variable names must be explicitly declared at the beginning using **IMPLICIT NONE** statement to prevent possible problems arising from unwanted or mistyped variables. This statement, however, can be turned off by simply commenting it.

2. Though the implicit typing is disabled by the statement **IMPLICIT NONE**, when explicitly declaring a variable, the conventional default real (names starting with letters *A-H* and *O-Z*) and default integer (names starting with letters *I-N*) are still employed, except for cases in which the variable name is identical to its actual name, but begins with a letter different from the default implicit rule.

3. The FORTRAN names for the variables and subroutines are usually identical to the actual names, which generally represent corresponding physical nature.

4. In most of the subroutines, the arguments are explained in detail with their updating attribute denoted. [in] means that the dummy argument is expected to have a value when the subroutine is referenced, and this value is not updated by the subroutine. [out] means that the dummy argument has no value when the subroutine is referenced, it will be given one before the subroutine finishes. [in/out] means that the dummy argument has an initial value that will be updated by the subroutine. [in|out] means that the array length of the argument is increased with more array elements, while the values of the existed array elements are not changed. This treatment is somewhat similar to the INTENT attribute in FORTRAN 90.

5. Different materials and particle types can be employed in the programs. For real particles, the particle type is required to be positive. For example, the real idea gas particle is denoted as type 1 and the real water particle type 2. For the corresponding virtual particles, the particle type is the corresponding negative counterpart. For example, the virtual idea gas particle type is -1; the virtual water particle type is -2.

10.4.3 Description of the SPH code

Structure of the code

Except for the files related to time calculation, in the sample SPH code, 17 FORTRAN 77 source files and 1 including header file are include in the entire package (Table 10.3). The subroutines related to the initial configuration, boundary treatment and virtual particle generation are quite problem dependent. The corresponding function descriptions are listed in Table 10.5. The relationship of these FORTRAN source files is schematically shown in Figure 10.4. The files for time calculation are written in FORTRAN 90, and can be removed from the entire package or can be replaced by other user added subroutines for calculating the elapsed CPU time. These FORTRAN 90 files for time calculation are not shown in Figure 10.4.

Table 10.3 Files contained in the 3D SPH code package

File type	File name		
Including file	param.inc		
F90 source files	time_elapsed.f90		Time_print.f90
F77 source files	art_heat.f	art_visc.f	av_vel.f
	density.f	direct_find.f	eos.f
	external_force.f	hsml.f	input.f
	internal_force.f	kernel.f	output.f
	single_step.f	sph.f	time_integration.f
	virt_part.f	viscosity.f	

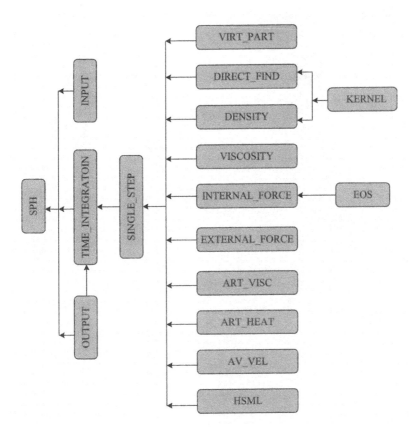

Figure 10.4 Relationship of the source files used in the attached SPH code.

Key variables used in the code

Some key variables are listed in Table 10.4 where I, DB and L represent integer, double precision and logical, respectively.

Table 10.4 Key variables and corresponding description

Variable name	Type	Description
General variables		
maxn	I	Maximum number of particles (real + virtual)
ntotal	I	Total number of real particles used in the simulation
nvirt	I	Number of virtual particles
itype	I	Particle type
x	DB	Particle position
vx	DB	Particle velocity
dvx	DB	Particle acceleration
mass	DB	Particle mass
rho	DB	Particle density
p	DB	Pressure
u	DB	Internal energy
c	DB	Sound speed
hsml	DB	Particle smoothing length
eta	DB	Shear viscosity
$t_{\beta\alpha}$	DB	Viscous shear stress in α direction exerting on a plane perpendicular to β direction
Time integration parameters		
dt	DB	Time step used in the time integration scheme

time	DB	Current time instant
maxtimestep	I	Maximum number of time steps to be run
itimestep	I	Current number of time step

Particle interaction information

max_interation	I	Maximum allowed number of interaction pairs
niac	I	Number of interaction pairs
pair_i	I	List of the first partner of interaction pair
pair_j	I	List of the second partner of interaction pair
countiac	I	Number of neighboring particles for a specified particle
w	DB	Smoothing kernel for a given pair of interaction
dwdx	DB	First derivative for a given pair of interaction

Indices for various SPH implementations

pa_sph	I	Index for particle approximation algorithm
nnps	I	Index for nearest neighboring particle searching (nnps) method
sle	I	Index for smoothing length evolution (sle) algorithm
skf	I	Index for smoothing kernel function

Switches for various SPH implementations

summation_density	L	Switch for density approximation by the summation approach
average_velocity	L	Switch for XSPH technique (equation (4.92))
config_input	L	Switch for initial configuration input
virtual_part	L	Switch for virtual particle consideration
vp_input	L	Switch for virtual particle input
visc	L	Switch for viscosity consideration

ex_force	L	Switch for external force consideration
visc_artificial	L	Switch for artificial viscosity consideration
heat_artificial	L	Switch for artificial heat consideration
self_gravity	L	Switch for gravity consideration
nor_density	L	Switch for density normalization

Control parameter for particle information output

int_stat	L	Switch for print statistics about SPH particle interactions.
print_step	I	Time steps for printing particle information on screen
save_step	I	Time steps for saving particle information to external disk
moni_particle	I	The particle number to be monitored

Parameters employed for linked list algorithm

x_maxgeom	DB	Upper limit of allowed x-regime
x_mingeom	DB	Lower limit of allowed x-regime
y_maxgeom	DB	Upper limit of allowed y-regime
y_mingeom	DB	Lower limit of allowed y-regime
z_maxgeom	DB	Upper limit of allowed z-regime
z_mingeom	DB	Lower limit of allowed z-regime

Description of the program units

The function descriptions for the source files in the attached SPH code package are concisely shown in the following table.

Table 10.5 Description of the source files

File name	Descriptions
param.inc	File containing parameters and constants to be included into other F77 programs as the including header file. The parameters are explained in detail in the file.
art_heat.f	Subroutine to calculate the artificial heat. Details can be seen from Monaghan (1992), Fulk (1994) or the descriptions in chapter 4.
art_visc.f	Subroutine to calculate the artificial viscosity. Details can be seen from Monaghan (1992), Hernquist and Katz (1989) or the descriptions in Chapter 4.
av_vel.f	Subroutine to calculate correcting average velocity. Details can be seen from Monaghan (1992) (XSPH) and the descriptions in Chapter 4.
density.f	File for updating density by using summation density approach or continuity density approach. Details can be seen from Chapter 4.
direct_find.f	Subroutine to search nearest neighboring particles by simply paring the particle spacing with smoothing length.
eos.f	File for calculating pressure through density and energy with an equation of state.
external_force.f	Subroutine to calculate the external forces, e.g. gravitational force. The forces from the interactions with boundary virtual particles are also calculated here as external forces.
hsml.f	Subroutine for updating smoothing length, Related algorithm could be seen from Chapter 4. More complicated algorithms to keep the neighboring particles roughly constant are not given herein.

input.f	File for loading or generating initial configuration. The initial particle information for 1D shock tube and 2D shear driven cavity are given. Initial particle configuration for other cases can be loaded from external disk files or generated by the users.
internal_force.f	Subroutine for calculating SPH interaction force. Details can bee seen from Chapter 4. The internal force is calculated by using nested SPH approximations on the shear stress. Two types of SPH particle approximation formulations are realized in this subroutine. Other related references are (Riffert et al., 1995), (Flebbe et al., 1994).
kernel.f	Subroutine for calculating smoothing function and its derivatives. The W4 cublic spline (Monaghan, 1992), Gauss function (Gingold and Monaghan, 1982) and Quintic smoothing function (Morris, 1997) are implemented. Detailed descriptions of these smoothing kernels can be seen in Chapter 3.
output.f	Subroutine for saving particle configurations to disk file.
single_step.f	Subroutine to carry out a single step for the time integration algorithm.
sph.f	Main program of the sample SPH program
time_integration.f	Subroutine for Leapfrog time integration, in which the particle velocities and positions are offset by a half time step when integrating the equation of motion. For simplicity, all particles are advanced with the same time step, and the time step keeps unchanged in the system evolution. However, the time step can be temporally or spatially (to each particle) variable. Related references are (Hernquist and Katz, 1989), (Simpson, 1995), (Monaghan, 1992) and etc.
virt_part.f	Subroutine to load virtual particles or generate virtual particle information according the problem geometry. The type I virtual particles for the 2D shear driven cavity problem are given as an example. More details about virtual or ghost

	particle generation can be referred to Monaghan (1994), Libersky et al. (1993), Morris (1997), Dilts and Haque (2000) and Chapter 4
viscosity.f	Subroutine to define fluid particle viscosity.
time_print.f90	F90 Subroutine for printing out the current date and time. It can be removed from the package if conflicting the system.
time_elapsed.f90	F90 subroutine for calculating elapsed CPU time. It can be removed from the package.

10.4.4 Two benchmark problems

Two benchmark problems are provided for test runs of the SPH code, one is the 1 D shock tube problem described in Chapter 3 and Chapter 5; another is the 2 D shear driven cavity problem described in Chapter 4. These two benchmark problems belong to two typical classes of problems as the first 1 D shock tube involves compressible gas flow with shock wave; while the second 2 D shear cavity problem involves viscous incompressible flow with low Reynolds number. Through varying the simulation control parameters, different SPH algorithms such as different SPH approximation schemes, different smoothing functions, and different smoothing length evolution models etc. can be investigated. Observations from these two benchmark problems can be extended to many other applications.

1 D shock tube

For the problem of 1D shear driven cavity, the initial particle information is generated in the subroutine *shock_tube* in *input.f*. The time step is set to 0.005 s in *sph.f* with 30 steps in the run. The initial particle information is the same as that described in Chapter 3. No virtual boundary particles are used in this case since the simulation is only run to the instant of 0.15 s. Before this instant, the boundary effect has not propagated to influence the interested shock phenomena. There is no external force and no viscosity in the simulation. The gamma law equation of state for idea gas is used in *eos.f*.

To obtain results comparable with those by Hernquist and Katz (1989), the control parameter in *param.inc* are shown in Table 10.6 through varying the control parameters in *param.inc*, different SPH algorithms can be investigated.

Table 10.6 Control parameters used in the 1 D shock tube problem

Control parameters	Value
dim	1
pa_sph	2
sle	0
skf	1
summation_density	.true.
average_velocity	.false.
config_input	.false.
virtual_part	.false.
vp_input	.false.
visc	.false.
ex_force	.false.
visc_artificial	.true.
heat_artificial	.false.
self_gravity	.false.
nor_density	.false.
shocktube	.true.
shearcavity	.false.

2 D Shear driven cavity

For the problem of 2 D shear driven cavity, the initial particle information is generated in the subroutine *shear_cavity* in *input.f.* As can be seen from subroutine *shear_cavity*, the side of the square domain is $l = 10^{-3}$ m; the density of the fluid is $\rho = 10^3$ kg/m³. A total of 1600 (40×40) real particles are placed in the square region. The smoothing length is equal to the initial particle spacing and keeps constant in the simulation. The dynamic viscosity of the fluid is given in *viscosity.f* as 1.0×10^{-3} Pa·s . Since the density is $\rho = 10^3$ kg/m³, the corresponding kinetic viscosity is $\upsilon = 10^{-6}$ m²/s.

In ***virt_part.f***, the virtual boundary particles of type I (see, Chapter 4) are generated. 320 (81 on each side of the square) virtual particles are placed right on the four edges. The velocities of the particles located on the upper edge are 1.0×10^{-3} m/s; the velocities for other virtual particles are set to 0. This treatment actually gives the velocity boundary conditions for the shear cavity problem. When the SPH run begins, virtual particles interact with inner real particles while the variation of the field properties propagates over the domain. Except for the SPH type interaction with neighbor real particles, the virtual particles also serve to exert a repulsive force to push back those particles that tend to penetrate the boundary. The repulsive force is of the form in equation (4.93), where the parameter D, n_1, n_2 are taken as 0.01, 12 and 4 respectively. The cutoff distance r_0 is the same as the virtual particle spacing, i.e. 1.25×10^{-5} m/s. The repulsive force is implemented in ***external_force.f*** and is regarded as a special kind of external force.

In the subroutine ***p_art_water*** of ***eos.f***, the pressure-density relation is defined according to equation (3.89). The sound speed is 0.01 m/s, which corresponds to a Mach number of 0.1.

To obtain the results presented in Chapter 4, the time step is set to 5×10^{-5} s in ***sph.f*** with around 3000 steps in the run. The control parameters in ***param.inc*** are shown in Table 10.7. Fine tune of the control parameters for different SPH approaches (e.g. different particle approximation approaches, different smoothing functions, considering artificial viscosity, artificial heat, smoothing length evolution, etc.) may result in better results.

Table 10.7 Control parameters used in the shear cavity simulation.

Control parameters	Value
dim	2
pa_sph	2
sle	0
skf	1
summation_density	.true.
average_velocity	.true.
config_input	.false.
virtual_part	.true.

vp_input	.false.
visc	.true.
ex_force	.true.
visc_artificial	.false.
heat_artificial	.false.
self_gravity	.false.
nor_density	.true.
shocktube	.false.
shearcavity	.true.

10.4.5 List of the FORTRAN source files

1. param.inc

```
c-------------------------------------------------------------
c     Including file for parameters and constants used
c     in the entire SPH software packages.
c-------------------------------------------------------------

c     dim : Dimension of the problem (1, 2 or 3)
      integer dim
      parameter ( dim = 1)

c     maxn     : Maximum number of particles
c     max_interation : Maximum number of interaction pairs
      integer maxn,max_interaction
      parameter ( maxn    = 12000    ,
     &            max_interaction = 100 * maxn )

c     Parameters for the computational geometry,
c     x_maxgeom : Upper limit of allowed x-regime
c     x_mingeom : Lower limit of allowed x-regime
c     y_maxgeom : Upper limit of allowed y-regime
c     y_mingeom : Lower limit of allowed y-regime
c     z_maxgeom : Upper limit of allowed z-regime
c     z_mingeom : Lower limit of allowed z-regime
      double precision x_maxgeom,x_mingeom,y_maxgeom,
     &                 y_mingeom,z_maxgeom,z_mingeom
      parameter ( x_maxgeom =  10.e0    ,
     &            x_mingeom = -10.e0    ,
     &            y_maxgeom =  10.e0    ,
     &            y_mingeom = -10.e0    ,
     &            z_maxgeom =  10.e0    ,
     &            z_mingeom = -10.e0    )

c     SPH algorithm for particle approximation (pa_sph)
c     pa_sph = 1 : (e.g. (p(i)+p(j))/(rho(i)*rho(j))
c              2 : (e.g. (p(i)/rho(i)**2+p(j)/rho(j)**2)
      integer pa_sph
      parameter(pa_sph = 2)

c     Nearest neighboring particle searching (nnps) method
c     nnps = 1 : Simplest and direct searching
c            2 : Sorting grid linked list
c            3 : Tree algorithm
      integer nnps
      parameter(nnps = 1 )

c     Smoothing length evolution (sle) algorithm
c     sle = 0 : Keep unchanged,
c           1 : h = fac * (m/rho)^(1/dim)
c           2 : dh/dt = (-1/dim)*(h/rho)*(drho/dt)
c           3 : Other approaches (e.g. h = h_0 * (rho_0/rho)**(1/dim) )

      integer sle
      parameter(sle = 0)

c     Smoothing kernel function
c     skf = 1, cubic spline kernel by W4 - Spline (Monaghan 1985)
c         = 2, Gauss kernel   (Gingold and Monaghan 1981)
c         = 3, Quintic kernel (Morris 1997)
      integer skf
      parameter(skf = 1)

c     Switches for different senarios

c     summation_density = .TRUE. : Use density summation model in the code,
```

```
c                       .FALSE.: Use continuiity equation
c       average_velocity = .TRUE. : Monaghan treatment on average velocity,
c                       .FALSE.: No average treatment.
c       config_input = .TRUE. : Load initial configuration data,
c                    .FALSE.: Generate initial configuration.
c       virtual_part = .TRUE. : Use vritual particle,
c                    .FALSE.: No use of vritual particle.
c       vp_input = .TRUE. : Load virtual particle information,
c                  .FALSE.: Generate virtual particle information.
c       visc = .true. : Consider viscosity,
c              .false.: No viscosity.
c       ex_force =.true. : Consider external force,
c                .false.: No external force.
c       visc_artificial = .true. : Consider artificial viscosity,
c                        .false.: No considering of artificial viscosity.
c       heat_artificial = .true. : Consider artificial heating,
c                        .false.: No considering of artificial heating.
c       self_gravity = .true. : Considering self_gravity,
c                    .false.: No considering of self_gravity
c       nor_density =  .true. : Density normalization by using CSPM,
c                     .false.: No normalization.
        logical summation_density, average_velocity, config_input,
     &          virtual_part, vp_input, visc, ex_force, heat_artificial,
     &          visc_artificial, self_gravity, nor_density
        parameter ( summation_density = .true. )
        parameter ( average_velocity = .false. )
        parameter ( config_input = .false. )
        parameter ( virtual_part = .false. )
        parameter ( vp_input = .false. )
        parameter ( visc = .false. )
        parameter ( ex_force = .false.)
        parameter ( visc_artificial = .true. )
        parameter ( heat_artificial = .false. )
        parameter ( self_gravity = .false. )
        parameter ( nor_density = .false. )

c       Symmetry of the problem
c       nsym = 0 : no symmetry,
c            = 1 : axis symmetry,
c            = 2 : center symmetry.
        integer   nsym
        parameter ( nsym = 0)

c       Control parameters for output
c       int_stat = .true. : Print statistics about SPH particle interactions.
c                    including virtual particle information.
c       print_step: Print Timestep (On Screen)
c       save_step : Save Timestep   (To Disk File)
c       moni_particle: The particle number for information monitoring.
        logical int_stat
        parameter ( int_stat = .true. )
        integer print_step, save_step, moni_particle
        parameter ( print_step = 100 ,
     &             save_step = 500,
     &             moni_particle = 1600   )

        double precision pi
        parameter ( pi = 3.14159265358979323846 )

c       Simulation cases
c       shocktube = .true. : carry out shock tube simulation
c       shearcavity = .true. : carry out shear cavity simulation
        logical shocktube, shearcavity
        parameter ( shocktube = .true. )
        parameter ( shearcavity = .false. )
```

2. art_heat.f

```
      subroutine art_heat(ntotal,hsml,mass,x,vx,niac,rho,u,
     &           c,pair_i,pair_j,w,dwdx,dedt)

c-------------------------------------------------------------------
c     Subroutine to calculate the artificial heat(Fulk, 1994, p, a-17)

c     ntotal : Number of particles                              [in]
c     hsml   : Smoothing Length                                 [in]
c     mass   : Particle masses                                  [in]
c     x      : Coordinates of all particles                     [in]
c     vx     : Velocities of all particles                      [in]
c     rho    : Density                                          [in]
c     u      : specific internal energy                         [in]
c     c      : Sound veolcity                                   [in]
c     niac   : Number of interaction pairs                      [in]
c     pair_i : List of first partner of interaction pair        [in]
c     pair_j : List of second partner of interaction pair       [in]
c     w      : Kernel for all interaction pairs                 [in]
c     dwdx   : Derivative of kernel with respect to x, y and z   [in]
c     dedt   : produced artificial heat, adding to energy Eq.   [out]

      implicit none
      include 'param.inc'

      integer ntotal,niac,pair_i(max_interaction),
     &        pair_j(max_interaction)
      double precision hsml(maxn), mass(maxn), x(dim,maxn),vx(dim,maxn),
     &                 rho(maxn), u(maxn), c(maxn),w(max_interaction),
     &                 dwdx(dim,max_interaction), dedt(maxn)
      integer i,j,k,d
      double precision dx, dvx(dim), vr, rr, h, mc, mrho, mhsml,
     &                 vcc(maxn), hvcc, mui, muj, muij, rdwdx, g1,g2

c---  Parameter for the artificial heat conduction:

      g1=0.1
      g2=1.0
      do i=1,ntotal
        vcc(i) = 0.e0
        dedt(i) = 0.e0
      enddo

      do k=1,niac
        i = pair_i(k)
        j = pair_j(k)
        do d=1,dim
          dvx(d) = vx(d,j) - vx(d,i)
        enddo
        hvcc = dvx(1)*dwdx(1,k)
        do d=2,dim
          hvcc = hvcc + dvx(d)*dwdx(d,k)
        enddo
        vcc(i) = vcc(i) + mass(j)*hvcc/rho(j)
        vcc(j) = vcc(j) + mass(i)*hvcc/rho(i)
      enddo

      do k=1,niac
        i = pair_i(k)
        j = pair_j(k)
        mhsml= (hsml(i)+hsml(j))/2.
        mrho = 0.5e0*(rho(i) + rho(j))
        rr = 0.e0
        rdwdx = 0.e0
        do d=1,dim
          dx = x(d,i) - x(d,j)
          rr = rr + dx*dx
          rdwdx = rdwdx + dx*dwdx(d,k)
```

```
      enddo
      mui=g1*hsml(i)*c(i)  + g2*hsml(i)**2*(abs(vcc(i))-vcc(i))
      muj=g1*hsml(j)*c(j)  + g2*hsml(j)**2*(abs(vcc(j))-vcc(j))
      muij= 0.5*(mui+muj)
      h = muij/(mrho*(rr+0.01*mhsml**2))*rdwdx
      dedt(i) = dedt(i) + mass(j)*h*(u(i)-u(j))
      dedt(j) = dedt(j) + mass(i)*h*(u(j)-u(i))
   enddo

   do i=1,ntotal
      dedt(i) = 2.0e0*dedt(i)
   enddo

   end
```

3. art_visc.f

```
      subroutine art_visc(ntotal,hsml,mass,x,vx,niac,rho,c,
     &           pair_i,pair_j,w,dwdx,dvxdt,dedt)

c-----------------------------------------------------------------------
c     Subroutine to calculate the artificial viscosity (Monaghan, 1992)

c     ntotal : Number of particles (including virtual particles) [in]
c     hsml   : Smoothing Length                                  [in]
c     mass   : Particle masses                                   [in]
c     x      : Coordinates of all particles                      [in]
c     vx     : Velocities of all particles                       [in]
c     niac   : Number of interaction pairs                       [in]
c     rho    : Density                                           [in]
c     c      : Temperature                                       [in]
c     pair_i : List of first partner of interaction pair         [in]
c     pair_j : List of second partner of interaction pair        [in]
c     w      : Kernel for all interaction pairs                  [in]
c     dwdx   : Derivative of kernel with respect to x, y and z   [in]
c     dvxdt  : Acceleration with respect to x, y and z           [out]
c     dedt   : Change of specific internal energy               [out]

      implicit none
      include 'param.inc'

      integer ntotal, niac, pair_i(max_interaction),
     &        pair_j(max_interaction)
      double precision hsml(maxn), mass(maxn), x(dim,maxn),vx(dim,maxn),
     &     rho(maxn), c(maxn), w(max_interaction),
     &     dwdx(dim,max_interaction), dvxdt(dim,maxn), dedt(maxn)
      integer i,j,k,d
      double precision dx, dvx(dim), alpha, beta, etq, piv,
     &        muv, vr, rr, h, mc, mrho, mhsml

c     Parameter for the artificial viscosity:
c     Shear viscosity
      parameter( alpha = 1.e0  )

c     Bulk viscosity
      parameter( beta  = 1.e0  )

c     Parameter to avoid singularities
      parameter( etq   = 0.1e0 )

      do i=1,ntotal
        do d=1,dim
          dvxdt(d,i) = 0.e0
        enddo
        dedt(i) = 0.e0
      enddo

c     Calculate SPH sum for artificial viscosity

      do k=1,niac
        i = pair_i(k)
        j = pair_j(k)
        mhsml= (hsml(i)+hsml(j))/2.
        vr = 0.e0
        rr = 0.e0
        do d=1,dim
          dvx(d) = vx(d,i) - vx(d,j)
          dx     = x(d,i) -  x(d,j)
          vr     = vr + dvx(d)*dx
          rr     = rr + dx*dx
        enddo

c     Artificial viscous force only if v_ij * r_ij < 0

        if (vr.lt.0.e0) then
```

```
c       Calculate muv_ij = hsml v_ij * r_ij / ( r_ij^2 + hsml^2 etq^2 )

            muv = mhsml*vr/(rr + mhsml*mhsml*etq*etq)

c       Calculate PIv_ij = (-alpha muv_ij c_ij + beta muv_ij^2) / rho_ij

            mc   = 0.5e0*(c(i) + c(j))
            mrho = 0.5e0*(rho(i) + rho(j))
            piv  = (beta*muv - alpha*mc)*muv/mrho

c       Calculate SPH sum for artificial viscous force

            do d=1,dim
              h = -piv*dwdx(d,k)
              dvxdt(d,i) = dvxdt(d,i) + mass(j)*h
              dvxdt(d,j) = dvxdt(d,j) - mass(i)*h
              dedt(i) = dedt(i) - mass(j)*dvx(d)*h
              dedt(j) = dedt(j) - mass(i)*dvx(d)*h
            enddo
          endif
        enddo

c    Change of specific internal energy:

        do i=1,ntotal
          dedt(i) = 0.5e0*dedt(i)
        enddo

        end
```

4. av_vel.f

```
      subroutine av_vel(ntotal,mass,niac,pair_i,pair_j,
     &           w, vx, rho, av)

c-----------------------------------------------------------------------
c     Subroutine to calculate the average velocity to correct velocity
c     for preventing.penetration (monaghan, 1992)

c     ntotal : Number of particles                              [in]
c     mass   : Particle masses                                  [in]
c     niac   : Number of interaction pairs                      [in]
c     pair_i : List of first partner of interaction pair        [in]
c     pair_j : List of second partner of interaction pair       [in]
c     w      : Kernel for all interaction pairs                 [in]
c     vx     : Velocity of each particle                        [in]
c     rho    : Density of each particle                         [in]
c     av     : Average velocityof each particle                 [out]

      implicit none
      include 'param.inc'

      integer ntotal, niac, pair_i(max_interaction),
     &        pair_j(max_interaction)
      double precision  mass(maxn),w(max_interaction),
     &        vx(dim,maxn), rho(maxn), av(dim, maxn)
      integer i,j,k,d
      double precision   vcc, dvx(dim), epsilon

c     epsilon --- a small constants chosen by experence,
c                 may lead to instability.
c     for example, for the 1 dimensional shock tube problem, the E <= 0.3

      epsilon = 0.3

      do i = 1, ntotal
        do d = 1, dim
          av(d,i) = 0.
        enddo
      enddo

      do k=1,niac
        i = pair_i(k)
        j = pair_j(k)
        do d=1,dim
          dvx(d) = vx(d,i) - vx(d,j)
          av(d, i) = av(d,i) - 2*mass(j)*dvx(d)/(rho(i)+rho(j))*w(k)
          av(d, j) = av(d,j) + 2*mass(i)*dvx(d)/(rho(i)+rho(j))*w(k)
        enddo
      enddo

      do i = 1, ntotal
        do d = 1, dim
          av(d,i) = epsilon * av(d,i)
        enddo
      enddo

      end
```

5. density.f

```
      subroutine sum_density(ntotal,hsml,mass,niac,pair_i,pair_j,w,
     &           itype,rho)

C-------------------------------------------------------------------------
C    Subroutine to calculate the density with SPH summation algorithm.
C    ntotal : Number of particles                                     [in]
C    hsml   : Smoothing Length                                        [in]
C    mass   : Particle masses                                         [in]
C    niac   : Number of interaction pairs                             [in]
C    pair_i : List of first partner of interaction pair              [in]
C    pair_j : List of second partner of interaction pair             [in]
C    w      : Kernel for all interaction pairs                        [in]
c    itype  : type of particles                                      [in]
c    x      : Coordinates of all particles                           [in]
c    rho    : Density                                               [out]

      implicit none
      include 'param.inc'

      integer ntotal, niac, pair_i(max_interaction),
     &        pair_j(max_interaction), itype(maxn)
      double precision hsml(maxn),mass(maxn), w(max_interaction),
     &        rho(maxn)
      integer i, j, k, d
      double precision selfdens, hv(dim), r, wi(maxn)

c     wi(maxn)---integration of the kernel itself
      do d=1,dim
        hv(d) = 0.e0
      enddo

c     Self density of each particle: Wii (Kernel for distance 0)
c     and take contribution of particle itself:

      r=0.

c     Firstly calculate the integration of the kernel over the space

      do i=1,ntotal
        call kernel(r,hv,hsml(i),selfdens,hv)
        wi(i)=selfdens*mass(i)/rho(i)
      enddo

      do k=1,niac
        i = pair_i(k)
        j = pair_j(k)
        wi(i) = wi(i) + mass(j)/rho(j)*w(k)
        wi(j) = wi(j) + mass(i)/rho(i)*w(k)
      enddo

c     Secondly calculate the rho integration over the space

      do i=1,ntotal
        call kernel(r,hv,hsml(i),selfdens,hv)
        rho(i) = selfdens*mass(i)
      enddo

c     Calculate SPH sum for rho:
      do k=1,niac
        i = pair_i(k)
        j = pair_j(k)
        rho(i) = rho(i) + mass(j)*w(k)
        rho(j) = rho(j) + mass(i)*w(k)
      enddo

c     Thirdly, calculate the normalized rho, rho=sum(rho)/sum(w)

      if (nor_density) then
```

```
      do i=1, ntotal
        rho(i)=rho(i)/wi(i)
      enddo
    endif

    end

    subroutine con_density(ntotal,mass,niac,pair_i,pair_j,
   &                dwdx,vx,itype,x,rho, drhodt)

c---------------------------------------------------------------------
c     Subroutine to calculate the density with SPH continuiity approach.

c     ntotal : Number of particles                                [in]
c     mass   : Particle masses                                    [in]
c     niac   : Number of interaction pairs                        [in]
c     pair_i : List of first partner of interaction pair          [in]
c     pair_j : List of second partner of interaction pair         [in]
c     dwdx   : derivation of Kernel for all interaction pairs      [in]
c     vx     : Velocities of all particles                        [in]
c     itype   : type of particles                                 [in]
c     x      : Coordinates of all particles                       [in]
c     rho    : Density                                            [in]
c     drhodt : Density change rate of each particle               [out]

    implicit none
    include 'param.inc'

    integer ntotal,niac,pair_i(max_interaction),
   &        pair_j(max_interaction), itype(maxn)
    double precision mass(maxn), dwdx(dim, max_interaction),
   &        vx(dim,maxn), x(dim,maxn), rho(maxn), drhodt(maxn)
    integer i,j,k,d
    double precision     vcc, dvx(dim)

    do i = 1, ntotal
      drhodt(i) = 0.
    enddo

    do k=1,niac
      i = pair_i(k)
      j = pair_j(k)
      do d=1,dim
        dvx(d) = vx(d,i) - vx(d,j)
      enddo
      vcc = dvx(1)*dwdx(1,k)
      do d=2,dim
        vcc = vcc + dvx(d)*dwdx(d,k)
      enddo
      drhodt(i) = drhodt(i) + mass(j)*vcc
      drhodt(j) = drhodt(j) + mass(i)*vcc
    enddo

    end
```

6. direct_find.f

```
      subroutine direct_find(itimestep, ntotal,hsml,x,niac,pair_i,
     &           pair_j,w,dwdx,countiac)

c----------------------------------------------------------------
c    Subroutine to calculate the smoothing funciton for each particle and
c    the interaction parameters used by the SPH algorithm. Interaction
c    pairs are determined by directly comparing the particle distance
c    with the corresponding smoothing length.

c    itimestep : Current time step                                [in]
c    ntotal    : Number of particles                              [in]
c    hsml      : Smoothing Length                                 [in]
c    x         : Coordinates of all particles                     [in]
c    niac      : Number of interaction pairs                      [out]
c    pair_i    : List of first partner of interaction pair        [out]
c    pair_j    : List of second partner of interaction pair       [out]
c    w         : Kernel for all interaction pairs                 [out]
c    dwdx      : Derivative of kernel with respect to x, y and z  [out]
c    countiac  : Number of neighboring particles                  [out]

      implicit none
      include 'param.inc'

      integer itimestep, ntotal,niac,pair_i(max_interaction),
     &        pair_j(max_interaction), countiac(maxn)
      double precision hsml(maxn), x(dim,maxn), w(max_interaction),
     &        dwdx(dim,max_interaction)
      integer i, j, d, sumiac, maxiac, miniac, noiac,
     &        maxp, minp, scale_k
      double precision dxiac(dim), driac, r, mhsml, tdwdx(dim)

      if (skf.eq.1) then
        scale_k = 2
      else if (skf.eq.2) then
        scale_k = 3
      else if (skf.eq.3) then
        scale_k = 3
      endif

      do i=1,ntotal
        countiac(i) = 0
      enddo

      niac = 0

      do i=1,ntotal-1
        do j = i+1, ntotal
          dxiac(1) = x(1,i) - x(1,j)
          driac    = dxiac(1)*dxiac(1)
          do d=2,dim
            dxiac(d) = x(d,i) - x(d,j)
            driac    = driac + dxiac(d)*dxiac(d)
          enddo
          mhsml = (hsml(i)+hsml(j))/2.
          if (sqrt(driac).lt.scale_k*mhsml) then
            if (niac.lt.max_interaction) then

c    Neighboring pair list, and totalinteraction number and
c    the interaction number for each particle

              niac = niac + 1
              pair_i(niac) = i
              pair_j(niac) = j
              r = sqrt(driac)
              countiac(i) = countiac(i) + 1
              countiac(j) = countiac(j) + 1

c    Kernel and derivations of kernel
```

```
            call kernel(r,dxiac,mhsml,w(niac),tdwdx)
            do d=1,dim
               dwdx(d,niac) = tdwdx(d)
            enddo
         else
            print *,
     &      ' >>> ERROR <<< : Too many interactions'
            stop
         endif
      endif
    enddo
  enddo

c   Statistics for the interaction

    sumiac = 0
    maxiac = 0
    miniac = 1000
    noiac  = 0
    do i=1,ntotal
      sumiac = sumiac + countiac(i)
      if (countiac(i).gt.maxiac) then
        maxiac = countiac(i)
        maxp = i
      endif
      if (countiac(i).lt.miniac) then
        miniac = countiac(i)
        minp = i
      endif
      if (countiac(i).eq.0)        noiac  = noiac + 1
    enddo

    if (mod(itimestep,print_step).eq.0) then
      if (int_stat) then
        print *,' >> Statistics: interactions per particle:'
        print *,'**** Particle:',maxp, ' maximal interactions:',maxiac
        print *,'**** Particle:',minp, ' minimal interactions:',miniac
        print *,'**** Average :',real(sumiac)/real(ntotal)
        print *,'**** Total pairs : ',niac
        print *,'**** Particles with no interactions:',noiac
      endif
    endif

    end
```

7. eos.f

```fortran
      subroutine p_gas(rho, u, p, c)

c-------------------------------------------------------------------------
c  Gamma law EOS: subroutine to calculate the pressure and sound

c     rho    : Density                                              [in]
c     u      : Internal energy                                      [in]
c     p      : Pressure                                             [out]
c     c      : sound velocity                                       [out]

      implicit none
      double precision rho, u, p, c
      double precision gamma

c     For air (idea gas)

      gamma=1.4
      p = (gamma-1) * rho * u
      c = sqrt((gamma-1) * u)

      end

      subroutine p_art_water(rho, p, c)

c-------------------------------------------------------------------------
c  Artificial equation of state for the artificial compressibility

c     rho    : Density                                              [in]
c     u      : Internal energy                                      [in]
c     p      : Pressure                                             [out]
c     c      : sound velocity                                       [out]
c     Equation of state for artificial compressibility

      implicit none
      double precision rho, u, p, c
      double precision gamma, rho0

c     Artificial EOS, Form 1 (Monaghan, 1994)
c       gamma=7.
c       rho0=1000.
c       b = 1.013e5
c       p = b*((rho/rho0)**gamma-1)
c       c = 1480.

c     Artificial EOS, Form 2 (Morris, 1997)
      c = 0.01
      p = c**2 * rho

      end
```

8. external_force.f

```
      subroutine ext_force(ntotal,mass,x,niac,pair_i,pair_j,
     &           itype,hsml,dvxdt)

c-----------------------------------------------------------------------
c     Subroutine to calculate the external forces, e.g. gravitational forces.
c     The forces from the interactions with boundary virtual particles
c     are also calculated here as external forces.

c     here as the external force.
c     ntotal  : Number of particles                        [in]
c     mass    : Particle masses                             [in]
c     x       : Coordinates of all particles                [in]
c     pair_i  : List of first partner of interaction pair   [in]
c     pair_j  : List of second partner of interaction pair  [in]
c     itype   : type of particles                           [in]
c     hsml    : Smoothing Length                            [in]
c     dvxdt   : Acceleration with respect to x, y and z     [out]

      implicit none
      include 'param.inc'

      integer ntotal, itype(maxn), niac,
     &        pair_i(max_interaction), pair_j(max_interaction)
      double precision mass(maxn), x(dim,maxn), hsml(maxn),
     &        dvxdt(dim,maxn)
      integer i, j, k, d
      double precision dx(dim), rr, f, rr0, dd, p1, p2

      do i = 1, ntotal
        do d = 1, dim
          dvxdt(d, i) = 0.
        enddo
      enddo

c     Consider self-gravity or not ?
      if (self_gravity) then
        do i = 1, ntotal
          dvxdt(dim, i) = -9.8
        enddo
      endif

c     Boundary particle force and penalty anti-penetration force.
      rr0 = 1.25e-5
      dd = 1.e-2
      p1 = 12
      p2 = 4

      do  k=1,niac
        i = pair_i(k)
        j = pair_j(k)
        if(itype(i).gt.0.and.itype(j).lt.0) then
          rr = 0.
          do d=1,dim
            dx(d) = x(d,i) - x(d,j)
            rr = rr + dx(d)*dx(d)
          enddo
          rr = sqrt(rr)
          if(rr.lt.rr0) then
            f = ((rr0/rr)**p1-(rr0/rr)**p2)/rr**2
            do d = 1, dim
              dvxdt(d, i) = dvxdt(d, i) + dd*dx(d)*f
            enddo
          endif
        endif
      enddo

      end
```

9. hsml.f

```
      subroutine h_upgrade(dt,ntotal, mass, vx, rho, niac,
     &           pair_i, pair_j, dwdx, hsml)

c----------------------------------------------------------------------
c     Subroutine to evolve smoothing length

c     dt     : time step                                      [in]
c     ntotal : Number of particles                            [in]
c     mass   : Particle masses                                [in]
c     vx     : Velocities of all particles                    [in]
c     rho    : Density                                        [in]
c     niac   : Number of interaction pairs                    [in]
c     pair_i : List of first partner of interaction pair      [in]
c     pair_j : List of second partner of interaction pair     [in]
c     dwdx   : Derivative of kernel with respect to x, y and z [in]
c     hsml   : Smoothing Length                            [in/out]

      implicit none
      include 'param.inc'

      integer ntotal, niac, pair_i(max_interaction),
     &        pair_j(max_interaction)
      double precision mass(maxn), vx(dim, maxn), rho(maxn),
     &        dwdx(dim, max_interaction), hsml(maxn)
      integer i,j,k,d
      double precision dt, fac, dvx(dim), hvcc, vcc(maxn), dhsml(maxn)

      if (sle.eq.0 ) then

c---  Keep smoothing length unchanged.
        return
      else if (sle.eq.2) then

c---  dh/dt = (-1/dim)*(h/rho)*(drho/dt).
        do i=1,ntotal
          vcc(i) = 0.e0
        enddo

        do k=1,niac
          i = pair_i(k)
          j = pair_j(k)
          do d=1,dim
            dvx(d) = vx(d,j) - vx(d,i)
          enddo
          hvcc = dvx(1)*dwdx(1,k)
          do d=2,dim
            hvcc = hvcc + dvx(d)*dwdx(d,k)
          enddo
          vcc(i) = vcc(i) + mass(j)*hvcc/rho(j)
          vcc(j) = vcc(j) + mass(i)*hvcc/rho(i)
        enddo

        do i = 1, ntotal
          dhsml(i) = (hsml(i)/dim)*vcc(i)
          hsml(i) = hsml(i) + dt*dhsml(i)
          if (hsml(i).le.0) hsml(i) = hsml(i) - dt*dhsml(i)
        enddo
      else if(sle.eq.1) then
        fac = 2.0
        do i = 1, ntotal
          hsml(i) = fac * (mass(i)/rho(i))**(1./dim)
        enddo

      endif

      end
```

10. input.f

```
      subroutine input(x, vx, mass, rho, p, u, itype, hsml, ntotal)

c-----------------------------------------------------------------------
c     Subroutine for loading or generating initial particle information

c     x-- coordinates of particles                           [out]
c     vx-- velocities of particles                           [out]
c     mass-- mass of particles                               [out]
c     rho-- dnesities of particles                           [out]
c     p-- pressure  of particles                             [out]
c     u-- internal energy of particles                       [out]
c     itype-- types of particles                             [out]
c     hsml-- smoothing lengths of particles                  [out]
c     ntotal-- total particle number                         [out]

      implicit none
      include 'param.inc'

      integer itype(maxn), ntotal
      double precision x(dim, maxn), vx(dim, maxn), mass(maxn),
     &                 p(maxn), u(maxn), hsml(maxn), rho(maxn)
      integer i, d, im

c     load initial particle information from external disk file

      if(config_input) then

         open(1,file="../data/f_xv.dat")
         open(2,file="../data/f_state.dat")
         open(3,file="../data/f_other.dat")

         write(*,*)'  **************************************************'
         write(*,*)'      Loading initial particle configuration...   '
         read (1,*) ntotal
         write(*,*)'      Total number of particles   ', ntotal
         write(*,*)'  **************************************************'
         do i = 1, ntotal
           read(1,*)im, (x(d, i),d = 1, dim), (vx(d, i),d = 1, dim)
           read(2,*)im, mass(i), rho(i), p(i), u(i)
           read(3,*)im, itype(i), hsml(i)
         enddo

      else

         open(1,file="../data/ini_xv.dat")
         open(2,file="../data/ini_state.dat")
         open(3,file="../data/ini_other.dat")

      if (shocktube) call shock_tube(x, vx, mass, rho, p, u,
     &                itype, hsml, ntotal)

      if (shearcavity) call shear_cavity(x, vx, mass, rho, p, u,
     &                itype, hsml, ntotal)
         do i = 1, ntotal
           write(1,1001) i, (x(d, i),d = 1, dim), (vx(d, i),d = 1, dim)
           write(2,1002) i, mass(i), rho(i), p(i), u(i)
           write(3,1003) i, itype(i), hsml(i)
         enddo
1001     format(1x, I5, 6(2x, e14.8))
1002     format(1x, I5, 7(2x, e14.8))
1003     format(1x, I5, 2x, I2, 2x, e14.8)
         write(*,*)'  **************************************************'
         write(*,*)'      Initial particle configuration generated   '
         write(*,*)'      Total number of particles   ', ntotal   ..........
         write(*,*)'  **************************************************'

      endif
```

```
          close(1)
          close(2)
          close(3)

          end


          subroutine shock_tube(x, vx, mass, rho, p, u,
        &                      itype, hsml, ntotal)
c-------------------------------------------------------------------
c     This subroutine is used to generate initial data for the
c     1 d noh shock tube problem
c     x-- coordinates of particles                          [out]
c     vx-- velocities of particles                          [out]
c     mass-- mass of particles                              [out]
c     rho-- dnesities of particles                          [out]
c     p-- pressure  of particles                            [out]
c     u-- internal energy of particles                      [out]
c     itype-- types of particles                            [out]
c         =1    ideal gas
c     hsml-- smoothing lengths of particles                 [out]
c     ntotal-- total particle number                        [out]

          implicit none
          include 'param.inc'

          integer itype(maxn), ntotal
          double precision x(dim, maxn), vx(dim, maxn), mass(maxn),
        &     rho(maxn), p(maxn), u(maxn), hsml(maxn)
          integer i, d
          double precision space_x

          ntotal=400
          space_x=0.6/80.

          do i=1,ntotal
            mass(i)=0.75/400.
            hsml(i)=0.015
            itype(i)=1
            do d = 1, dim
              x(d,i) = 0.
              vx(d,i) = 0.
            enddo
          enddo

          do i=1,320
            x(1,i)=-0.6+space_x/4.*(i-1)
          enddo

          do i=320+1,ntotal
            x(1,i)=0.+space_x*(i-320)
          enddo

          do i=1,ntotal
            if (x(1,i).le.1.e-8) then
              u(i)=2.5
              rho(i)=1.
              p(i)=1.
            endif
            if (x(1,i).gt.1.e-8)   then
              u(i)=1.795
              rho(i)=0.25
              p(i)=0.1795
            endif
          enddo

          end

          subroutine shear_cavity(x, vx, mass, rho, p, u,
        &                      itype, hsml, ntotal)
```

```
c-------------------------------------------------------------------------
c       This subroutine is used to generate initial data for the
c       2 d shear driven cavity probem with Re = 1
c       x-- coordinates of particles                              [out]
c       vx-- velocities of particles                              [out]
c       mass-- mass of particles                                  [out]
c       rho-- dnesities of particles                              [out]
c       p-- pressure  of particles                                [out]
c       u-- internal energy of particles                          [out]
c       itype-- types of particles                                [out]
c          =2    water
c       h-- smoothing lengths of particles                        [out]
c       ntotal-- total particle number                            [out]

        implicit none
        include 'param.inc'

        integer itype(maxn), ntotal
        double precision x(dim, maxn), vx(dim, maxn), mass(maxn),
     &      rho(maxn), p(maxn), u(maxn), hsml(maxn)
        integer i, j, d, m, n, mp, np, k
        double precision xl, yl, dx, dy

c       Giving mass and smoothing length as well as other data.

        m = 41
        n = 41
        mp = m-1
        np = n-1
        ntotal = mp * np
        xl = 1.e-3
        yl = 1.e-3
        dx = xl/mp
        dy = yl/np

        do i = 1, mp
          do j = 1, np
            k = j + (i-1)*np
            x(1, k) = (i-1)*dx + dx/2.
            x(2, k) = (j-1)*dy + dy/2.
          enddo
        enddo

        do i = 1, mp*np
          vx(1, i) = 0.
          vx(2, i) = 0.
          rho (i) = 1000.
          mass(i) = dx*dy*rho(i)
          p(i)= 0.
          u(i)=357.1
          itype(i) = 2
          hsml(i) = dx
        enddo

        end
```

11. internal_force.f

```
      subroutine int_force(itimestep,dt,ntotal,hsml,mass,vx,niac,rho,
     &          eta,pair_i,pair_j,dwdx,u,itype,x,t,c,p,dvxdt,tdsdt,dedt)

c-------------------------------------------------------------------------
c     Subroutine to calculate the internal forces on the right hand side
c     of the Navier-Stokes equations, i.e. the pressure gradient and the
c     gradient of the viscous stress tensor, used by the time integration.
c     Moreover the entropy production due to viscous dissipation, tds/dt,
c     and the change of internal energy per mass, de/dt, are calculated.

c     itimestep: Current timestep number                              [in]
c     dt       :   Time step                                          [in]
c     ntotal : Number of particles                                   [in]
c     hsml   : Smoothing Length                                      [in]
c     mass   : Particle masses                                       [in]
c     vx     : Velocities of all particles                           [in]
c     niac   : Number of interaction pairs                           [in]
c     rho    : Density                                               [in]
c     eta    : Dynamic viscosity                                     [in]
c     pair_i : List of first partner of interaction pair             [in]
c     pair_j : List of second partner of interaction pair            [in]
c     dwdx   : Derivative of kernel with respect to x, y and z        [in]
c     itype  : Type of particle (material types)                     [in]
c     u      : Particle internal energy                              [in]
c     x      : Particle coordinates                                  [in]
c     itype  : Particle type                                         [in]
c     t      : Particle temperature                              [in/out]
c     c      : Particle sound speed                                 [out]
c     p      : Particle pressure                                    [out]
c     dvxdt  : Acceleration with respect to x, y and z              [out]
c     tdsdt  : Production of viscous entropy                        [out]
c     dedt   : Change of specific internal energy                   [out]

      implicit none
      include 'param.inc'

      integer itimestep, ntotal,niac,pair_i(max_interaction),
     &        pair_j(max_interaction), itype(maxn)
      double precision dt, hsml(maxn), mass(maxn), vx(dim,maxn),
     &     rho(maxn), eta(maxn), dwdx(dim,max_interaction), u(maxn),
     &     x(dim,maxn), t(maxn), c(maxn), p(maxn), dvxdt(dim,maxn),
     &     tdsdt(maxn),dedt(maxn)
      integer i, j, k, d
      double precision  dvx(dim), txx(maxn), tyy(maxn),
     &     tzz(maxn), txy(maxn), txz(maxn), tyz(maxn), vcc(maxn),
     &     hxx, hyy, hzz, hxy, hxz, hyz, h, hvcc, he, rhoij

c     Initialization of shear tensor, velocity divergence,
c     viscous energy, internal energy, acceleration

      do i=1,ntotal
        txx(i) = 0.e0
        tyy(i) = 0.e0
        tzz(i) = 0.e0
        txy(i) = 0.e0
        txz(i) = 0.e0
        tyz(i) = 0.e0
        vcc(i) = 0.e0
        tdsdt(i) = 0.e0
        dedt(i) = 0.e0
        do d=1,dim
          dvxdt(d,i) = 0.e0
        enddo
      enddo

c     Calculate SPH sum for shear tensor Tab = va,b + vb,a - 2/3 delta_ab
vc,c
```

```
      if (visc) then
        do k=1,niac
          i = pair_i(k)
          j = pair_j(k)
          do d=1,dim
            dvx(d) = vx(d,j) - vx(d,i)
          enddo
          if (dim.eq.1) then
            hxx = 2.e0*dvx(1)*dwdx(1,k)
          else if (dim.eq.2) then
            hxx = 2.e0*dvx(1)*dwdx(1,k) -  dvx(2)*dwdx(2,k)
            hxy = dvx(1)*dwdx(2,k) + dvx(2)*dwdx(1,k)
            hyy = 2.e0*dvx(2)*dwdx(2,k) - dvx(1)*dwdx(1,k)
          else if (dim.eq.3) then
            hxx = 2.e0*dvx(1)*dwdx(1,k) - dvx(2)*dwdx(2,k)
     &                                  - dvx(3)*dwdx(3,k)
            hxy = dvx(1)*dwdx(2,k) + dvx(2)*dwdx(1,k)
            hxz = dvx(1)*dwdx(3,k) + dvx(3)*dwdx(1,k)
            hyy = 2.e0*dvx(2)*dwdx(2,k) - dvx(1)*dwdx(1,k)
     &                                  - dvx(3)*dwdx(3,k)
            hyz = dvx(2)*dwdx(3,k) + dvx(3)*dwdx(2,k)
            hzz = 2.e0*dvx(3)*dwdx(3,k) - dvx(1)*dwdx(1,k)
     &                                  - dvx(2)*dwdx(2,k)
          endif
          hxx = 2.e0/3.e0*hxx
          hyy = 2.e0/3.e0*hyy
          hzz = 2.e0/3.e0*hzz
          if (dim.eq.1) then
            txx(i) = txx(i) + mass(j)*hxx/rho(j)
            txx(j) = txx(j) + mass(i)*hxx/rho(i)
          else if (dim.eq.2) then
            txx(i) = txx(i) + mass(j)*hxx/rho(j)
            txx(j) = txx(j) + mass(i)*hxx/rho(i)
            txy(i) = txy(i) + mass(j)*hxy/rho(j)
            txy(j) = txy(j) + mass(i)*hxy/rho(i)
            tyy(i) = tyy(i) + mass(j)*hyy/rho(j)
            tyy(j) = tyy(j) + mass(i)*hyy/rho(i)
          else if (dim.eq.3) then
            txx(i) = txx(i) + mass(j)*hxx/rho(j)
            txx(j) = txx(j) + mass(i)*hxx/rho(i)
            txy(i) = txy(i) + mass(j)*hxy/rho(j)
            txy(j) = txy(j) + mass(i)*hxy/rho(i)
            txz(i) = txz(i) + mass(j)*hxz/rho(j)
            txz(j) = txz(j) + mass(i)*hxz/rho(i)
            tyy(i) = tyy(i) + mass(j)*hyy/rho(j)
            tyy(j) = tyy(j) + mass(i)*hyy/rho(i)
            tyz(i) = tyz(i) + mass(j)*hyz/rho(j)
            tyz(j) = tyz(j) + mass(i)*hyz/rho(i)
            tzz(i) = tzz(i) + mass(j)*hzz/rho(j)
            tzz(j) = tzz(j) + mass(i)*hzz/rho(i)
          endif

c     Calculate SPH sum for vc,c = dvx/dx + dvy/dy + dvz/dz:

          hvcc = 0.
          do d=1,dim
            hvcc = hvcc + dvx(d)*dwdx(d,k)
          enddo
          vcc(i) = vcc(i) + mass(j)*hvcc/rho(j)
          vcc(j) = vcc(j) + mass(i)*hvcc/rho(i)
        enddo
      endif

      do i=1,ntotal

c     Viscous entropy Tds/dt = 1/2 eta/rho Tab Tab

        if (visc) then
          if (dim.eq.1) then
            tdsdt(i) = txx(i)*txx(i)
          else if (dim.eq.2) then
```

```
                tdsdt(i) = txx(i)*txx(i)  + 2.e0*txy(i)*txy(i)
       &                              + tyy(i)*tyy(i)
            else if (dim.eq.3) then
                tdsdt(i) = txx(i)*txx(i)  + 2.e0*txy(i)*txy(i)
       &                              + 2.e0*txz(i)*txz(i)
       &                  + tyy(i)*tyy(i)  + 2.e0*tyz(i)*tyz(i)
       &                              + tzz(i)*tzz(i)
            endif
            tdsdt(i) = 0.5e0*eta(i)/rho(i)*tdsdt(i)
          endif

c      Pressure from equation of state

          if (abs(itype(i)).eq.1) then
            call p_gas(rho(i), u(i), p(i),c(i))
          else if (abs(itype(i)).eq.2) then
            call p_art_water(rho(i), p(i), c(i))
          endif

      enddo

c      Calculate SPH sum for pressure force -p,a/rho
c      and viscous force (eta Tab),b/rho
c      and the internal energy change de/dt due to -p/rho vc,c

      do k=1,niac
        i = pair_i(k)
        j = pair_j(k)
        he = 0.e0

c      For SPH algorithm 1

        rhoij = 1.e0/(rho(i)*rho(j))
        if(pa_sph.eq.1) then
          do d=1,dim

c      Pressure part

            h = -(p(i) + p(j))*dwdx(d,k)
            he = he + (vx(d,j) - vx(d,i))*h

c      Viscous force

            if (visc) then

              if (d.eq.1) then

c      x-coordinate of acceleration

                h = h + (eta(i)*txx(i) + eta(j)*txx(j))*dwdx(1,k)
                if (dim.ge.2) then
                  h = h + (eta(i)*txy(i) + eta(j)*txy(j))*dwdx(2,k)
                  if (dim.eq.3) then
                    h = h + (eta(i)*txz(i) + eta(j)*txz(j))*dwdx(3,k)
                  endif
                endif
              elseif (d.eq.2) then

c      y-coordinate of acceleration

                h = h + (eta(i)*txy(i) + eta(j)*txy(j))*dwdx(1,k)
       &                  + (eta(i)*tyy(i) + eta(j)*tyy(j))*dwdx(2,k)
                if (dim.eq.3) then
                  h = h + (eta(i)*tyz(i) + eta(j)*tyz(j))*dwdx(3,k)
                endif
              elseif (d.eq.3) then

c      z-coordinate of acceleration

                h = h + (eta(i)*txz(i) + eta(j)*txz(j))*dwdx(1,k)
       &                  + (eta(i)*tyz(i) + eta(j)*tyz(j))*dwdx(2,k)
```

```
      &                      + (eta(i)*tzz(i) + eta(j)*tzz(j))*dwdx(3,k)
                  endif
                endif
                h = h*rhoij
                dvxdt(d,i) = dvxdt(d,i) + mass(j)*h
                dvxdt(d,j) = dvxdt(d,j) - mass(i)*h
              enddo
              he = he*rhoij
              dedt(i) = dedt(i) + mass(j)*he
              dedt(j) = dedt(j) + mass(i)*he

c     For SPH algorithm 2
            else if (pa_sph.eq.2) then
              do d=1,dim
                h = -(p(i)/rho(i)**2 + p(j)/rho(j)**2)*dwdx(d,k)
                he = he + (vx(d,j) - vx(d,i))*h

c     Viscous force

                if (visc) then
                  if (d.eq.1) then

c     x-coordinate of acceleration
                    h = h + (eta(i)*txx(i)/rho(i)**2 +
      &                      eta(j)*txx(j)/rho(j)**2)*dwdx(1,k)
                    if (dim.ge.2) then
                      h = h + (eta(i)*txy(i)/rho(i)**2 +
      &                        eta(j)*txy(j)/rho(j)**2)*dwdx(2,k)
                      if (dim.eq.3) then
                        h = h + (eta(i)*txz(i)/rho(i)**2 +
      &                          eta(j)*txz(j)/rho(j)**2)*dwdx(3,k)
                      endif
                    endif
                  elseif (d.eq.2) then

c     y-coordinate of acceleration

                    h = h + (eta(i)*txy(i)/rho(i)**2
      &                    + eta(j)*txy(j)/rho(j)**2)*dwdx(1,k)
      &                    + (eta(i)*tyy(i)/rho(i)**2
      &                    + eta(j)*tyy(j)/rho(j)**2)*dwdx(2,k)
                    if (dim.eq.3) then
                      h = h + (eta(i)*tyz(i)/rho(i)**2
      &                      + eta(j)*tyz(j)/rho(j)**2)*dwdx(3,k)
                    endif
                  elseif (d.eq.3) then

c     z-coordinate of acceleration

                    h = h + (eta(i)*txz(i)/rho(i)**2 +
      &                      eta(j)*txz(j)/rho(j)**2)*dwdx(1,k)
      &                    + (eta(i)*tyz(i)/rho(i)**2 +
      &                      eta(j)*tyz(j)/rho(j)**2)*dwdx(2,k)
      &                    + (eta(i)*tzz(i)/rho(i)**2 +
      &                      eta(j)*tzz(j)/rho(j)**2)*dwdx(3,k)
                  endif
                endif
                dvxdt(d,i) = dvxdt(d,i) + mass(j)*h
                dvxdt(d,j) = dvxdt(d,j) - mass(i)*h
              enddo
              dedt(i) = dedt(i) + mass(j)*he
              dedt(j) = dedt(j) + mass(i)*he
          endif
        enddo

c     Change of specific internal energy de/dt = T ds/dt - p/rho vc,c:
        do i=1,ntotal
            dedt(i) = tdsdt(i) + 0.5e0*dedt(i)
        enddo

        end
```

12. kernel.f

```fortran
subroutine kernel(r,dx,hsml,w,dwdx)

c------------------------------------------------------------------------
c   Subroutine to calculate the smoothing kernel wij and its
c   derivatives dwdxij.
c     if skf = 1, cubic spline kernel by W4 - Spline (Monaghan 1985)
c         = 2, Gauss kernel   (Gingold and Monaghan 1981)
c         = 3, Quintic kernel (Morris 1997)

c     r    : Distance between particles i and j            [in]
c     dx   : x-, y- and z-distance between i and j         [in]
c     hsml : Smoothing length                              [in]
c     w    : Kernel for all interaction pairs              [out]
c     dwdx : Derivative of kernel with respect to x, y and z  [out]

      implicit none
      include 'param.inc'

      double precision r, dx(dim), hsml, w, dwdx(dim)
      integer i, j, d
      double precision q, dw, factor

      q = r/hsml
      w = 0.e0
      do d=1,dim
        dwdx(d) = 0.e0
      enddo

      if (skf.eq.1) then

        if (dim.eq.1) then
          factor = 1.e0/hsml
        elseif (dim.eq.2) then
          factor = 15.e0/(7.e0*pi*hsml*hsml)
        elseif (dim.eq.3) then
          factor = 3.e0/(2.e0*pi*hsml*hsml*hsml)
        else
         print *,' >>> Error <<< : Wrong dimension: Dim =',dim
         stop
        endif
        if (q.ge.0.and.q.le.1.e0) then
          w = factor * (2./3. - q*q + q**3 / 2.)
          do d = 1, dim
            dwdx(d) = factor * (-2.+3./2.*q)/hsml**2 * dx(d)
          enddo
        else if (q.gt.1.e0.and.q.le.2) then
          w = factor * 1.e0/6.e0 * (2.-q)**3
          do d = 1, dim
            dwdx(d) =-factor * 1.e0/6.e0 * 3.*(2.-q)**2/hsml * (dx(d)/r)
          enddo
        else
          w=0.
          do d= 1, dim
            dwdx(d) = 0.
          enddo
        endif

      else if (skf.eq.2) then

        factor = 1.e0 / (hsml**dim * pi**(dim/2.))
        if(q.ge.0.and.q.le.3) then
          w = factor * exp(-q*q)
          do d = 1, dim
            dwdx(d) = w * ( -2.* dx(d)/hsml/hsml)
          enddo
        else
          w = 0.
          do d = 1, dim
```

```
              dwdx(d) = 0.
          enddo
        endif

    else if (skf.eq.3) then

        if (dim.eq.1) then
          factor = 1.e0 / (120.e0*hsml)
        elseif (dim.eq.2) then
          factor = 7.e0 / (478.e0*pi*hsml*hsml)
        elseif (dim.eq.3) then
          factor = 1.e0 / (120.e0*pi*hsml*hsml*hsml)
        else
          print *,' >>> Error <<< : Wrong dimension: Dim =',dim
          stop
        endif
        if(q.ge.0.and.q.le.1) then
          w = factor * ( (3-q)**5 - 6*(2-q)**5 + 15*(1-q)**5 )
          do d= 1, dim
            dwdx(d) = factor * ( (-120 + 120*q - 50*q**2)
  &                              / hsml**2 * dx(d) )
          enddo
        else if(q.gt.1.and.q.le.2) then
          w = factor * ( (3-q)**5 - 6*(2-q)**5 )
          do d= 1, dim
            dwdx(d) = factor * (-5*(3-q)**4 + 30*(2-q)**4)
  &                              / hsml * (dx(d)/r)
          enddo
        else if(q.gt.2.and.q.le.3) then
          w = factor * (3-q)**5
          do d= 1, dim
            dwdx(d) = factor * (-5*(3-q)**4) / hsml * (dx(d)/r)
          enddo
        else
          w = 0.
          do d = 1, dim
            dwdx(d) = 0.
          enddo
        endif

    endif

    end
```

13. output.f

```
      subroutine output(x, vx, mass, rho, p, u, c, itype, hsml, ntotal)

c----------------------------------------------------------------------
c     Subroutine for saving particle information to external disk file

c     x-- coordinates of particles                                [in]
c     vx-- velocities of particles                                [in]
c     mass-- mass of particles                                    [in]
c     rho-- dnesities of particles                                [in]
c     p-- pressure  of particles                                  [in]
c     u-- internal energy of particles                            [in]
c     c-- sound velocity of particles                             [in]
c     itype-- types of particles                                  [in]
c     hsml-- smoothing lengths of particles                       [in]
c     ntotal-- total particle number                              [in]

      implicit none
      include 'param.inc'

      integer itype(maxn), ntotal
      double precision x(dim, maxn), vx(dim, maxn), mass(maxn),
     &        rho(maxn),p(maxn), u(maxn), c(maxn), hsml(maxn)
      integer i, d, npart

      open(1,file="../data/f_xv.dat")
      open(2,file="../data/f_state.dat")
      open(3,file="../data/f_other.dat")

      write(1,*) ntotal
      do i = 1, ntotal
        write(1,1001) i, (x(d, i), d=1,dim), (vx(d, i), d = 1, dim)
        write(2,1002) i, mass(i), rho(i), p(i), u(i)
        write(3,1003) i, itype(i), hsml(i)
      enddo

1001  format(1x, I6, 6(2x, e14.8))
1002  format(1x, I6, 7(2x, e14.8))
1003  format(1x, I6, 2x, I4, 2x, e14.8)

      close(1)
      close(2)
      close(3)

      end
```

14. single_step.f

```
      subroutine single_step(itimestep, dt, ntotal, hsml, mass, x, vx,
     &                u, s,rho, p, t, tdsdt, dx, dvx, du, ds, drho,itype, av)

c-----------------------------------------------------------------------
c    Subroutine to determine the right hand side of a differential
c    equation in a single step for performing time integration

c    In this routine and its subroutines the SPH algorithms are performed.
c        itimestep: Current timestep number                        [in]
c        dt       :  Timestep                                       [in]
c        ntotal   :  Number of particles                           [in]
c        hsml     :  Smoothing Length                              [in]
c        mass     :  Particle masses                               [in]
c        x        :  Particle position                            [in]
c        vx       :  Particle velocity                            [in]
c        u        :  Particle internal energy                     [in]
c        s        :  Particle entropy (not used here)             [in]
c        rho      :  Density                                   [in/out]
c        p        :  Pressure                                    [out]
c        t        :  Temperature                               [in/out]
c        tdsdt    :  Production of viscous entropy t*ds/dt        [out]
c        dx       :  dx = vx = dx/dt                              [out]
c        dvx      :  dvx = dvx/dt, force per unit mass            [out]
c        du       :  du  = du/dt                                  [out]
c        ds       :  ds  = ds/dt                                  [out]
c        drho     :  drho =  drho/dt                              [out]
c        itype    :  Type of particle                             [in]
c        av       :  Monaghan average velocity                    [out]

      implicit none
      include 'param.inc'

      integer itimestep, ntotal, itype(maxn)
      double precision dt, hsml(maxn),  mass(maxn), x(dim,maxn),
     &         vx(dim,maxn), u(maxn), s(maxn), rho(maxn), p(maxn),
     &         t(maxn), tdsdt(maxn), dx(dim,maxn), dvx(dim,maxn),
     &         du(maxn), ds(maxn), drho(maxn), av(dim, maxn)
      integer i, d, nvirt, niac, pair_i(max_interaction),
     &         pair_j(max_interaction), ns(maxn)
      double precision w(max_interaction), dwdx(dim,max_interaction),
     &         indvxdt(dim,maxn),exdvxdt(dim,maxn),ardvxdt(dim,maxn),
     &         avdudt(maxn), ahdudt(maxn), c(maxn), eta(maxn)

      do  i=1,ntotal
        avdudt(i) = 0.
        ahdudt(i) = 0.
        do   d=1,dim
          indvxdt(d,i) = 0.
          ardvxdt(d,i) = 0.
          exdvxdt(d,i) = 0.
        enddo
      enddo

c---  Positions of virtual (boundary) particles:

      nvirt = 0
      if (virtual_part) then
        call virt_part(itimestep, ntotal,nvirt,hsml,mass,x,vx,
     &        rho,u,p,itype)
      endif

c---  Interaction parameters, calculating neighboring particles
c     and optimzing smoothing length

      if (nnps.eq.1) then
        call direct_find(itimestep, ntotal+nvirt,hsml,x,niac,pair_i,
     &        pair_j,w,dwdx,ns)
      else if (nnps.eq.2) then
```

```
         call link_list(itimestep, ntotal+nvirt,hsml(1),x,niac,pair_i,
     &        pair_j,w,dwdx,ns)
       else if (nnps.eq.3) then
         call tree_search(itimestep, ntotal+nvirt,hsml,x,niac,pair_i,
     &        pair_j,w,dwdx,ns)
       endif

c---   Density approximation or change rate
       if (summation_density) then
         call sum_density(ntotal+nvirt,hsml,mass,niac,pair_i,pair_j,w,
     &        itype,rho)
       else
         call con_density(ntotal+nvirt,mass,niac,pair_i,pair_j,
     &        dwdx,vx, itype,x,rho, drho)
       endif

c---   Dynamic viscosity:
       if (visc) call viscosity(ntotal+nvirt,itype,x,rho,eta)

c---   Internal forces:

       call int_force(itimestep,dt,ntotal+nvirt,hsml,mass,vx,niac,rho,
     &      eta, pair_i,pair_j,dwdx,u,itype,x,t,c,p,indvxdt,tdsdt,du)

c---   Artificial viscosity:

       if (visc_artificial) call art_visc(ntotal+nvirt,hsml,
     &        mass,x,vx,niac,rho,c,pair_i,pair_j,w,dwdx,ardvxdt,avdudt)

c---   External forces:

       if (ex_force) call ext_force(ntotal+nvirt,mass,x,niac,
     &                  pair_i,pair_j,itype, hsml, exdvxdt)

c      Calculating the neighboring particles and undating HSML

       if (sle.ne.0) call h_upgrade(dt,ntotal, mass, vx, rho, niac,
     &                  pair_i, pair_j, dwdx, hsml)

       if (heat_artificial) call art_heat(ntotal+nvirt,hsml,
     &         mass,x,vx,niac,rho,u, c,pair_i,pair_j,w,dwdx,ahdudt)

c      Calculating average velocity of each partile for avoiding penetration
       if (average_velocity) call av_vel(ntotal,mass,niac,pair_i,
     &                  pair_j, w, vx, rho, av)

c---   Convert velocity, force, and energy to f and dfdt

       do i=1,ntotal
         do d=1,dim
           dvx(d,i) = indvxdt(d,i) + exdvxdt(d,i) + ardvxdt(d,i)
         enddo
         du(i) = du(i) + avdudt(i) + ahdudt(i)
       enddo

       if (mod(itimestep,print_step).eq.0) then
         write(*,*)
         write(*,*) '**** Information for particle ****',
     &                  moni_particle
         write(*,101)'internal a ','artifical a=',
     &                  'external a ','total a '
         write(*,100)indvxdt(1,moni_particle),ardvxdt(1,moni_particle),
     &                  exdvxdt(1,moni_particle),dvx(1,moni_particle)
       endif
101    format(1x,4(2x,a12))
100    format(1x,4(2x,e12.6))

       end
```

15. sph.f

```
      program SPH

c-------------------------------------------------------------------------
c     This is a three dimensional SPH code. the followings are the
c     basic parameters needed in this codeor calculated by this code

c     mass-- mass of particles                                     [in]
c     ntotal-- total particle number ues                          [in]
c     dt--- Time step used in the time integration                [in]
c     itype-- types of particles                                  [in]
c     x-- coordinates of particles                            [in/out]
c     vx-- velocities of particles                            [in/out]
c     rho-- dnesities of particles                            [in/out]
c     p-- pressure  of particles                              [in/out]
c     u-- internal energy of particles                       [in/out]
c     hsml-- smoothing lengths of particles                   [in/out]
c     c-- sound velocity of particles                            [out]
c     s-- entropy of particles                                   [out]
c     e-- total energy of particles                              [out]

      implicit none
      include 'param.inc'

      integer ntotal, itype(maxn), maxtimestep, d, m, i, yesorno
      double precision x(dim,maxn), vx(dim, maxn), mass(maxn),rho(maxn),
     &       p(maxn), u(maxn), c(maxn), s(maxn), e(maxn), hsml(maxn), dt
      double precision s1, s2

      call time_print
      call time_elapsed(s1)

      if (shocktube)  dt = 0.005
      if (shearcavity) dt = 5.e-5
      call input(x, vx, mass, rho, p, u, itype, hsml, ntotal)
1     write(*,*)' ***************************************************'
      write(*,*)'            Please input the maximal time steps '
      write(*,*)' ***************************************************'
      read(*,*) maxtimestep
      call time_integration(x, vx, mass, rho, p, u, c, s, e, itype,
     &     hsml, ntotal, maxtimestep, dt )
      call output(x, vx, mass, rho, p, u, c, itype, hsml, ntotal)
      write(*,*)' ***************************************************'
      write(*,*) 'Are you going to run more time steps ? (0=No, 1=yes)'
      write(*,*)' ***************************************************'
      read (*,*) yesorno
      if (yesorno.ne.0) go to 1
      call time_print
      call time_elapsed(s2)
      write (*,*)'            Elapsed CPU time = ', s2-s1

      end
```

16. time_integration.f

```
      subroutine time_integration(x,vx, mass, rho, p, u, c, s, e, itype,
     &            hsml, ntotal, maxtimestep, dt )

c-------------------------------------------------------------------
c     x-- coordinates of particles                    [input/output]
c     vx-- velocities of particles                    [input/output]
c     mass-- mass of particles                               [input]
c     rho-- dnesities of particles                    [input/output]
c     p-- pressure  of particles                      [input/output]
c     u-- internal energy of particles                [input/output]
c     c-- sound velocity of particles                       [output]
c     s-- entropy of particles, not used here               [output]
c     e-- total energy of particles                         [output]
c     itype-- types of particles                             [input]
c           =1    ideal gas
c           =2    water
c           =3    tnt
c     hsml-- smoothing lengths of particles           [input/output]
c     ntotal-- total particle number                         [input]
c     maxtimestep-- maximum timesteps                        [input]
c     dt-- timestep                                          [input]


      implicit none
      include 'param.inc'

      integer itype(maxn), ntotal, maxtimestep
      double precision x(dim, maxn), vx(dim, maxn), mass(maxn),
     &       rho(maxn), p(maxn), u(maxn), c(maxn), s(maxn), e(maxn),
     &       hsml(maxn), dt
      integer i, j, k, itimestep, d, current_ts, nstart
      double precision  x_min(dim, maxn), v_min(dim, maxn), u_min(maxn),
     &       rho_min(maxn), dx(dim,maxn), dvx(dim, maxn), du(maxn),
     &       drho(maxn), av(dim, maxn), ds(maxn),
     &       t(maxn), tdsdt(maxn)
      double precision  time, temp_rho, temp_u

      do i = 1, ntotal
        do d = 1, dim
          av(d, i) = 0.
        enddo
      enddo

      do itimestep = nstart+1, nstart+maxtimestep
        current_ts=current_ts+1
        if (mod(itimestep,print_step).eq.0) then
          write(*,*)'_____'
          write(*,*)'  current number of time step =',
     &               itimestep,'   current time=', real(time+dt)
          write(*,*)'_____'
        endif

c     If not first time step, then update thermal energy, density and
c     velocity half a time step

        if (itimestep .ne. 1) then
          do i = 1, ntotal
            u_min(i) = u(i)
            temp_u=0.
            if (dim.eq.1) temp_u=-nsym*p(i)*vx(1,i)/x(1,i)/rho(i)
            u(i) = u(i) + (dt/2.)*(du(i)+temp_u)
            if(u(i).lt.0)  u(i) = 0.
            if (.not.summation_density) then
              rho_min(i) = rho(i)
              temp_rho=0.
              if (dim.eq.1) temp_rho=-nsym*rho(i)*vx(1,i)/x(1,i)
              rho(i) = rho(i) +(dt/2.)*( drho(i)+ temp_rho)
            endif
```

```
          do    1, dim
            v   ,(d, i) = vx(d, i)
            \        i) = vx(d, i) + (dt/2.)*dvx(d, i)
            en
          enddo
        endif

c---  Definition of variables out of the function vector:
        call single_step(itimestep, dt, ntotal, hsml, mass, x, vx, u, s,
     &       rho, p, t, tdsdt, dx, dvx, du, ds, drho,itype, av)
        if (itimestep .eq. 1) then
          do i=1,ntotal
            temp_u=0.
            if (dim.eq.1) temp_u=-nsym*p(i)*vx(1,i)/x(1,i)/rho(i)
            u(i) = u(i) + (dt/2.)*(du(i) + temp_u)
            if(u(i).lt.0)  u(i) = 0.
            if (.not.summation_density ) then
              temp_rho=0.
              if (dim.eq.1) temp_rho=-nsym*rho(i)*vx(1,i)/x(1,i)
              rho(i) = rho(i) + (dt/2.)* (drho(i)+temp_rho)
            endif
            do d = 1, dim
              vx(d, i) = vx(d, i) + (dt/2.) * dvx(d, i) + av(d, i)
              x(d, i) = x(d, i) + dt * vx(d, i)
            enddo
          enddo
        else
          do i=1,ntotal
            temp_u=0.
            if (dim.eq.1) temp_u=-nsym*p(i)*vx(1,i)/x(1,i)/rho(i)
            u(i) = u_min(i) + dt*(du(i)+temp_u)
            if(u(i).lt.0)  u(i) = 0.

            if (.not.summation_density ) then
              temp_rho=0.
              if (dim.eq.1) temp_rho=-nsym*rho(i)*vx(1,i)/x(1,i)
              rho(i) = rho_min(i) + dt*(drho(i)+temp_rho)
            endif

            do d = 1, dim
              vx(d, i) = v_min(d, i) + dt * dvx(d, i) + av(d, i)
              x(d, i) = x(d, i) + dt * vx(d, i)
            enddo
          enddo

        endif

        time = time + dt

        if (mod(itimestep,save_step).eq.0) then
          call output(x, vx, mass, rho, p, u, c, itype, hsml, ntotal)
        endif

        if (mod(itimestep,print_step).eq.0) then
          write(*,*)
          write(*,101)'x','velocity', 'dvx'
          write(*,100)x(1,moni_particle), vx(1,moni_particle),
     &                dvx(1,moni_particle)
        endif

101     format(1x,3(2x,a12))
100     format(1x,3(2x,e12.6))

      enddo

      nstart=current_ts

      end
```

17. virt_part.f

```fortran
      subroutine virt_part(itimestep, ntotal,nvirt,hsml,mass,x,vx,
     &           rho,u,p,itype)

c------------------------------------------------------------------------
c     Subroutine to determine the information of virtual particles
c     Here only the Monaghan type virtual particles for the 2D shear
c     cavity driven problem are generated.
c       itimestep : Current time step                         [in]
c       ntotal : Number of particles                          [in]
c       nvirt  : Number of virtual particles                  [out]
c       hsml   : Smoothing Length                             [in|out]
c       mass   : Particle masses                              [in|out]
c       x      : Coordinates of all particles                 [in|out]
c       vx     : Velocities of all particles                  [in|out]
c       rho    : Density                                      [in|out]
c       u      : internal energy                              [in|out]
c       itype  : type of particles                            [in|out]

      implicit none
      include 'param.inc'
      integer itimestep, ntotal, nvirt, itype(maxn)
      double precision hsml(maxn),mass(maxn),x(dim,maxn),vx(dim,maxn),
     &                 rho(maxn), u(maxn), p(maxn)
      integer i, j, d, im, mp
      double precision xl, dx, v_inf

      if (vp_input) then

        open(1,file="../data/xv_vp.dat")
        open(2,file="../data/state_vp.dat")
        open(3,file="../data/other_vp.dat")
        read(1,*) nvirt
        do j = 1, nvirt
          i = ntotal + j
          read(1,*)im, (x(d, i),d = 1, dim), (vx(d, i),d = 1, dim)
          read(2,*)im, mass(i), rho(i), p(i), u(i)
          read(3,*)im, itype(i), hsml(i)
        enddo
        close(1)
        close(2)
        close(3)

        else

        nvirt = 0
        mp = 40
        xl = 1.0e-3
        dx = xl / mp
        v_inf = 1.e-3

c     Monaghan type virtual particle on the Upper side

        do i = 1, 2*mp+1
          nvirt = nvirt + 1
          x(1, ntotal + nvirt) = (i-1)*dx/2
          x(2, ntotal + nvirt) = xl
          vx(1, ntotal + nvirt) = v_inf
          vx(2, ntotal + nvirt) = 0.
        enddo

c     Monaghan type virtual particle on the Lower side

        do i = 1, 2*mp+1
          nvirt = nvirt + 1
          x(1, ntotal + nvirt) = (i-1)*dx/2
          x(2, ntotal + nvirt) = 0.
          vx(1, ntotal + nvirt) = 0.
          vx(2, ntotal + nvirt) = 0.
```

```
            enddo

c     Monaghan type virtual particle on the Left side

            do i = 1, 2*mp-1
              nvirt = nvirt + 1
              x(1, ntotal + nvirt) = 0.
              x(2, ntotal + nvirt) = i*dx/2
              vx(1, ntotal + nvirt) = 0.
              vx(2, ntotal + nvirt) = 0.
            enddo

c     Monaghan type virtual particle on the Right side

            do i = 1, 2*mp-1
              nvirt = nvirt + 1
              x(1, ntotal + nvirt) = xl
              x(2, ntotal + nvirt) = i*dx/2
              vx(1, ntotal + nvirt) = 0.
              vx(2, ntotal + nvirt) = 0.
            enddo

            do i = 1, nvirt
              rho (ntotal + i) = 1000.
              mass(ntotal + i) = rho (ntotal + i) * dx * dx
              p(ntotal + i) = 0.
              u(ntotal + i) = 357.1
              itype(ntotal + i) = -2
              hsml(ntotal + i) = dx
            enddo

        endif

        if (mod(itimestep,save_step).eq.0) then
          open(1,file="../data/xv_vp.dat")
          open(2,file="../data/state_vp.dat")
          open(3,file="../data/other_vp.dat")
          write(1,*) nvirt
          do i = ntotal + 1, ntotal + nvirt
            write(1,1001) i, (x(d, i), d=1,dim), (vx(d, i), d = 1, dim)
            write(2,1002) i, mass(i), rho(i), p(i), u(i)
            write(3,1003) i, itype(i), hsml(i)
          enddo
1001      format(1x, I6, 6(2x, e14.8))
1002      format(1x, I6, 7(2x, e14.8))
1003      format(1x, I6, 2x, I4, 2x, e14.8)
          close(1)
          close(2)
          close(3)
        endif

        if (mod(itimestep,print_step).eq.0) then
          if (int_stat) then
          print *,' >> Statistics: Virtual boundary particles:'
          print *,'            Number of virtual particles:',NVIRT
          endif
        endif

        end
```

18. viscosity.f

```fortran
      subroutine viscosity(ntotal,itype,x,rho,eta)

c-------------------------------------------------------------------------
c   Subroutine to define the fluid particle viscosity

c     ntotal  : Number of particles                              [in]
c     itype   : Type of particle                                 [in]
c     x       : Coordinates of all particles                     [in]
c     rho     : Density                                          [in]
c     eta     : Dynamic viscosity                                [out]

      implicit none
      include 'param.inc'

      integer ntotal,i,itype(maxn)
      double precision x(dim,maxn),rho(maxn),eta(maxn)

      do i=1,ntotal
        if (abs(itype(i)).eq.1) then
          eta(i)=0.
        else if (abs(itype(i)).eq.2) then
          eta(i)=1.0e-3
        endif
      enddo

      end
```

19. time_elapsed.f90

```
   subroutine time_elapsed(s)

!=====================================================================
!   The standard Fortran 90 routine RTC is used to calculate the elapsed CPU
!=====================================================================

   use dfport
   implicit none

   integer, parameter :: output = 6
   real(8) :: s

   s = rtc()

   end subroutine time_elapsed
```

20. time_print.f90

```
   subroutine time_print

!=================================================================
!   TIME_PRINT                      Print out the current date and time.
!
!   Notes:
!
!   The standard Fortran 90 routine DATE_AND_TIME is used to get
!   the current date and time strings.
!
!=================================================================

   implicit none
   integer, parameter :: output = 6

   ! . local scalars.
   character ( len =  8 ) :: datstr
   character ( len = 10 ) :: timstr

   ! . Get the current date and time.
   call date_and_time ( datstr, timstr )

   ! . Write out the date and time.
   write ( output, "(/A)"  ) "                        Date = " // datstr(7:8) //
"/" // &
                             datstr(5:6) // "/" // &
                             datstr(1:4)
   write ( output, "(A)"   ) "                        Time = " // timstr(1:2) //
":" // &
                             timstr(3:4) // ":" // &
                             timstr(5:10)

   write ( output, *)

   end subroutine time_print
```

Bibliography

Anderson, J. D. (1995), Computational fluid dynamics: the basics with applications, Mc Graw-Hill.

Aktas O. and Aluru N. R. (2002), A combined continuum/DSMC technique for multiscale analysis of microfluidc filters, Journal of Computational Physics, 178:342-372.

Alder B. J. and Wainright T. E. (1957), Phase transition for a hard sphere system, Journal of Chemical Physics, 27:1208-1209.

Allen M. P. and Tildesley D. J. (1987), Computer simulation of liquids, Clarendon Press, Oxford.

Allen M. P. and Tildesley D. J. (1993), Computer simulation in chemical physics, Kluwer, Dordrecht.

Asphaug E. and Olson K. (1998), Impacts into asteroids and planets: parallel tree code simulations using smooth particle hydrodynamics, WWW address: http://sdcd.gsfc.nasa.gov/ESS/annual.reports/ess98/asphaug.html.

Atluri S. N. and Zhu T. (1998), A new meshless local Petrov-Galerkin (MPLG) approach in computational mechanics, Computational Mechanics, 22:117-127.

Atluri S. N. and Shen S. P. (2002), The meshless local Petrov-Galerkin (MLPG) method. Tech Science Pree, USA.

Atluri S. N., Cho J. Y. and Kim H. G. (1999), Analysis of thin beams, using the meshless local Petrov-Galerkin (MLPG) method, with generalized moving least squares interpolation. Computational Mechanics, 24:334-347.

Attaway S. W., Heinstein M. W., and Swegle J. W. (1994), Coupling of smooth particle hydrodynamics with the finite element method, Nuclear Engineering and Design, 150(2/3): 199–205.

Balsara D. S. (1995), von Neumann stability analysis of smoothed particle hydrodynamics-suggestions for optimal algorithms, Journal of

Computational Physics, 121:357-372.

Bangash M. Y. H. (1993), Impact and explosion, Blackwell Scientific Publications, Oxford.

Belytschko T., Krongauz Y., Dolbow J. and Gerlach, C. (1998), On the completeness of the meshfree particle methods, International Journal for Numerical Methods in Engineering, 43(5):785-819.

Belytschko T., Krongauz Y.; Organ D., Fleming M. and Krysl P. (1996), Meshless methods: an overview and recently developments, Computer Methods in Applied Mechanics and Engineering, 139:3-47.

Belytschko T., Liu W. K. and Moran B. (2000), Nonlinear finite elements for continua and structures. John Wiley and Sons, New York.

Belytschko T., Lu Y. Y. and Gu L. (1994), Element-Free Galerkin methods, International Journal for Numerical Methods in Engineering, 37:229-256.

Ben Moussa B. and Vila J. P. (1996a), Convergence of SPH method for scalar nonlinear conservation laws, preprint MIP.

Ben Moussa B. and Vila J. P. (1996b), Convergence of SPH method for scalar nonlinear conservation laws on boundary domains, preprint MIP.

Benson D. J. (1992), Computational methods in Lagrangian and Eulerian hydrocodes, Computer Methods in Applied Mechanics and Engineering, 99:235-394.

Benz W. (1988), Applications of smoothed particle hydrodynamics (SPH) to astrophysical problems, Computer Physics Communications, 48:97-105.

Benz W. (1989), Smoothed particle hydrodynamics: a review, NATO Workshop, Les; Arcs, France.

Benz W. (1990), Smoothed particle hydrodynamics: a review, In Numerical Modeling of Non-linear Stellar Pulsation: Problems and Prospects, Kluwer Academic, Boston.

Benz W. and Asphaug E. (1993), Explicit 3d continuum fracture modeling with smoothed particle hydrodynamics, Proceedings of 24th Lunar and Planetary Science Conference in Lunar and Planetary Institute, pp 99-100.

Benz W. and Asphaug E. (1994), Impact simulations with fracture, I methods and tests, ICARUS, 107:98-116.

Benz W. and Asphaug E. (1995), Simulations of brittle solids using Smoothed Particle Hydrodynamics, Computer Physics Communications, 87:253-265.

Berczik P. (2000), Modeling the star formation in galaxies using the chemo-dynamical SPH code, Astronomy and Astrophysics, 360:76-84.

Berczik P. and Kolesnik I. G. (1993), Smoothed particle hydrodynamics and its applications to astrophysical problems, Kinematics and Physics of Celestial Bodies, 9:1-11.

Berczik P. and Kolesnik I. G. (1998), Gas dynamical model of the triaxial protogalaxy collapse, Astronomy and Astrophysical transactions, 16:163-185.

Beson D. J. (1992), Computational methods in Lagrangian and Eulerian hydrocodes, Computer Methods in Applied Mechanics and Engineering, 99:235-394.

Beissel S. and Belytschko T. (1996), Nodal integration of the element-free Galerkin method, Computer Methods in Applied Mechanics and Engineering, 139: 49-74.

Bicknell G. V. and Gingold R. A. (1983), On tidal detonation of stars by massive black holes, Astrophysical Journal, 273: 749-760.

Binder K. (1988), The Monte Carlo method in condensed matter physics, Springer Berlin.

Binder K. (1992), The Monte Carlo method in statistical physics, Springer Berlin.

Bird G. A. (1994), Molecular gas dynamics and the direct simulation of gas flow, Oxford University Press, Oxford, U. K.

Birkhoff G., MacDougall D., Pugh E. and Taylor G. (1948), Explosives with lined cavities. Journal of Applied Physics, 19(6).

Bonet J. and Kulasegaram S. (2000), Correction and stabilization of smoothed particle hydrodynamics method with applications in metal forming simulations, International Journal for Numerical Methods in Engineering, 47:1189-1214.

Brackbill J. U., Kothe D. B. and Ruppel H. M. (1988), FLIP: a low-dissipation, particle-in-cell method for fluid flow, Computer Physics Communications, 48:25-38.

Brett J. M. (1997), Numerical modeling of shock wave and pressure pulse generation by underwater explosions, DSTO-TR-0677, Austrian.

Brinkley S. R. and Kirkwood J. G. (1947), Theory of the propagation of shock waves, Physics Review, 71:606.

Broughton J. Q., Abraham F. F., Bernstein N. and Kaxiras E. (1999), Concurrent coupling of length scale: methodology and application, Physical Reivew B, 60(4): 2391-2403.

Brown K., Attaway S., Plimpton S. J. and Hendrickson B. (2000), Parallel strategies for crash and impact simulations, Computer Methods in Applied Mechanics and Engineering, 184:375-390.

Campbell J., Vignjevic R. and Libersky L. D. (2000), A contact algorithm for smoothed particle hydrodynamics, Computer Methods in Applied Mechanics and Engineering, 184:49-65.

Campbell P. M. (1989), Some new algorithms for boundary values problems in smoothed particle hydrodynamics, DNA Report, DNA-88-286.

Century Dynamics Incorporated (1997), AUTODYN Release Notes Version 3.1, AUTODYN™ Interactive Non-Linear Dynamic Analysis Software.

Charles E. Anderson Jr. (1987), An overview of the theory of hydrocodes, International Journal of Impact Engineering, 5:33-59.

Chen J. K. and Beraun J. E. (2000), A generalized smoothed particle hydrodynamics method for nonlinear dynamic problems, Computer Methods in Applied Mechanics and Engineering, 190:225-239.

Chen J. K., Beraun J. E. and Carney T. C. (1999a), A corrective smoothed particle method for boundary value problems in heat conduction, Computer Methods in Applied Mechanics and Engineering, 46:231-252.

Chen J. K., Beraun J. E. and Jih C. J. (1999b), An improvement for tensile instability in smoothed particle hydrodynamics, Computational Mechanics, 23:279-287.

Chen J. K., Beraun J. E. and Jih C. J. (1999c), Completeness of corrective smoothed particle method for linear elastodynamics, Computational Mechanics, 24:273-285.

Chen J. S. Yoon S., Wang H. P. and Liu W. K. (2000), An improved reproducing kernel particle method for nearly incompressible hyperelastic solids, Computer Methods in Applied Mechanics and Engineering, 181:117-145.

Chen J. S., Pan C. and Wu C. T. (1997), Large deformation analysis of rubber based on a reproducing kernel particle method, Computational Mechanics, 19:153-168.

Chen J. S., Pan C. and Wu C. T. (1998), Application of reproducing kernel particle methods to large deformations and contact analysis of elastomers, Rubber Chemistry and Technology, 7:191-213.

Chen J. S., Pan C., Wu C. T. and Liu W. K. (1996), Reproducing kernel particle methods for large deformation analysis of nonlinear structures, Computer Methods in Applied Mechanics and Engineering, 139:195-227.

Chen S. and Doolen G. D. (1998), Lattice Boltzmann method for fluid flows, Annual Review of Fluid Mechanics, 30:329-346.

Chiesum J. E. and Shin Y. S. (1997), Explosion gas bubbles near simple boundaries, Shock and Vibration, 4 (1): 11-25.

Chin G. L. (2001), Smoothed particle hydrodynamics and adaptive smoothed particle hydrodynamics with strength of materials, Master thesis, National University of Singapore, Singapore.

Chin G. L., Lam K. Y. and Liu G. R. (2002a), Adaptive smoothed particle hydrodynamics with strength of material, Part I, Proceedings of the 2nd International Conference on Structural Stability and Dynamics, Singapore, December.

Chin G. L., Lam K. Y. and Liu G. R. (2002b), Adaptive smoothed particle hydrodynamics with strength of material, Part II, Proceedings of the 2nd International Conference on Structural Stability and Dynamics, Singapore.

Chong W. K., Lam K. Y., Yeo K. S., Khoo B. C., Liu G. R. and Chong O. Y. (1998a), Computational Study of water mitigation effects on an explosion inside a vented tunnel system. Explosive Safety.

Chong W. K., Lam K. Y., Yeo K. S., Khoo B. C., Liu G. R. and Chong O. Y. (1998b), Blast Suppression using water in a tunnel system. Proceedings of

HPC ASIA'98 3rd High performance Computing Asia Conference, 2: 968-982, Singapore.

Chong W. K., Lam K. Y., Yeo K. S., Liu G. R. and Chong O. Y. (1999), A comparison of simulation's results with experiment on water mitigation of an explosion. Shock and Vibration, 6:73-80.

Chorin A. J. (1973), Discretization of a vertex sheet, with an example of roll-up, Journal of Computational Physics, 13:423-429.

Chou P. C., Ciccarelli R. D. and Walters W. P. (1983), The Formation of jets from hemispherical-liner warheads. Proc. 7th Intl Symp On Ballistics, The Hague, Netherlands.

Ciccootti G., Frenkel D. and McDonald I. R. (1987), Simulation of liquids and solids, North-Holland.

Cleary P. W. (1998), Modeling Confined Multi-material Heat and Mass Flows Using SPH, Applied Mathematical Modeling, 22-981-993.

Cole R. H. (1948), Underwater explosions, Princeton University Press.

Cowler M. S. (1987), AUTODYN-an interactive non-linear analysis program for microcomputers though supercomputers, Proc 9th Intl Conf SMIRT.

Crepeau J. and Needham C. (1998), The effects of water on blast from the simultaneous detonation of 180 152 mm shells, 28th DDESB Seminar, Orlando, Florida.

Cundall P. A. (1987), Distinct element models of rock and soil structure, In Brown (Ed.) Analytical and Computational Methods in Engineering Rock Mechanics, London, 129-163.

Cushman-Roision B., Esenkow O. E. and Mathias B. J. (2000), A particle-in-cell method for the solution of two-layer shallow-water equations. International Journal for Numerical Methods in Fluids, 32:515-543.

Dai K. Y., Liu G. R., Lim K. M. and Chen X. L. (2002), An element-free Galerkin method for static and free vibration analysis of shear-deformable laminated composite plates, Journal of Sound and Vibration, in press.

Dave R., Dubinski J. and Hernquist L. (1997), Parallel TreeSPH, New Astron. 2, 227-297.

Dilts G. A. (1999), Moving least square particle hydrodynamics i: consistency and stability, International Journal for Numerical Methods in Engineering, 44:1115-1155.

Dilts G. A. (2000), Moving least square particle hydrodynamics ii: conservation and boundaries, International Journal for Numerical Methods in Engineering, 48:1503-1524.

Dilts G. A. and Haque A. I. (2000), Three-dimensional boundary detection for smooth particle methods, LA-UR-00-4419, Los Alamos National Laboratory.

Dobratz B. M. (1981), LLNL Explosive Handbook, UCRL-52997, Lawrence Livermore National Laboratory, Livermore, CA.

Duarte C. A. and Oden J. T. (1996), An HP adaptive method using clouds,

Computer Methods in Applied Mechanics and Engineering, 139:237-262.

Dyka C. T. and Ingel R. P. (1995), An approach for tension instability in smoothed particle hydrodynamics (SPH), Computers & Structures, 57:573-580.

Dyka C. T., Randles P. W., and Ingel R. P. (1997), Stress points for tension instability in smoothed particle hydrodynamics, International Journal for Numerical Methods in Engrgineering, 40:2325-2341.

Español P. (1998), Fluid particle model, Physical Review E, 57/3, 2930-2948.

Evans M. W. and Harlow F. H. (1957), The particle-in-cell method for hydrodynamic calculations, Los Alamos National Laboratory Report LA-2139.

Evrard, August E. (1988), Beyond N-Body: 3D cosmological gas dynamics, Monthly Notices of the Royal Astronomical Society, 235:911.

Fickett W. and Wood W. W. (1966), Flow calculations for pulsating one-dimensional detonations, The Physics of Fluids, 9:903-916.

Flebbe O., Muzei S., Herold H., Riffert H. and Ruder H. (1994), Smoothed particle hydrodynamics-physical viscosity and the simulation of accretion disks, The Astrophysical Journal, 431:754-760.

Flekkoy E. G., Wagner G. and Feder J. (2000),Hybrid model for combined particle and continuum dynamics, Europhysics Letter, 52(3): 271.

Forsen R., Carlberg A. and Eriksson S. (1996), Small scale tests on mitigation effects of water in a model of the KLOTZ Club Tunnel in Alvdalen. 27th DDESB Seminar, Las Vegas, NV.

Forsen R., Hansson H. and Carlberg A. (1997), Large scale test on mitigation effects of water in the KLOTZ Club installation in Alvdalen. FOA Report R-97-00470-311-SE, Defense Research Establishment, Sweden.

Fraga S., Parher J. M. R. and Pocock J. M. (1995), Computer simulations of protein structures and interactions, Springer, Berlin.

Franz J. V. (2001), Computational physics, an introduction. Kluwer Academic/Plenum Publishers, New York.

Frederic A. R. and James C. L. (1999), Smoothed particle hydrodynamics calculations of stellar interactions, Journal of Computational and Applied Mathematics, 109:213-230.

Fromm J. and Harlow F. H. (1963), Numerical solution of the problem of Vortex Sheet development, Physics of Fluids, 6:975.

Fulk D.A. (1994), A numerical analysis of smoothed particle hydrodynamics, PhD. Thesis, Air Force Institute of Technology.

Garcia A. L., Bell J. B. (1999), Crutchfield W. Y. and Alder B. J., Adaptive mesh and algorithm refinement using direct simulation Monte Carlo, Journal of Computational Physics, 154:121-134.

Geers T. L. (1971), Residual potential and approximate methods for three-dimensional fluid structure interaction problems, Journal of the Acoustical Society of America, 49:1505-1510.

Geers T. L. (1978), Doubly asymptotic approximations for transient motions of submerged structures, Journal of the Acoustical Society of America, 64:1152-1159.

Gentry R. A., Martin R. E. and Daly B. J. (1966), An Eulerian differencing method for unsteady compressible flow problems. Journal of Computational Physics, 1:87-118.

Gibson J. B., Goland A. N. (1960), Milgram M. and Vineyard G. H., Dynamics of radiation damage, Physics Review 120:1229-1253.

Gingold R. A. and Monaghan J. J. (1977), Smoothed Particle Hydrodynamics: Theory and Application to Non-spherical stars, Monthly Notices of the Royal Astronomical Society, 181:375-389.

Gingold R. A. and Monaghan J. J. (1982), Kernel estimates as a basis for general particle method in hydrodynamics, Journal of Computational Physics, 46:429-453.

Gong S. W., Lam K. Y. and Liu G. R. (1999), Computational simulation of ship section subjected to underwater shock, In Shim et al. (eds.) Impact Response of Materials and Structures, pp 413-419, Oxford.

Gray J. P., Monaghan J. J. and Swift R. P. (2001) SPH elastic dynamics, Computer Methods in Applied Mechanics and Engineering, 190: 6641-6662.

Grest G. S. and Kremer K. (1986), Molecular dynamics simulation for polymers in the presence of a heat bath, Physics Review A, 33, 3628.

Gu Y. T. and Liu G. R. (2001a), A boundary point interpolation method (BPIM) using radial function basis, First MIT Conference on Computational Fluid and Solid Mechanics, pp. 1590-1592, June 2001, MIT.

Gu Y. T. and Liu G. R. (2001b), A coupled element free Galerkin/Boundary element method for stress analysis of two-dimensional solids, Computer Methods in Applied Mechanics and Engineering, 190(34): 4405-4419.

Gu Y. T. and Liu G. R. (2001c), A local point interpolation method for static and dynamic analysis of thin beams, Computer Methods in Applied Mechanics and Engineering, 190: 5515-5528.

Gu Y. T. and Liu G. R. (2001d), A meshless local Petrov-Galerkin (MLPG) method for free and forced vibration analyses for solids, Computational Mechanics 27(3): 188-198.

Gu YT and Liu GR(2001e), A meshless Local Petrov-Galerkin (MLPG) formulation for static and free vibration analyses of thin plates. Computer Modeling in Engineering & Sciences, 2(4), 463-476.

Gu Y. T. and Liu G. R. (2002), A boundary point interpolation method for stress analysis of solids, Computational Mechanics, 28(1):47-54.

Gu Y. T. and Liu G. R. (2003), A boundary radial point interpolation method (BRPIM) for 2-D structural analyses, *Structural Engineering and Mechanics, An International Journal.* 15 (5).

Gutfraind R. and Savage S. B. (1998), Flow of fractured ice through wedge-shaped channels: smoothed particle hydrodynamics and discrete-element simulations, Mechanics of Materials, 29:1-17.

Hadjiconstantinou N. G. (1999a), Combining atomistic and continuum simulations of contact-line motion, Physical Review E, 59(2):2475-2479.

Hadjiconstantinou N. G. (1999b), Hybrid atomistic-continuum formulations and the moving contact line problem, Journal of Computational Physics, 154:245-265.

Hadjiconstantinou N. G. and Patera A. T. (1997), Heterogneous atomistic-continuum representations for dense fluid systems, International Journal of Modern Physics C, 8(4): 967-976.

Hageman, L. J. (1975), HELP, a multi-material Eulerian program for two compressible fluid and elastic-plastic flows in two space dimensions and time, Systems, Science and Software, SSS-R75-2.

Hallquist J. O. (1988), User's manual for DYNA2D – an explicit two-dimensional hydrodynamic finite element code with interactive rezoning and graphic display, Lawrence Livermore National Laboratory report UCID – 18756, Rev. 3.

Hallquist J. O. (1998), LS-DYNA THEORETICAL MANUAL. Livermore Software Technology Corporation, 2876 Waverley Way, Livermore, California 94550-1740.

Hallquist J. Q. (1986), DYNA3D User's manual (nonlinear dynamic analysis of solids in three dimensions). Lawrence Livermore National Laboratory, UCID-19592.

Hans U Mair (1999), Review: hydrocodes for structure response to underwater explosions, Shock and Vibration, 6(2): 81-96.

Hansson H. and Forsen R. (1997), Mitigation effects of water on ground shock: large scale testing in alvdalen. FOA-R-97-311, Defense Research Establishment, Weapons and Protection Division, S-17290, Stockholm, Sweden.

Harlow F. H. (1957), Hydrodynamic problems involving large fluid distortion, Journal of the Association for Computing Machinery, 4: 137.

Harlow F. H. (1963), The particle-in-cell method for numerical solution of problems in fluid dynamics, Proceedings of Symposia in Applied Mathematics.

Harlow F. H. (1964), The particle-in-cell computing method for fluid dynamics, In Methods in Computational Physics, Fundamental Methods in Hydrodynamics, vol. 3, Academic Press, New York.

Harlow F. H. and Welch E. (1965), Numerical calculation of time-dependent viscous incompressible flow of fluids with free surface, Phys. Fluids, Vol. 8, p. 2182.

Hernquist L. and Katz N. (1989), TreeSPH- A unification of SPH with the hierarchical tree method, The Astrophysical Journal Supplement Series, 70: 419-446.

Hicks D. L. and Liebrock L. M. (2000), Lanczo's generalized derivative: insights and applications, Applied Mathematics and Computations, 112:63-73.

Hicks D. L., Swegle J. W. and Attaway S. W. (1997), Conservative smoothing stabilizes discrete-numerical instabilities in SPH material dynamics computations, Applied Mathematics and Computation, 85:205-226.

Hiermaier S., Konke D., Stilp A. J. and Thoma K. (1997), Computational simulation of the hypervelocity impact of Al-sphere on thin plates of different materials, International Journal of Impact Engineering, 20(1-5):363-374.

Hirsch C. (1988), Numerical Computation of Internal and External Flows, Volume 1, Wiley-Interscience publication.

Hirt C. W., Amsden A. A. and Cook J. L. (1974), An arbitrary Lagrangian-Eulerian computing method for all flow speeds, Journal of Computational Physics, 14:227-253.

Hockney R. W. and Eastwood J. W. (1988), Computer simulations using particles, Adamhilger, New York.

Hoogerbrugge P., J. and Koelman J. M. V. A. (1992), Simulating microscopic hydrodynamic phenomena with dissipative particle dynamics, Europhysics Letters, 19(3), 155-160.

Hoover W. G. (1991), Computational statistical mechanics, Elsevier, Amsterdam.

Hultman J. and Källander D. (1997), Tests of a galaxy formation code, Merging of particles', Astronomy and Astrophysics, 324:534.

Hultman J. and Pharayn A. (1999), Hierarchical, dissipative formation of elliptical galaxies: is thermal instability the key mechanism ? hydrodynamic simulations including supernova feedback multi-phase gas and metal enrichment in cdm: structure and dynamics of elliptical galaxies, Astronomy and Astrophysics, 347:769-798.

Lacome Jean Luc and Gallet Céline (2002), SPH: a solution to avoid using erosion criterion, Livermore Software Technology Corporation.

Johnson G. R. (1981), Recent developments and analyses associated with the EPIC-2 and EPIC-3 codes. In: Wang S. S. and Renton W. J. (eds.) Advances in Aerospace Structures and Materials, AD-01, New York, ASME.

Johnson G. R. (1994), Linking of Lagrangian particle methods to standard finite element methods for high velocity impact computations. Nuclear Engineering and Design, 150(2/3): 265–274.

Johnson G. R. and Beissel S. R. (1996), Normalized smoothed functions for SPH impact computations. International Journal for Numerical Methods in Engineering, 39:2725-2741.

Johnson G. R. and Hollquist W. H. (1998), Evaluation of cylinder impact test data for constitutive model constants, Journal of Applied Physics, 64(8):3901-3910.

Johnson G. R. Petersen E. H. and Stryk R. A. (1993), Incorporation of an SPH option into the EPIC code for a wide range of high velocity impact computations, International Journal of Impact Engineering, 14: 385-394.

Johnson G. R. Stryk R.A. and Beissel S.R. (1996a), Interface effects for SPH computations, In Jones N. et al. (eds.) Structures Under Shock and Impact, 285-294.

Johnson G. R., Stryk R.A. and Beissel S.R. (1996b), SPH for high velocity impact computations, Computer Methods in Applied Mechanics and Engineering, 139:347-373.

Johnson N. L. (1996), The legacy and future of CFD at Los Alamos, Proceedings of the 1996 Canadian CFD Conference, Ottawa, Canada June 3-4.

Jun S. Liu W. K. and Belytschko T. (1998), Explicit reproducing kernel particle methods for large deformation problems, International Journal for Numerical Methods in Engineering, 41:137-166.

Kandanoff L., McNamara G. R. and Zanetti G. (1989), From Automata to fluid flow: comparisons of simulation and theory, Physics Review A, 40:4527.

Keenan W. A. and Wager P. C. (1992), Mitigation of confined explosion effects by placing water in proximity of explosions. 25th DDESB Explosives Safety Seminar, Anaheim, CA.

Kirkwood J. G. and Bethe H. A. (1942), Basic propagation theory. -OSRD 558.

Koshizuka S., Nobe A. and Oka Y. (1998), Numerical analysis of breaking waves using the moving particle semi-implicit method, International Journal for Numerical Methods in Fluids, 26:751-769.

Krysl, P. and Belytschko, T., Analysis of Thin Plates by the Element-Free Galerkin Method, Comput. Mech., 17,26-35, 1996a.

Krysl, P., and Belytschko,T., Analysis of thin shells by the element-free Galerkin method, International Journal of Solids and Structures, 33,3057-3080,1996b

Kum O., Hoover W. G. and Posch H. A. (1995), Viscous conducting flows with smooth-particle applied mechanics, Physics Review E, 109:67-75.

Laguna Pablo (1995), Smoothed particle interpolation, The Astrophysical Journal, 439:814-821.

Lam K. Y., Gong S. W. and Liu G. R. (1998), Underwater shock analysis capability for FRP marine structures, Proceedings of Naval Platform Technology Seminar 1998, Advancing the Frontiers of Technology, 17:1-11, Singapore.

Lam, K Y, Liu G. R. and T H Lim, Applications of arbitrary-lagrangian-eulerian (ALE) techniques for shock response of vehicle and related problems. In Proceedings of the DTG Technology Seminar Series 96: Land Platform Technology, 8 November 1996, Singapore, pp. 56-67.

Lam K. Y., Liu G. R., Liu M. B. and Zong Z. (2000), Smoothed particle hydrodynamics for fluid dynamic problems, Proceedings of International Symposium on Supercomputing and Fluid Science, pp. 1-16, Institute of fluid Science, Tohoku University, Sendai, Japan.

Lattanzio, J. C., Monaghan J. J., Pongracic H. and Schwartz M. P. (1986), Controlling penetration, SIAM Journal on Scientific and Statistical Computing, 7(2):591-598.

Lee W. H. (1998), Newtonian hydrodynamics of the coalescence of black holes with neutron stars ii, tidally locked binaries with a soft equation of state, Monthly Notices of the Royal Astronomical Society, 308:780-794.

Lee W. H. (2000), Newtonian hydrodynamics of the coalescence of black holes with neutron stars iii, irrotational binaries with a stiff equation of state, Monthly Notices of the Royal Astronomical Society, 318:606-624.

Leonard A. (1980), Vortex methods for flow simulation, Journal of Computational Mechanics, 37:289-335.

Li S. and Liu W. K. (2002), Meshfree and particle methods and their applications, Applied Mechanics Review, 55(1):1-34.

Li S., Hao W. and Liu W. K. (2000a), Meshfree simulation of shear banding in large deformation, International Journal of Solids and Structures, 37:7185-7206.

Li S., Hao W. and Liu W. K. (2000b), Numerical simulation of large deformation of thin shell structures using meshfree methods, Computational Mechanics, 25:102-116.

Li S., Liu W. K., Rosakis A. J., Ted Belytschko and Hao W. (2002), Mesh-free Galerkin simulations of dynamic shear band propagation and failure mode transition, International Journal of Solids and Structures, 39(5):1213-1240.

Lia C. and Carraro G. (2000), A parallel tree SPH code for galaxy formation, Mon. Not. R. Astron. Soc., 314, 145-161.

Libersky L. D. and Petscheck A. G. (1991), Smoothed particle hydrodynamics with strength of materials, in H. Trease, J. Fritts and W. Crowley (ed.): Proceedings of The Next Free Lagrange Conference, Springer-Verlag, NY, 395:248-257.

Libersky L. D., Petscheck A. G., Carney T. C, Hipp J. R. and Allahdadi F. A. (1993), High strain Lagrangian hydrodynamics-a three-dimensional SPH code for dynamic material response, Journal of Computational Physics, 109: 67-75.

Libersky L. D., Randles P. W. and Carney T. C. (1995), SPH calculations of fragmentation, Proceedings of the 3rd U. S. Congress on Computational Mechanics, Dallas, TX.

Lin H and Atluri SN (2001), Analysis of incompressible Navier-Stokes flows by the meshless MLPG method. Computer Modeling in Engineering & Sciences, 2(2), 117-142.

Liszka T. and Orkisz J. (1980), The finite difference method at arbitrary irregular grids and its applications in applied mechanics, Computers and Structures, 11:83-95.

Liu G. R. (1999), A Point Assembly Method for Stress Analysis for Solid, in V. P.

W. Shim et al. (Eds.): Impact Response of Materials & Structures, pp. 475-480, Oxford, 1999.

Liu G. R. (2002), Mesh Free Methods: moving beyond the finite element method, CRC Press, Boca Raton.

Liu G. R. and Chen X. L. (2001), A mesh-free method for static and free vibration analyses of thin plates of complicated shape, Journal of Sound and Vibration. 241 (5): 839-855.

Liu, G.R. and Chen, X.L., Buckling of symmetrically laminated composite plates using the element-free Galerkin method, International Journal of Structural Stability and Dynamics, 2(3), 2002, pp. 281-294

Liu G. R. and Gu Y. T. (1999), A point interpolation method, Proceedings of 4th Asia-Pacific Conference on Computational Mechanics, pp 1009-1014, December 1999, Singapore.

Liu G. R. and Gu Y. T. (2000a) Vibration analyses of 2-D solids by the local point interpolation method (LPIM), Proceedings of 1st international conference on structural stability and dynamics, pp. 411-416, December, Taipei, Taiwan.

Liu G. R. and Gu Y. T. (2000b), Coupling of element free Galerkin and hybrid boundary element methods using modified variational formulation, Computational Mechanics, 26(2): 166-173.

Liu G. R. and Gu Y. T. (2000c), Coupling of Element Free Galerkin method with Boundary Point Interpolation Method, In Atluri SN and Brust FW (Eds.): Advances in computational engineering & science 2000, pp. 1427-1432, Los Angles.

Liu G. R. and Gu Y. T. (2000d), Meshless local Petrov-Galerkin (MLPG) method in combination with finite element and boundary element approaches, Computational Mechanics, 26:536-546.

Liu G. R. and Gu Y. T. (2001a), A local point interpolation method for stress analysis of two-dimensional solids, Structural Engineering and Mechanics, 11(2): 221-236.

Liu G. R. and Gu Y. T. (2001b), A local radial point interpolation method (LR-PIM) for free vibration analyses of 2-D solids. Journal of Sound and Vibration, 246(1): 29-46.

Liu G. R. and Gu Y. T. (2001c), A point interpolation method for two-dimensional solids. International Journal for Numerical Methods in Engineering, 50, 937-951.

Liu G. R. and Gu Y. T. (2001d), On formulation and application of local point interpolation methods for computational mechanics. Computational Mechanics of Proceeding of First Asia-Pacific Congress on Computational Mechanics, pp. 97-106, Sydney, Australia (Invited paper).

Liu G. R. and Gu Y. T. (2002), A truly meshless method based on the strong-weak from, in Liu G. R. (Ed.) Advances in Meshfree and X-FEM Methods, pp. 259-261.

Liu G. R. and Gu Y. T. (2003a), A meshfree method: meshfree weak-strong (MWS) form method for 2-D solids (submitted).

Liu G. R. and Gu Y. T. (2003b), Boundary meshfree methods based on the boundary point interpolation methods. Engineering Analysis with Boundary Elements, Special Issue (in press)

Liu G. R. and Gu Y. T. (2003c), A Meshfree Weak-Strong (MWS) form method, 25th World Conference on Boundary Element Methods, 8 -10 September 2003, Split, Croatia. (Accepted).

Liu G. R. and Han X. Computational inverse techniques in nondestructive evaluation., (in press, to appear in June 2003), ~560 pages, CRC Press. ISBN: 0849315239.

Liu G. R. and Quek (2003), The finite element method: a practical course, Butterworth Heinemann.

Liu G. R. and Tu Z. H. (2001), MFree2D©: an adaptive stress analysis package based on mesh-free technology, First MIT Conference on Computational Fluid and Solid Mechanics, pp. 327-329, June 2001, MIT.

Liu, G. R. and Tu Z. H. (2002), An Adaptive Procedure Based on Background Cells for Meshless Methods, Computer Methods in Applied Mechanics and Engineering., 191:1923-1943.

Liu G. R., Dai K. Y., Lim K. M. and Gu Y. T. (2002a), A point interpolation mesh free method for static and frequency analysis of two-dimensional piezoelectric structures, Computational Mechanics, 29(6): 510-519.

Liu G. R., Dai K. Y., Lim K. M. and Gu Y. T. (2002b), A radial point interpolation method for simulation of two-dimensional piezoelectric structures, Smart Materials and Structures. 12(2): 171-180.

Liu G. R., Lam K. Y. and Lu C. (1998), Computational simulation of sympathetic explosion in a high performance magazine. Proc. of 3rd Weapon Effects Seminar, Singapore.

Liu G. R., Liu M. B. and Lam K. Y. (2002), A general approach for constructing smoothing function for meshfree methods, Proceedings of the 9th International Conference on Computing in Civil and Building Engineering, pp. 431-436, Taipei, Taiwan.

Liu L. and Tan V. B. C. (2002), A meshfree method for dynamics analysis of thin shells, In Liu G. R. (Ed.): Advances in Meshfree and X-FEM methods, pp. 90-95.

Liu L., Liu G. R., and Tan V. B. C. (2002), Element free method for static and free vibration analysis of spatial thin shell structures. Computer Methods in Applied Mechanics and Engineering. 2002(191), 5923-5942.

Liu M. B., Liu G. R., Zong Z. and Lam K. Y. (2000), Numerical simulation of underwater explosion by SPH. In Atluri SN & Brust FW (Eds.): Advances in Computational Engineering & Science 2000, pp. 1475-1480.

Liu M. B., Liu G. R. and Lam K. Y. (2001a), A new technique to treat material interfaces for smoothed particle hydrodynamics, Proceedings of 1st Asia-pacific Congress on Computational Mechanics (APCOM'01), pp. 977-982, Sydney.

Liu M. B., Liu G. R., Zong Z. and Lam K. Y. (2001b), Numerical simulation of incompressible flows by SPH, International Conference on Scientific & Engineering Computing, Beijing.

Liu M. B., Liu G. R. and Lam K. Y. (2002a), Investigations into water mitigations using a meshless particle method, Shock Waves 12(3):181-195.

Liu M. B., Liu G. R. and Lam K. Y. (2002b), Coupling meshfree particle method with molecular dynamics—a novel approach for multi-scale simulations, in Liu G. R. (Ed.) Advances in Meshfree and X-FEM Methods, pp. 211-216.

Liu M. B., Liu G. R. and Lam K. Y. (2003a), Comparative study of the real and artificial detonation models in underwater explosions, Engineering Simulation, 25(1).

Liu M. B., Liu G. R. and Lam K. Y. (2003b), Constructing smoothing functions in smoothed particle hydrodynamics with applications, Journal of Computational and Applied Mathematics (In press).

Liu M. B., Liu G. R. and Lam K. Y. (2003c), Meshfree particle simulation of the explosion process for high explosive in shaped charge, Shock Waves (In press).

Liu M. B., Liu G. R., Zong Z. and Lam K. Y. (2003d), Computer simulation of the high explosive explosion using smoothed particle hydrodynamics methodology, Computers & Fluids, 32(3): 305-322.

Liu M. B., Liu G. R. and Lam K. Y. (2003e), A one dimensional meshfree particle formulation for simulating shock waves, Shock Waves (Revised).

Liu M. B., Liu G. R., Zong Z. and Lam K. Y. (2003f), Smoothed particle hydrodynamics for numerical simulation of underwater explosions, Computational Mechanics, 30(2):106-118.

Liu W. K. and Chen Y. (1995), Wavelet and multiple-scale reproducing kernel methods, International Journal for Numerical Methods in Fluids, 21:901-931.

Liu W. K., Jun S. and Zhang Y. F. (1995a), Reproducing Kernel Particle Methods, International Journal for Numerical Methods in Fluids, 20:1081-1106.

Liu W. K., Jun S., Li S., Adee J. and Belytschko T. (1995b), Reproducing kernel particle methods for structural dynamics, International Journal for Numerical Methods in Engineering, 38:1655-1679.

Liu W. K., Chen Y., Chang C. T. and Belytschko T. (1996a), Advances in multiple scale kernel particle methods, Computational Mechanics, 18(2):73-111.

Liu W. K., Chen Y., Jun S., Chen J. S., Belyschko T., Pan C., Uras R. A. and Chang C. T. (1996b), Overview and applications of the reproducing kernel particle methods, Archives of Computational Methods in Engineering State of the Art, Reviews, 3: 3-80.

Liu W. K., Chen Y., Uras R. A. and Chang C. T. (1996c), Generalized multiple scale reproducing kernel particle methods, Computer Methods in Applied Mechanics and Engineering, 139:91-157.

Liu W. K., Jun S., Sihling D. T., Chen Y. J. and Hao W. (1997), Multiresolution reproducing kernel particle method for computational fluid dynamics, International Journal for Numerical Methods in Fluids, 24:1391-1415.

Liu X., Lee C. K. and Fan S. C. (2002), On using enriched cover function in the Partition-of-unity method for singular boundary-value problems, Computational Mechanics, 29:212-225.

Long S. Y. and Atluri S. N. (2002), A meshless local Petrov-Galerkin (MLPG) method for solving the bending problem of a thin plate. CMES, Vol. 3, No. 1, pp. 53-64.

Lucy L. B. (1977), Numerical approach to testing the fission hypothesis, Astronomical Journal, 82:1013-1024.

Mader C. L. (1979), Numerical modeling of detonations, University of California Press.

Mader C. L. (1998), Numerical modeling of explosives and propellants, CRC Press, New York.

Malvar L. J. and Tancreto J. E. (1998), Analytical and test results for water mitigation of explosion effects. 28th DDESB Seminar, Orlando, FL.

Marchand K. A., Oswald C. J. and Plocyn M. A. (1996), Testing and snalysis done in support of the development of a container for on-site weapon demilitarization. 27th DDESB Seminar, Las Vegas, NV.

Masson C. and Baliga B. R. (1992), A control volume finite element bethod for two- dimensional dilute fluid-solid particle flows, Numerical Methods in Engng'92, In Hirsch et al. (eds), Elsevier Science Publishers B.V., 1: 541-548.

Meglicki Z., Wickramasinghe D. and Dewar R. L. (1995), Gravitational collapse of a magnetized vortex: Applications to the galactic center, Monthly Notices of the Royal Astronomical Society, 272:717-729.

Metropolis N. and Ulam S. (1949), The Monte Carlo method, Journal of American Statistical Association, 44:335-341.

Meyer M. and Pontikis V. (1991), Computer simulation in materials science, interatomic potentials, simulation techniques and applications, Kluwer, Dordrecht.

Molyneaux T. C. K., Li L. Y. and Firth N. (1994), Numerical simulation of underwater explosions, Computers and Fluids, 23:903-911.

Monaghan J. J. (1982), Why particle methods work (hydrodynamics), SIAM Journal on Scientific and Statistical Computing, 3:422-433.

Monaghan J. J. (1985), Particle methods for hydrodynamics, Computer Physics Report, 3:71-124.

Monaghan J. J. (1987), SPH meets the Shocks of Noh, Monash University Paper.

Monaghan J. J. (1988), An introduction to SPH, Computer Physics Communications, 48:89-96.

Monaghan J. J. (1989), On the problem of penetration in particle methods, Journal of Computational physics, 82:1-15.

Monaghan J. J. (1990), Modeling the universe; Proceedings of the Astronomical Society of Australia, 18:233-237.

Monaghan J. J. (1992), Smoothed particle hydrodynamics, Annual Review of Astronomical and Astrophysics, 30:543-574.

Monaghan J. J. (1994), Simulating free surface flow with SPH, Journal of Computational Physics, 110:399-406.

Monaghan J. J. (1995a), Simulating gravity currents with SPH lock gates, Applied Mathematics Reports and Preprints, Monash University.

Monaghan J. J. (1995b), Heat conduction with discontinuous conductivity, Applied Mathematics Reports and Preprints, Monash University, (95/18).

Monaghan J. J. (2000), SPH without a tensile instability, Journal of Computational Physics, 159:290-311.

Monaghan J. J. and Gingold R. A. (1983), Shock simulation by the particle method SPH, Journal of Computational Physics, 52:374-389.

Monaghan J. J. and Kocharyan A. (1995), SPH simulation of multi-phase flow, Computer Physics Communication, 87:225-235.

Monaghan J. J. and Lattanzio J. C. (1985), A refined particle method for astrophysical problems, Astronomy and Astrophysics, 149:135-143.

Monaghan J. J. and Lattanzio J. C. (1991), A simulation of the collapse and fragmentation of cooling molecular clouds, Astrophysical Journal, 375:177-189.

Monaghan J. J. and Poinracic J. (1985), Artificial viscosity for particle methods, Applied Numerical Mathematics, 1:187-194.

Monaghan J. J. and Gingold R. A. (1983), Shock simulation by the particle method of SPH, Journal of Computational Physics, 52:374-381.

Morris J. P. (1994), A study of the stability properties of SPH, Applied Mathematics Reports and Preprints, Monash University.

Morris J. P. (1996), Analysis of smoothed particle hydrodynamics with applications, Ph. D. thesis, Monash University.

Morris J. P. and Monaghan J. J. (1997), A switch to reduce SPH viscosity, Journal of Computational Physics, 136:41-50.

Morris J. P., Fox P. J. and Zhu Y. (1997), Modeling low Reynolds number incompressible flows using SPH, Journal of Computational Physics, 136:214-226.

Morris J. P., Zhu Y. and Fox P. J. (1999), Parallel simulation of pore-scale flow though porous media, Computers and Geotechnics, 25:227-246.

MSC/Dyna (1991) User's manual, Los Angeles, CA.

MSC/Dytran (1997), User's manual, version 4: The MacNeal-Schwndler

Corporation, USA.

Mukherjee, Y. X. and Mukherjee S. (1997a), Boundary node method for potential problems. International Journal for Numerical Methods in Engineering, 40:797-815.

Mukherjee, Y. X. and Mukherjee S. (1997b), On boundary conditions in the element-fee Galerkin method, Computational Mechanics, 19:264-270.

Munz C. D., Schneider R. and Voss U. (1999), A finite-volume particle-in-cell method for the numerical simulation of devices in pulsed-power technology. Survey on Mathematics for Industry, 8:243-257.

Nayroles B., Touzot G. and Villon P. (1992), Generalizing the finite element methods: diffuse approximation and diffuse elements, Computational Mechanics, 10:307-318.

Nelson R. P. and Papaloizou John C. B. (1994), Variable smoothing lengths and energy conservation in smoothed particle hydrodynamics, Monthly Notices of the Royal Astronomical Society, 270:1-20.

Nitsche L. C. and Zhang W. D. (2002), Atomistic SPH and a link between diffusion and interfacial tension, Fluid Mechanics and Transport Phenomena, 48(2):201-211.

Noguchi, H. T. Kawashima and T. Miyamura, (2000) Element free analyses of shell and spatial structures, International Journal for Numerical Methods in Engineering. 47, 1215-1240.

Noh W. F. (1987), Errors for calculations of strong shocks using an artificial viscosity and an artificial heat flux, Journal of Computational Physics, 72:78-120.

O'Connell S. T. and Thompson P. A. (1995), Molecular dynamics-continuum hybrid computations: A tool for studying complex fluid flows, Physical Review E, 52 (6) 5792-5795.

Oger L. and Savage S. B. (1999), Smoothed particle hydrodynamics for cohesive grains, Computer Methods in Applied Mechanics and Engineering, 180:169-183.

Onate E., Idelsohn S., Zienkiewicz, O. C. and Taylor R. L. (1996), A finite point method in computational mechanics applications to convective transport and fluid flow, International Journal for Numerical methods in Engineering, 39(22):3839-3866.

Owen J. M., Villumsen J. V., Shapiro P. R. and Martel H. (1998), Adaptive smoothed particle hydrodynamics methodology ii, Astrophysical Journal Supplement Series, 116:155-209.

Owen, D. R. J. N. Petrinic and J. P. Macedo, (1996), Finite/discrete element models for masonry structures. Proc. First South African Conference on Applied Mechanics, 1-5 July

Pan L. S., Liu G. R. and Lam K. Y. (1999), Determination of Slip Coefficient for Rarefied Gas Flows Using Direct Simulation Monte Carlo, Journal of Micromechanics and Microengineering, 9:89-96.

Pan L. S., Liu, G. R., Khoo B. C. and Song, B. (2000), A Modified Direct Simulation Monte Carlo Method for Low-speed Microflows, Journal of Micromechanics and Microengineering, 10:21-27.

Pan L. S., Ng T. Y., Xu D., Liu G. R. and Lam K. Y. (2002), Coefficient using the direct simulation monte carlo method, Journal of Micromechanics and microengineering, 12:41-52.

Penney and Dasgupta H. K. (1942), British Report, RC 333.

Plimpton S. J., Attaway S., Hendrickson B. and Swegle J., Vaughan C. and Gardner D. (1998), Parallel transient dynamics simulations: algorithms for contact detection and smoothed particle hydrodynamics, Journal of Parallel and Distributed Computing, 50:104-122.

Posch H. A., Hoover W. G. and Kum O. (1995), Steady-state shear flows via nonequilibrium molecular dynamics and smoothed-particle applied mechanics, Physics Review E, 52:1711-1719.

Pugh E., Eichelberger R. and Rostoker N. (1952), Theory of jet formation by charges with lined conical cavities. Journal of Applied Physics, 23(5).

Qian Y. H., Succi S. and Orszag S. A. (2000), Recent advances in lattice Boltzmann computing, In Stauffer (Ed.): Annual Reviews of Computational Physics, Volume III, World Scientific, Singapore.

Qian D., Liu W. K. and Ruoff R. S. (2001), Mechanics of C60 in Nanotubes, Journal of Physical Chemistry B, 105:10753-10758.

Rahman A. (1964), Correlations in the motion of atoms in liquid Argon, Physics Review, 136:405-411.

Randles P. W. and Libersky L. D. (1996), Smoothed particle hydrodynamics some recent improvements and applications, Computer Methods in Applied Mechanics and Engineering, 138:375-408.

Randles P. W. Carney T. C., Libersky L. D., Renick J. R. and Petschek A. G. (1995a), Calculation of oblique impact and fracture of tungsten cubes using smoothed particle hydrodynamics, International Journal of Impact Engineering, 17.

Randles P. W. et al. (1995b), SPH calculation of fragmentation in the MK82 bomb, Proceedings of APS Topical Conference on Shock Compression of Condensed Matter, Seattle, WA.

Randles P. W., and Libersky L. D. (2000), Normalized SPH with stress points, International Journal for Numerical Methods in Engineering, 47:1445-1462.

Rapaport D. C. (1995), The art of molecular dynamics simulation, Cambridge University Press.

Rhoades C E. (1992), A fast algorithm for calculating particle interactions in smooth particle hydrodynamic simulations, Computer Physics Communications, 70:478-482.

Riffert H., Herold H., Flebbe O. and Ruder H. (1995), Numerical aspects of the smoothed particle hydrodynamics method for simulating accretion disks,

Computer Physics Communications, 89:1-16.

Roberto C. D. and Roberto D. L. (2000), A criterion for the choice of the interpolation kernel in smoothed particle hydrodynamics, Applied Numerical Mathematics, 34:363-371.

Rudd R. E. and Broughton J. Q. (1999), Atomistic simulation of MEMS resonators through the coupling of length scales, Journal of Modeling and Simulation of Microsystem, 1(1): 29-38.

Ruth M. and Lynden-Bell (1994), Computer simulations of fracture at the atomic level, Science, 263 (5154):1704-1705.

Salter S. H., and Parkes J. H. (1994), The use of water-filled bags to reduce the effects of explosives. 26th DDESB Seminar, Miami, FL.

Schlatter B. (1999), Modeling fluid flow using smoothed particle hydrodynamics, Master Thesis, Oregon State University.

Senz D. G., Bravo E. and Woosley S. E. (1999), Single and multiple detonations in white dwarfs, Astronomy and Astrophysics, 349:177-188.

Shapiro P. R., Martel H., Villumsen J. V., and Owen, J. M. (1996), Adaptive smoothed particle hydrodynamics, with application to cosmology: methodology, Astrophysical Journal Supplement, 103:269-330.

Shin Y. S. and Chisum J. E. (1997), Modeling and simulation of underwater shock problems using a coupled Lagrangian-Eulerian analysis approach, Shock and Vibration, 4:1-10.

Shin Y. S., Lee M., Lam K. Y. and Yeo K. S. (1998), Modeling mitigation effects of watershield on shock waves, Shock and Vibration, 5: 225-234.

Simpson J. C. (1995), Numerical techniques for three-dimensional smoothed particle hydrodynamics simulations: applications to accretion disks, The Astrophysical Journal, 448:822-831.

Smirnova J. A., Zhigilei L. V. and Garisson B. J. (1999), A combined molecular dynamics and finite element method technique applied to laser induced pressure wave propagation, Computer Physics Communication, 118: 11-16.

Sod G. A. (1978), A survey of several finite difference methods for systems of hyperbolic conservation laws, Journal of Computational Physics, 27:1-31.

Speith R., Schnetter E., Kunze S., Riffert H. (2000), Distributed implementation of SPH for simulations of accretion disks. In: Esser R., Grassberger P., Grotendorst J. and Lewerenz M. (eds.): Molecular Dynamics on Parallel Computers, p. 276 – 285, World Scientific.

Steinberg D. J. (1987), Spherical explosions and the equation of state of water. Report UCID-20974, Lawrence Livermore National Laboratory, Livermore, CA.

Steinmetz M. and Muller E. (1993), On the capabilities and limits of smoothed particle hydrodynamics, Astronomy and Astrophysics268: 391-410.

Stellingwerf R. F. and Wingate C. A. (1993), Impact modeling with smoothed particle hydrodynamics, International Journal of Impact Engineering,

14:707-718.

Stillinger F. H. and Rahman A. (1974), Improved simulation of liquid water by molecular dynamics, Journal of Chemical Physics, 60:1545-1557.

Sulsky D., Zhou S. J. and Schreyer H. L. (1995), Application of the particle-in-cell method to solid mechanics, Computer Physics Communications, 87:236-252.

Swegle J. W. (1978), TOODY IV—a computer program for two-dimensional wave propagation, SAND-78-0552, Sandia National Laboratories, Albuquerque, New Mexico.

Swegle J. W. (1992), Report at Sandia National laboratories.

Swegle J. W. (1994), An analysis of smoothed particle hydrodynamics, SAND93-2513, Sandia National Laboratories, Albuquerque, NM.

Swegle J. W. and Attaway S. W. (1995), On the feasibility of using smoothed particle hydrodynamics for underwater explosion calculations, Computational Mechanics, 17: 151-168.

Swegle J. W., Hicks D. L. and Attaway S. W. (1995), Smoothed particle hydrodynamics stability analysis, Journal of Computational Physics, 116(1): 123-134.

Takeda H., Shoken M. Miyama and Minoru Sekiya (1994), Numerical simulation of viscous flow by smoothed particle hydrodynamics, Progress in Theoretical Physics, 92(5): 939-959.

Vignjevic R., Campbell J., and Libersky L. (2000), A treatment of zero-energy modes in the smoothed particle hydrodynamics, Computer Methods in Applied Mechanics and Engineering, 184:67-85.

von Neumann, J., and Richtmyer R.D. (1950), A method for the numerical calculation of hydrodynamic shocks, Journal of Applied Physics 21:232-247.

Walters W. P. and Zukas J. A. (1989), Fundamentals of shaped charges. John Wiley and Sons Inc.

Wang, J.G. and Liu, G. R., Radial point interpolation method for elastoplastic problems. Proc. of the 1st Int. Conf. On Structural Stability and Dynamics, Dec. 7-9, 2000, Taipei, Taiwan, 703-708, 2000.

Wang, J.G. and Liu, G. R., Radial point interpolation method for no-yielding surface models. Proc. Of the first M.I.T. Conference on Computational Fluid and Solid Mechanics, 12-14 June 2001, 538-540, 2001a.

Wang, J. G., Liu, G. R., and Wu, Y.G., A Point Interpolation Method for Simulating Dissipation Process of Consolidation, Comp. Meth. Appl. Mech. Eng., Vol.190, 5907-5922, 2001b.

Wang, J.G. and Liu, G. R., A Point Interpolation Meshless Method Based on Radial Basis Functions, Int. J Numer. Meth. Eng., 54, 2002, pp. 1623-1648.

Whitworth A. P. (1995), Estimating density in smoothed particle hydrodynamics, Astronomy and Astrophysics, 301:929-932.

Wilkins M. L. (1999) Computer simulation of dynamic phenomena, Springer-Verlag Berlin Heidelberg.

Wolfram S. (1983), Cellular Automata, Los Alamos Science, 9:2-21.

Wu H. A., Liu G. R. and Wang J. S. (2003), Surface effect on the mechanical property of metal nanowire, Sixth International Conference on Computer Methods and Experimental Measurements for Surface Treatment Effects, March, 2003, Greece.

Wu H. A., Liu G. R. and Wang X. X. (2002), A molecular dynamics simulation of the nano-void influence on the elastic property of metals, Modeling and Simulation in Materials Science and Engineering, (Submitted).

Wu Y. L. and Liu G. R. (2003a), A Meshfree Formulation of Local Radial Point Interpolation Method(LRPIM) for incompressible flow simulation, Computational Mechanics (in press).

Wu Y. L. and Liu G. R. (2003b), Meshfree weak-strong form (MWS) method for incompressible flows, the *Proceedings of 7th U.S. National Congress on Computational Mechanics*, Albuquerque, USA. July 28-30, 2003. (Accepted).

Xu Y. G and Liu G. R. (2002), Stepwise-equilibrium and Adaptive Molecular Dynamics Simulation for Fracture Toughness of Single Crystals With Defects, Nanotechnology (Submitted).

Xu Y. G and Liu G. R. (2003), Fitting interatomic potentials using molecular dynamics simulations and inter-generation projection genetic algorithm, Journal of Micro-mechanics and Micro-engineering, 13: 254-260.

Yagawa G. and Yamada T. (1996), Free mesh method: a new meshless finite element method, Computational Mechanics, 18:383-386.

Yagawa G. and Yamada T. (1998), Meshless method on massively parallel processor with application to fracture mechanics, Key Engineering Materials, 145-149:201-210.

Yukiya Aoyama and Jun Nakano (1999), RS/6000 SP: Practical MPI programming, International Technical Support Organization, IBM Corporation.

Zhang S. Z. (1976), Detonation and its applications. Press of National Defense Industry, Beijing.

Zhao H. Z. (1998), Water mitigation effects on the detonation in a confined chamber, proceedings of HPC Asia'98 3rd High performance Computing Asia Conference, 1:808-811, Singapore.

Zhao H. Z. (2001), Water effects on shock wave delay in free fields, Explosion and Shock Waves (in Chinese), 21 (1): 26-28.

Zhu Y., Fox P. J. and Morris J. P. (1999), A pore-scale numerical model for flow through porous media, International Journal for Numerical and Analytical methods in Geomechanics, 23:881-904.

Zienkiewicz O. C. and Taylor R. L. (2000), The Finite Element Method, 5th ed., Butterworth Herinemann.

Zong Z., Lam K. Y. and Liu G. R. (1998), Numerical investigations of damaging effects of underwater shock on submarine structures in the presence of free surfaces, Proceedings of the 69th shock and vibration symposium, Singpaore.

Zukas J. A. (1982), Impact dynamics, John Wiley & Sons, New York.

Zukas J. A. (1990), High velocity impact, John Wiley & Sons, New York.

Index